# PHARMACEUTICAL EQUIPMENT VALIDATION

## *The Ultimate Qualification Guidebook*

■■■

**Phil Cloud**

Interpharm/CRC
Boca Raton London New York Washington, D.C.

**Library of Congress Cataloging-in-Publication Data**

Cloud, Phil
Pharmaceutical equipment validation : the ultimate qualification guidebook / Phil Cloud
p. cm.
Includes bibliographical references and index.
ISBN 1-57491-079-5 (alk. paper)
1. Pharmaceutical technology—Quality control. 2. Pharmaceutical technology—standards. I. Title.
RS192.C57 1998
615′19—dc21 98-28032



Direct all inquiries to CRC Press LLC, 2000 N.W. Corporate Blvd., Boca Raton, Florida 33431.


Interpharm is an imprint of CRC Press LLC


International Standard Book Number 1-57491-079-5
Library of Congress Card Number 98-28032
Printed in the United States of America 4 5 6 7 8 9 0
Printed on acid-free paper

# Contents

# Preface

FDA regulations, such as Current Good Manufacturing Practice (cGMP) for pharmaceuticals, Good Laboratory Practice (GLP), Good Clinical Practice (GCP), and industry standard ISO 9000, require that documentation, such as protocols, be established and followed. These regulations do not provide guidelines on how specifically to produce the documentation. It is left up to individual companies to design their own validation documentation system. This book provides hands-on techniques for qualifying pharmaceutical equipment to achieve FDA compliance.

This book is designed for individuals responsible for writing and executing qualification protocols for drug products and for related industries, such as drug devices and diagnostics, pharmaceutical biotechnology, and bulk pharmaceutical chemicals. This book provides a complete set of 38 protocol templates. These templates are different because they are already filled in, thus providing answers to many of your protocol writing and testing questions. Simple diagrams and other graphics illustrate key ideas.

The author's personal experience with various regulators, such as FDA, FAA, NASA, and Explosives, Firearms & Tobacco, influenced the decision to write this book. The need came from always wanting to know exactly how to do something and the regulators' not providing it. The methods described in this book are those of the author and are not to be construed as the policy of his employer.

# Chapter 1

# Introduction

## Validation and Qualification

Validation and qualification—what is the difference? Well, there is a difference. According to *The American Heritage Dictionary:*

> Validation contains premises from which a conclusion can be logically derived.
>
> Qualification is a condition or circumstance that must be met or complied with.

This book covers the latter in great detail. The equipment qualification methods described in this book are designed to support the current Good Manufacturing Practices (cGMP) in a solid dosage form manufacturing environment. The mortar and pestle have been replaced by automated pharmaceutical processing equipment. Equipment is the main link between raw materials and the final product.

Raw Materials ➡ Equipment ➡ Final Product

Equipment must be installed, operated, and maintained within design specifications; therefore, the installation, operational, and performance qualifications (IQ, OQ, and PQ) of equipment for drug products are a crucial part of the validation process. This book provides a thorough understanding of how to prepare, test, and complete equipment qualification protocols. It also explains how to perform qualification testing and whether to test the equipment for worst case, the intended range of use, edge of failure, proven acceptable range, effective ranges for critical parameters, or range of experience.

Equipment qualification testing involves simple physics (i.e., electricity, air, water, steam, heat, vacuum, etc.). There is a bottom line to equipment testing. For example, you are watching an operation on television, and the people are speaking in a foreign language. You do not know what they are saying, but suddenly it clicks—they are doing the same thing that I do. The only difference is that I do not understand what they are saying. The same thing applies to equipment qualification; it does not matter if you work in drug devices and diagnostics, pharmaceutical biotechnology, or bulk pharmaceutical chemicals. The equipment that needs to be qualified is all controlled by the same physics. The qualification templates presented in this book can be tailored to fit any manufacturing discipline anywhere in the world.

A manufacturer wants to think that a safe product is being produced for the end user, and the best way to do that is to document and test the process every step of the way. In this way, a process can be repeated, with assurance that it is safe. This book covers the qualification portion of the documentation process. Documentation is not created just because it is a regulated requirement; it just makes good business sense.

## VALIDATION ORGANIZATION

Some companies are large enough to have each validation element performed by a different person or department; in some companies, one person performs all of the functions. In Figure 1.1, only the function that is bolded is covered in this book.

The following list assumes that a company is starting a validation department from scratch. If the validation department already exists, compare this list with your own to see if all of the bases have been covered. The complete validation structure has been listed to show the position of this book within the complete validation scheme.

- Setup a Validation Department
- Write a validation plan
- Write policies, SOPs (Standard Operating Procedures), and administrative procedures

***Figure 1.1. Validation Department Functions***

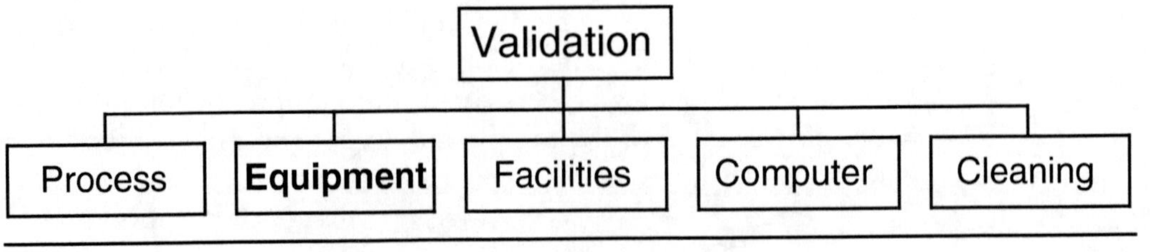

- Establish a change control board
- Establish document reviewers and approvers
- Establish document distribution
- Identify which equipment needs to be validated and establish a master list
- Assign protocols to process, equipment, facilities, computer, and cleaning validation
- Identify critical equipment to validate first, then most important to less important
- Assign equipment numbers
- Assign protocol numbers
- Assign change control numbers
- Change protocols
- Perform commissioning at the original equipment manufacturer (OEM)
- Write and execute qualification protocols
- Deficiencies
- Deviations
- Addendums
- Archiving
- Requalification
- Internal audits
- FDA (U.S. Food and Drug Administration) audits

## The Qualification Protocol

The qualification protocol is the center of gravity of the validation documentation process. This book shows examples of IQs, OQs, and PQs. The protocol is the industry tool used to document the qualification process, but it can take any form. A qualification protocol needs to be developed and approved before qualification testing can begin. During qualification testing, the performer and reviewer will sign the protocol at the bottom of each page where they make entries. The validation is complete when all acceptance criteria have been met. Protocols are test procedures; when they are executed, they become test reports. Test reports are snapshots in time and are archived this way, not to be changed. There are ways to make changes to the protocols after they are archived, but the original is never changed. Only the bolded areas in Figure 1.2 are covered in this book.

The major elements of the qualification protocol are given in Figure 1.3. The protocol examples in this book can be used as cGMP compliance guides, and they should be tailored to meet individual company requirements.

**Figure 1.2. Qualification Protocol Documentation**

**Figure 1.3. Qualification Protocol Elements**

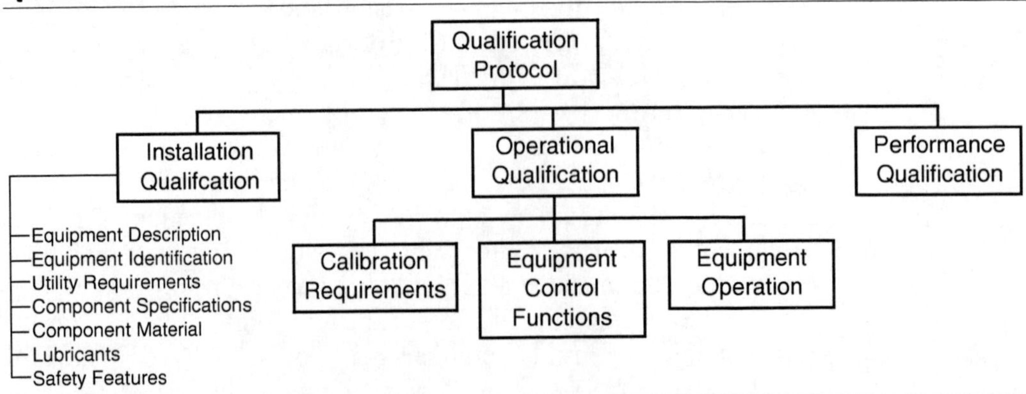

## Qualification Testing

It is important that written qualification protocols are prepared that specify the tests to be conducted and the data to be collected. Observations will be documented, and the reason for collecting test data must be clearly understood. The test data must reflect reality and be recorded accurately. The protocol should specify a sufficient number of tests to show repeatability and provide an accurate assessment of the variability of consecutive runs. The test conditions for these runs should impose upper and lower control limits. A thorough analysis will ensure that the control limits for the equipment are appropriate and that it will not fail if operated within those limits. Testing all functions at the extremes of operation and showing these limits to be acceptable will provide assurance that the intermediate levels are acceptable. Equipment is validated for its mechanical parameters, not how well it processes something. For example, a mixer will be validated for minimum and maximum speeds, loading, and so on, not how well it mixes powders. How well it mixes powders pertains to process validation, not equipment validation.

Qualification testing requires a knowledge of how to measure volts, amps, motor speeds, revolutions per minute, time, temperature, air pressure, vacuum, water pressure, steam pressure, flow rates, mechanical dimensions, volumes, and so on. Weights, volumes, and averages will be calculated, and tolerances will be established. The tester will be reading and interpreting gauges installed on the equipment and the test instruments used to take measurements. The tester also needs to know what type of control system is involved in the operation of the equipment—automatic, semiautomatic, or manual. Automatic means that the equipment is controlled by a microprocessor, PLC (programmable logic controller), or a PC (personal computer); manual indicates that adjustments are made by the operator.

## What Needs to Be Tested?

Everything on a piece of equipment needs to be validated in one way or another; how much and to what detail is another matter. If it moves, it must be tested. Use the following methods to obtain a list of components that need to be tested:

- Observe the equipment in operation.
- Obtain major components from the equipment manufacturers' installation and operational manual and spare parts list.

## What Level of Testing Is Required?

There have been several attempts throughout the pharmaceutical industry to establish a test standard for equipment challenge conditions. Before determining which test condition to use, find out what the normal operating conditions are for that piece of equipment from the equipment operator. Then pick a challenge condition from the following list. The objective is to gain by testing a level of confidence that the equipment is operating under a state of control.

- *Minimum/maximum operating ranges:* This applies to utilities, motor speeds, time, temperature, air pressure, vacuum, water pressure, steam pressure, flow rates, weights, volumes, and so on.
- *Minimum/maximum load evaluations:* This applies to mixers, blenders, and so on. Each company needs to have standards that state if the limits are exceeded or not met, good results will not be achieved. Challenge testing should go outside these limits.
- *The most appropriate challenge condition:* This is not an accurate description because there is room for various interpretations.
- *Effective ranges of critical parameters:* The problem with this one is the word *critical.* If several people are asked which components on a piece of equipment are the most critical, several different answers will be given.
- *Proven acceptable range (PAR) and range of experience:* These do not apply if it is a new piece of equipment. The only problem with these are that the ranges may not necessarily be the minimum and maximum ranges.
- *The intended range of use:* A frequently used condition. The drawback is that the ranges may not necessarily be the minimum and maximum ranges.
- *Worst case:* This is the one tested the most. It is similar to minimum/maximum load evaluations and minimum/maximum ranges.
- *Edge of failure:* This does not apply to equipment because something would have to be made to fail to obtain the edge of failure, although it could apply to a tablet. The point at which the tablet is destroyed is just beyond the edge of failure; therefore, a slight back-off would be its edge of failure.

## How Much Testing Is Enough?

What is an acceptable number of times a process step needs to be repeated successfully to say that an operation is validated—10, 100, or 1,000? The number of repetitions of a process step during qualification should be based on statistical significance. There is a conventional wisdom about validation that once is chance, twice is nice, and three times is validation. Validation is often performed over three trials. Three should sometimes be considered a minimum, because the experimental design may dictate that a higher level of confidence may be required to prove the process. The number of units or duration of trials used in validation is also intended to represent a typical production lot size or timeline. The key words are confidence level and statistical significance. *Juran's Quality Control Handbook* (Juran and Gryna 1988) is a study of statistical process control, which is beyond the scope of this book.

## How to Use This Book

This book covers all of the manufacturing equipment used from granulation to packaging in a solid dosage form pharmaceutical manufacturing company. This book can be used in several ways. (1) When preparing to validate a piece of equipment, refer to either the table of contents under the department or the index to find the section pertaining to a particular equipment type. Next, tailor these protocols to fit the situation. Because the templates in this book are already completed, the reader is given guidance on how to write qualification protocols and perform the testing. (2) This book can be used to compare different types of equipment qualifications for trending purposes. For example, investigate a specific subject in this book, such as volts, in all of the protocols. In the end, you will be left wondering why no standard has been established along the way. (3) This book can also be used for training new employees or retraining existing employees.

# Chapter 2

# Validation of Granulation Equipment

This chapter will cover the qualification testing of equipment used to perform operations in a granulation department (Figure 2.1).

## V-Shell Blender

The validation of a 218 L V-shell blender (Figure 2.2) will be described here. A motor drives the blender and operates at a fixed speed. As the blender rotates, the two inclined cylinders intermesh their flows, causing a radical blending action. In

***Figure 2.1. Granulation Equipment***

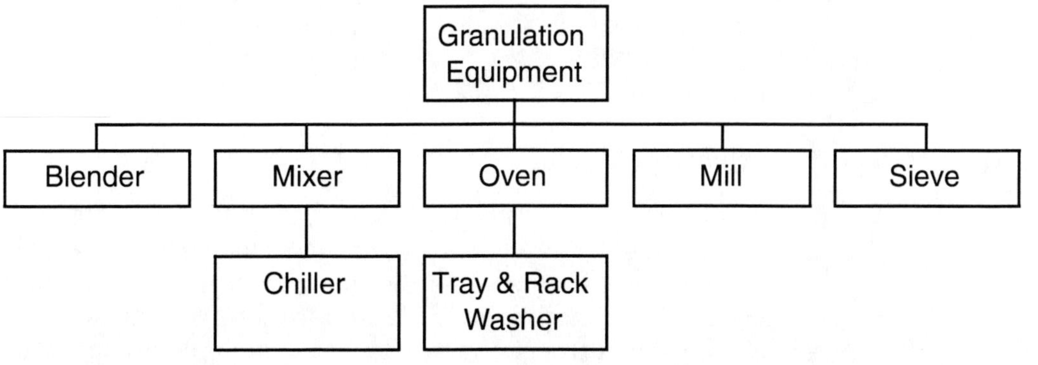

***Figure 2.2.*** ***Twin-Shell Blender (Courtesy of Patterson-Kelley Div. of Harsco Corp.)***

addition, an intensifier bar is located in the blender rotational axis. As it rotates, the solid particles are suspended so that a liquid spray can uniformly coat them. Thus, the three basic requirements for liquid-solids blending are achieved: the liquid and solids are finely divided, suspended in space, and kept in continual motion. The blender can be used in tablet formulations to make the powders more compressible to produce tablets that are resistant to breakage during handling.

The validation protocol for a blender is described in the following sections. This protocol example can be used as a cGMP compliance guide, and it should be tailored to meet individual company requirements. Italicized words in the tables represent examples of information that could be recorded in a validation protocol for a V-shell blender.

## Installation Qualification

An IQ evaluation establishes confidence that the equipment is properly installed (Figure 2.3). The installation must meet the manufacturer's specified guidelines, along with design changes at installation. Also, the supporting electrical utilities must meet all electrical codes. The information required for an IQ evaluation is equipment identification, required documentation, equipment utility requirements,

***Figure 2.3. IQ Elements of a V-Shell Blender***

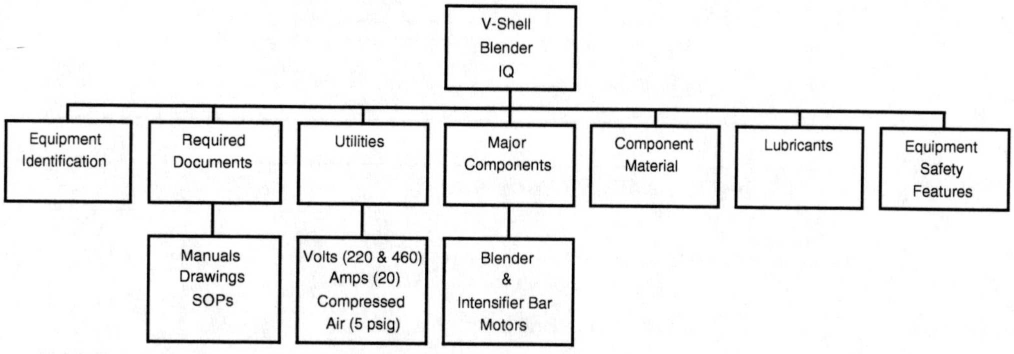


major component specifications, component material, lubricants, and equipment safety features.

### *Equipment Identification*

Record the equipment identification numbers in Table 2.1, along with the following information: equipment manufacturer's purchase order number, model number, serial number, company assigned equipment number, and the location of the equipment.

### *Required Documentation*

Record the equipment manufacturer's operation and maintenance manual and drawings in Table 2.2. Record the SOPs that cover the setup, operation, and cleaning of the blender in Table 2.3.

***Table 2.1. Equipment Identification***

| **Required Information** | **As-Found Conditions** |
|---|---|
| Manufacturer | *Any Blender Co.* |
| Purchase order number | *008649* |
| Model number | *PA4365X* |
| Serial number | *236778* |
| Equipment number | *1266* |
| Location | *Room 720* |

**Table 2.2. Manufacturers' Manuals and Drawings**

| Number | Description | Date |
|---|---|---|
| *1020* | *Blender Installation, Operation and Maintenance Manual* | *11/30/94* |
| *19357* | *Blender Assembly Drawings* | *11/30/94* |

**Table 2.3. Standard Operating Procedures**

| Number | Description | Release Date |
|---|---|---|
| *GRA021* | *V-Shell Blender Setup and Operation* | *10/23/92* |
| *GRA022* | *V-Shell Blender Cleaning Procedure* | *03/12/96* |

### *Equipment Utility Requirements*

Compare the manufacturer's specified volts (V), amps (A), and compressed air requirements to their as-found conditions at the time of qualification testing and record the results in Table 2.4. Also record the location of the power supply source. Record the instruments used to measure the volts, amps, and compressed air in Table 2.5.

***Volt calculation:***

Blender motor volt specification = 220 V ± 10%

±10% of 220 = ±22

220 + 22 = 242

220 – 22 = 198

The measured volts of 219 fall within ±10%.

**Table 2.4. Utilities**

| Utility | Specified | Measured Results | Acceptable (Yes/No) |
|---|---|---|---|
| Volts | Blender motor = 220 ± 10% | *219* | *Yes* |
| | Intensifier bar motor = 460 ± 10% | *A-B 469; A-C 466; B-C 470* | *Yes* |
| Amps | Blender motor = 1.8; Intensifier bar motor = 6.6 | 20 circuit rating | *Yes* |
| Compressed air | ≥ 5 psig | *36 psig* | *Yes* |

Power supply source: breaker box *BB18, wire number: 1 and BB23B, wire numbers: 1, 3, 5 respectively.*

**Table 2.5. Instruments Used**

| Test Instrument | Identification Number | Calibration Due Date |
|---|---|---|
| *Multimeter* | *ME-025* | *04/19/97* |
| *Air pressure gauge* | *P-102* | *02/10/97* |

Intensifier bar motor volt specification = 460 V ± 10%

±10% of 460 = ±46

460 + 46 = 506

460 – 46 = 414

The measured volts of 469/466/470 fall within ±10%.

***Amp calculation:***

Circuit rating = 20 A

Blender motor = 1.8 A

Intensifier bar drive motor = 6.6 A

Equipment current draw = blender motor A + intensifier bar motor A = 1.8 + 6.6 = 8.4 A

The circuit amp rating of 20 is greater than the maximum current draw of the equipment.

***Compressed air calculation:***

Compressed air specification ≥ 5 psig

The measured air pressure of 36 psig is greater than 5 psig.

### *Major Component Specifications*

The component specifications section of the protocol verifies that the blender components purchased were delivered and installed. Record the major components in Table 2.6.

### *Component Material*

Record the material of each component that contacts the product in Table 2.7.

### *Lubricants*

Record the lubricants used to operate the blender in Table 2.8 and indicate if they make contact with the product.

**Table 2.6. Major Components**

| Components | As-Found Conditions |
|---|---|
| Blending vessel | Volume: *218 L*<br>Material: *316 stainless steel* |
| Blender motor | Manufacturer: *Any Motor Co.*<br>Model number: *3BE07414*<br>Serial number: *LX*<br>Volts: *220*<br>Amperes: *1.8*<br>Phases: *1*<br>Cycles: *60 Hz*<br>hp: *1*<br>rpm: *1,740* |
| Charging hatches | Part number: *87-4170*<br>Type: *Full hinged, vented with a filter*<br>Material: *Stainless steel*<br>Gaskets: *White neoprene* |
| Discharge valve | Size: *15.24 cm*<br>Material: *Stainless steel*<br>Type: *Wafer* |
| Intensifier bar | Material: *Stainless steel*<br>Type: *High speed* |
| Intensifier bar motor | Manufacturer: *Any Motor Co.*<br>Model number: *80551137*<br>Serial number: *F594*<br>Volts: *460*<br>Amperes: *6.6*<br>Phases: *3*<br>Cycles: *60 Hz*<br>hp: *5*<br>rpm: *1,725* |

**Table 2.7. Component Material**

| Component | Material |
|---|---|
| Blending vessel | *316 stainless steel* |
| Charging hatches | *Stainless steel* |
| Discharge valve | *Stainless steel* |
| Intensifier bar | *Stainless steel* |

**Table 2.8. Lubricants**

| Where Used | Type | Manufacturer | Product Contact (Yes/No) |
|---|---|---|---|
| *Main drive reducer* | *APG 90 Oil* | *Any Oil Co.* | *No* |
| *Pillow block bearings and intensifier motor* | *Lubriplate 630-2 Grease* | *Any Oil Co.* | *No* |
| Is there a preventive maintenance procedure on file? (Yes/No) | | | *Yes* |

### *Equipment Safety Features*

There is an emergency stop button on the control panel. The operation of the emergency stop button is covered in "Equipment Control Functions."

Electronically interlocked barrier guards are the responsibility of the customer. However, it is the responsibility of the user to conduct a thorough hazards operation review of the installed blender. There are many considerations to be reviewed: barrier guarding to protect operators from injury by rotating equipment; electrical safety and grounding; exposure to chemical or biological hazards; safe handling of powders and liquids that may generate electrostatic hazards; and physical hazards related to cleaning, maintenance, or upset conditions. Specific hazards are identified by team review and or by using process hazards identification software.

Following the identification of process hazards, it is the user's responsibility to take appropriate action to ensure the continued safe operation of the process, including the blender and related equipment. Action may include design modifications of the equipment, guarding, protective apparel or appliances for operators, or specific training.

## Operational Qualification

An OQ evaluation should establish that the equipment can operate within specified tolerances and limits (Figure 2.4). The mechanical ranges of the blender are challenged, along with the basic blender operations. The blender will be validated for its operating ability, not how well it blends powders. Information required for the OQ evaluation is calibration of the instruments used to control the blender, equipment control functions (switches and push buttons), and equipment

***Figure 2.4. OQ Elements of a V-Shell Blender***

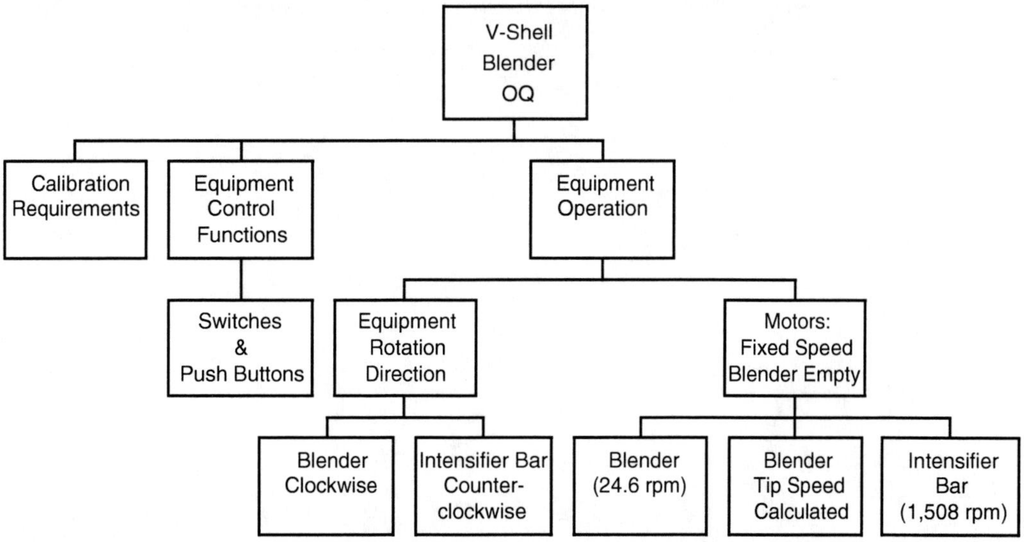

operation (blender motor and intensifier bar motor rotation direction, blender speed, blender tip speed, and intensifier bar speed).

### *Calibration Requirements*

Verify that all critical instruments on the equipment have been logged into the calibration system, have calibration procedures in place, and are in calibration at the time of qualification testing. Record all information for the calibrated instruments used to control the blender in Table 2.9.

### *Equipment Control Functions*

The objective of testing equipment control functions is to verify that the switches and push buttons on the blender operate according to manufacturers' specifications. Operate each control listed in Table 2.10 and verify its proper operation. Record the results in Table 2.10. The tests will be performed with the blender empty.

### *Equipment Operation: Blender and Intensifier Bar Rotation Direction Test*

The objective of the rotation direction test is to verify that the blender and intensifier bar rotate in the proper direction. The following are tests performed with the blender empty.

- Press the blender start push button, observe the direction of rotation of the blender as viewed from the drive side of the blender, and record the results in Table 2.11.

**Table 2.9. Calibrated and Noncalibrated Instruments**

*There were no noncalibrated instruments on the blender.*

| Calibrated Instrument | As-Found Conditions |
|---|---|
| Identification number | *AC23-14* |
| Type | *Timer, digital* |
| Manufacturer | *Any Gauge Co.* |
| Model number | *BL3409* |
| Serial number | *N033875* |
| Range | *0–999.9 min* |
| Scale division | *0–6 sec* |
| Location | *Mounted on the blender* |
| Use | *Sets the blending time* |
| Calibration due date | *07/08/97* |
| Critical or not critical | *Critical to the operation of the blender* |

**Table 2.10. *Equipment Control Function Test Results***

| Test Function | Expected Results | Acceptable (Yes/No) |
|---|---|---|
| Emergency stop push button | When the emergency stop push button is pressed (when the blender is running), the blender and intensifier bar stop operating. | *Yes* |
| Start push button | When the start push button is pressed (when the selector switch is in the run mode), the blender starts operating. | *Yes* |
| Start push-button operation in the jog mode | When the start push button is pressed (when the selector switch is in the jog mode), the blender operates while the push button is depressed. Blender rotation stops when the button is released. | *Yes* |
| Intensifier bar on push button | When the intensifier bar on push button is pressed (when the blender is operating), the intensifier bar starts. | *Yes* |
| Intensifier bar off push button | When the intensifier bar off push button is pressed, the intensifier bar stops. | *Yes* |
| Timer operation | When the timer is set to 5 min and the intensifier timer is set to 2 min, the timer counts down to 0:00 and the blender stops. | *Yes* |

- Press the intensifier bar on push button, observe the direction of rotation of the intensifier bar as viewed from the drive side of the blender, and record the results in Table 2.11.

### *Equipment Operation: Blender and Blender Tip Speed Test*

The objective of the tip speed test is to document the speed of the blender and its tip speed with the blender empty. The materials and instruments required for this test are a stopwatch and a tape measure. Perform the tip speed test as follows:

**Table 2.11. *Blender and Intensifier Bar Rotation Direction***

| Item | Expected Results | Results | Acceptable (Yes/No) |
|---|---|---|---|
| Blender rotation direction | Rotation should be clockwise as viewed from the drive side of the blender. | *Clockwise rotation was observed.* | *Yes* |
| Intensifier bar rotation direction | Rotation should be counter-clockwise as viewed from the drive side of the blender. | *Counterclockwise rotation was observed.* | *Yes* |

- Calculate the speed of the blender and verify that the calculated speed is within ±10% of the fixed speed of 24.6 rpm; then record the results in Table 2.12.
- Count the number of revolutions that the blender makes in 5 min. Use a calibrated stopwatch to measure the time and record the instrument used in Table 2.13.

  Number of revolutions = 125

  Number of minutes = 5

  Blender speed = number of revolutions/minutes = 125/5 = 25 rpm

  Blender speed specification = 24.6 rpm ± 10%

  ±10% of 24.6 = ±2.46

  24.6 + 2.46 = 27.06

  24.6 – 2.46 = 22.14

  The calculated speed of 25 rpm falls within ±10%.

- Calculate the blender's tip speed, then record the results in Table 2.12. Record the instrument used in Table 2.13.

  Approximate distance from the center of the shaft to the bottom of the blender (radius, $r$) is 25 in.

  Circumference of a circle is $2\pi r$

  2(3.1459)(25 in.) = 157 in. per revolution

  Tip speed = inches per revolution blender speed in rpm ÷ 12 = 157(25)/12 = 327 ft/min

**Table 2.12. Blender Speed Test Results**

| Item | Specification | Calculated Speed | Acceptable (Yes/No) |
|---|---|---|---|
| Blender speed | *24.6 rpm ± 10%* | *25 rpm* | *Yes* |
| Tip speed | *Information only* | *327 ft/min* | *Yes* |

**Table 2.13. Instruments Used**

| Test Instrument | Identification Number | Calibration Due Date |
|---|---|---|
| *Stopwatch* | *54039* | *02/16/97* |
| *Tape Measure* | *None* | *None* |

### *Equipment Operation: Intensifier Bar Speed Test*

The objective of the intensifier bar speed test is to document the speed of the intensifier bar with the blender empty. A tachometer is required for this test. Measure the speed of the intensifier bar with a calibrated tachometer. Verify that the measured speed is within ±10 percent of the fixed speed of 1,508 rpm; then record the results in Table 2.14. Record the instrument used to measure the speed in Table 2.15.

Intensifier bar speed specification = 1,508 rpm ± 10%

±10% of 1,508 = ±150.8

1,508 + 150.8 = 1,658.8

1,508 – 150.8 = 1,357.2

The measured rpm of 1,510 falls within ±10%.

## Performance Qualification

Once it has been established that the equipment is properly installed and functioning within specified operating parameters, it must be shown that the blender can operate reliably under routine, minimum, and maximum operating conditions. (Figure 2.5).

### *Blender Operation*

The blender operation test is designed to document the performance of the blender using a placebo and a colored dye. The placebo is used for maximum loading conditions. The speed of the blender will be calculated, and the speed of the intensifier bar will be measured and recorded.

The materials and instruments required for this test are a placebo, a dye, a scale, a stopwatch, and a tachometer. Use the following procedure to test blender operation:

***Table 2.14. Intensifier Bar Speed Test Results***

| Item | Specification (rpm) | Measured Speed (rpm) | Acceptable (Yes/No) |
|---|---|---|---|
| Intensifier bar speed | 1,508 ± 10% | *1,510* | *Yes* |

***Table 2.15. Instrument Used***

| Test Instrument | Identification Number | Calibration Due Date |
|---|---|---|
| *Tachometer* | *64020* | *06/21/97* |

**Figure 2.5. PQ Elements of a V-Shell Blender**

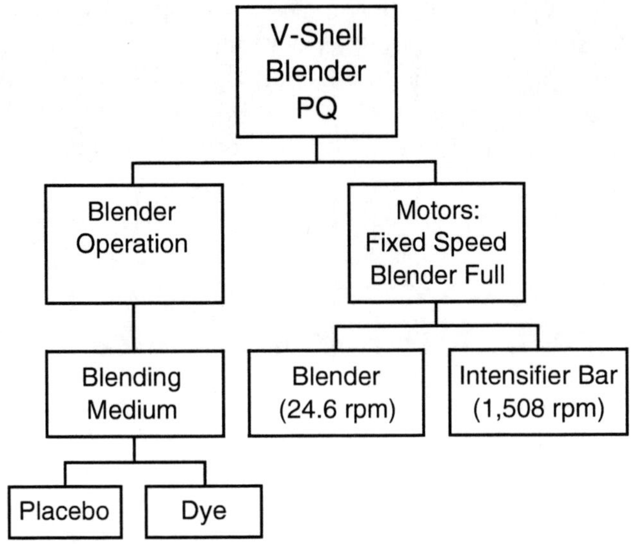


- Fill the blender with placebo to its maximum working volume of 142 L and record the amount used in Table 2.16.

  Total blender volume = 218 L

  Blending material volume = 65% of 218 = 142 L

- Record the weight of the placebo used in Table 2.16. Record the instrument used to weigh the placebo in Table 2.18.
- Add FDC blue #1 dye powder to the blender and record the weight of the dye used in Table 2.16. Record the instrument used to weigh the dye in Table 2.18.

**Table 2.16. Test Materials and Conditions**

| Item | Results |
|---|---|
| *Lactose hydrous* | *92.56 kg* |
| *Microcrystalline cellulose 102* | *0.40 kg* |
| *Magnesium stearate* | *1.04 kg* |
| *Weight of placebo* | *104.00 kg* |
| *FDC blue #1 dye* | *0.05 kg* |
| *Total weight blended* | *104.05 kg* |
| *Volume of blended material* | *142 L* |
| *Total blending time* | *5 min* |

- Blend the placebo for 5 min with the intensifier bar rotating and verify that the added dye is distributed uniformly throughout the container. Record the results in Table 2.17. Record the total blending time in Table 2.16. Record the instrument used to measure the time in Table 2.18.
- Calculate the speed of the blender and verify that the calculated speed is within ±10 percent of the fixed of 24.6 rpm, then record the results in Table 2.17.

  Count the number of revolutions that the blender makes in 5 min. Use a calibrated stopwatch to measure the time and record the instrument used in Table 2.18.

  Number of revolutions = 120

  Number of minutes = 5

  Blender speed = number of revolutions/minutes = 120/5 = 24 rpm

  Blender speed specification = 24.6 rpm ± 10%

  ±10% of 24.6 = ±2.46

  24.6 + 2.46 = 27.06

  24.6 – 2.46 = 22.14

  The calculated rpm of 24 falls within ±10%.

- Measure the speed of the intensifier bar with a calibrated tachometer. Verify that the measured speed is within ±10 percent of the fixed speed of 1,508 rpm;

**Table 2.17. Blender Performance Test Results**

| Test Function | Expected Results | Results | Acceptable (Yes/No) |
|---|---|---|---|
| Powder blending capabilities | The added dye should be distributed uniformly throughout the container. | *The dye was distributed uniformly throughout the container.** | *Yes* |
| Blender speed | 24.6 rpm ± 10% | *24 rpm* | *Yes* |
| Intensifier bar speed | 1,508 rpm ± 10% | *1,495 rpm* | *Yes* |

*The results of this test are qualitative only and are based on observation.

**Table 2.18. Instruments Used**

| Test Instrument | Identification Number | Calibration Due Date |
|---|---|---|
| *Scale* | *PE60* | *7/18/97* |
| *Stopwatch* | *54039* | *02/16/97* |
| *Tachometer* | *64020* | *06/21/97* |

then record the results in Table 2.17. Record the instrument used to measure the speed in Table 2.18.

Intensifier bar speed specification = 1,508 rpm ± 10%

±10% of 1,508 = ±150.8

1,508 + 150.8 = 1,658.8

1,508 – 150.8 = 1,357.2

The measured rpm of 1,495 falls within ±10%.

## Emulsifying Mixer

The validation of an emulsifying mixer (Figure 2.6), which is motor driven and operates at a fixed speed, will be described here. All mixing operations are operator controlled with respect to duration and endpoint determination. The mixer provides high-shear agitation to moderately low-viscosity fluids. As viscosity and/or density increases, the volume that the mixer will be able to handle effectively will decrease. The mixer is used to prepare solutions for use in a wet granulation process. The mixer cannot be operated without the impeller being submerged in water or another representative fluid or powder.

The validation protocol for an emulsifying mixer is described in the following sections. This protocol example can be used as a cGMP compliance guide, and it should be tailored to meet individual company requirements. Italicized words in the tables represent examples of information that could be recorded in a validation protocol for an emulsifying mixer.

### Installation Qualification

An IQ evaluation establishes confidence that the equipment is properly installed (Figure 2.7). The installation must meet the manufacturer's specified guidelines, along with design changes at installation. Also, the supporting electrical utilities must meet all electrical codes. The information required for an IQ evaluation is equipment identification, required documentation, equipment utility requirements, major component specifications, component material, lubricants, and equipment safety features.

#### *Equipment Identification*

Record the equipment identification numbers in Table 2.19, along with the following information: equipment manufacturer's purchase order number model number, serial number, company assigned equipment number, and the location of the equipment.

***Figure 2.6. Emulsifying Mixer (courtesy of Charles Ross & Son Company)***

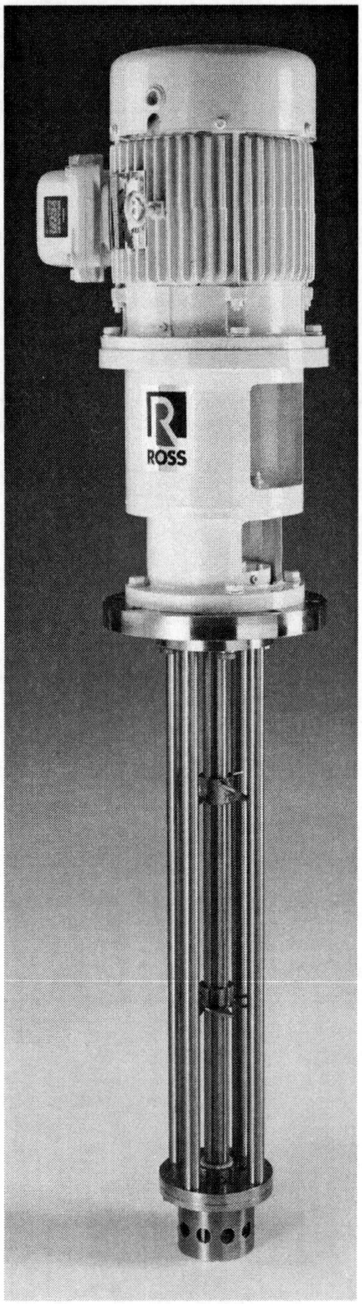


### *Required Documentation*

Record the equipment manufacturer's operation and maintenance manual and drawings in Table 2.20. Record the SOPs that cover the setup, operation, and cleaning of the mixer in Table 2.21.

**Figure 2.7. *IQ Elements of an Emulsifying Mixer***

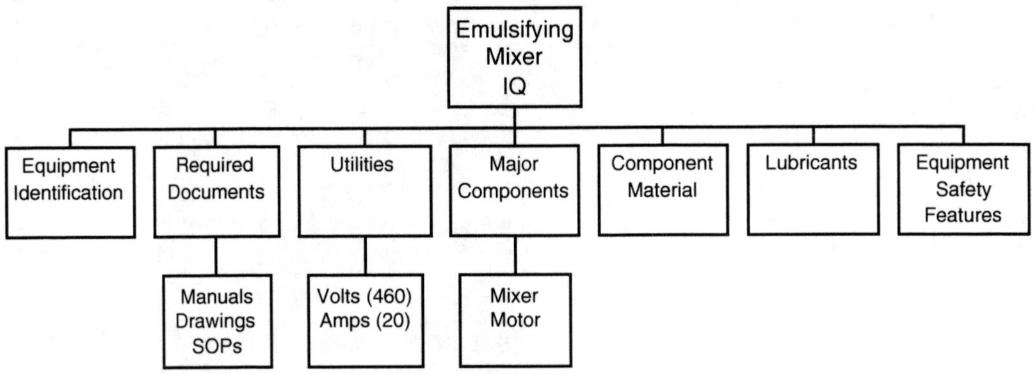


**Table 2.19. *Equipment Identification***

| Required Information | As-Found Conditions |
|---|---|
| Manufacturer | *Any Mixer Co.* |
| Purchase order number | *004482* |
| Model number | *Series 001, Model MC 501* |
| Serial number | *55356 B* |
| Equipment number | *2052* |
| Location | *Room 802 (This equipment is* portable.) |

**Table 2.20. *Manufacturers' Manuals and Drawings***

| Number | Description | Date |
|---|---|---|
| *None* | *Installation Operation Instructions* | *None* |
| *D-1346* | *Standard Production Mixer Emulsifier* | *1/4/83* |

**Table 2.21. *Standard Operating Procedures***

| Number | Description | Release Date |
|---|---|---|
| *GRA025* | *Granulation Equipment Setup* | *10/23/92* |
| *GRA026* | *Granulation Department Equipment Cleaning Procedure* | *03/12/96* |

### *Equipment Utility Requirements*

Compare the manufacturer's specified volt (V) and amp (A) requirements to their as-found condition at the time of qualification testing and record the results in Table 2.22. Also record the location of the power supply source. Record the instrument used to measure the volts and amps in Table 2.23.

***Volt calculation:***

Volt specification = 460 V ± 10%

±10% of 460 = ±46

460 + 46 = 506

460 – 46 = 414

The measured volts of 461/468/466 fall within ±10%.

***Amp calculation:***

Circuit rating = 20 A

Equipment current draw = 12.2 A

The circuit amp rating of 20 is greater than the maximum current draw of the equipment.

### *Major Component Specifications*

The component specifications section of the protocol verifies that the mixer components purchased were delivered and installed. Record the major components in Table 2.24.

### *Component Material*

Record the material of each component that contacts the product in Table 2.25.

### *Lubricants*

Record the lubricants used to operate the mixer in Table 2.26 and indicate if they make contact with the product.

***Table 2.22. Utilities***

| Utility | Specified | Measured Results | Acceptable (Yes/No) |
|---|---|---|---|
| Volts | 460 ± 10% | *A-B 461 A-C 468 B-C 466* | *Yes* |
| Amps | Motor = 12.2 | 20 circuit rating | *Yes* |

Power supply source, breaker box BB1, *wire numbers: 31, 33, 35*

***Table 2.23. Instrument Used***

| Test Instrument | Identification Number | Calibration Due Date |
|---|---|---|
| *Multimeter* | *ME-025* | *04/19/97* |

**Table 2.24. Major Components**

| Components | As-Found Conditions |
|---|---|
| Mixer motor | Manufacturer: *Any Motor Co.*<br>Model number: *ME 501*<br>Serial number: *55356 C*<br>Volts: *460*<br>Amperes: *12.2*<br>Phases: *3*<br>Cycles: *60 Hz*<br>hp: *5*<br>rpm: *3,480* |
| Shaft | Part number: *45267*<br>Size: *1/2 in. diameter by 36 in. long*<br>Material: *316 stainless steel* |
| Impeller | Part number: *45268*<br>Size: *1/2 in. bore by 6 in. diameter*<br>Material: *316 stainless steel* |

**Table 2.25. Component Material**

| Component | Material |
|---|---|
| Shaft | 316 stainless steel |
| Impeller | 316 stainless steel |
| Rotor shaft bushing | Teflon® |

**Table 2.26. Lubricants**

| Where Used | Type | Manufacturer | Product Contact (Yes/No) |
|---|---|---|---|
| *Motor bearings* | *Lubriplate 630 AA* | *Any Oil Co.* | *No* |
| Is there a preventive maintenance procedure on file? (Yes/No) | | | *Yes* |

### *Equipment Safety Features*

The mixer is used in an explosive atmosphere. There are no safety features on this equipment. Nevertheless, there are some general safety precautions that should be followed when operating the mixer:

- Never touch a mixer, which has an electric motor, or any part of an electrical service line cord with wet hands or wet feet or if standing on a wet surface.
- Never attempt to move or adjust a mixer while it is running.
- Never touch any rotating part of a mixer with bare hands, gloved hands, or any handheld object. Rotating parts include, but are not limited to, the mixer shaft, impeller(s), mechanical seals, and motor fans.

- Do not touch a mixer motor until it cools. The motor temperature may be high enough to cause burns.

## Operational Qualification

An OQ evaluation should establish that the equipment can operate within specified tolerances and limits (Figure 2.8). The mechanical ranges of the mixer are challenged, along with the basic mixer operations. The mixer will be validated for its operating ability, not how well it mixes liquids or powders. The information required for the OQ evaluation is calibration of the instrument used to control the mixer, equipment control functions (switches and push buttons), and equipment operation (mixer rotation direction and mixer speed).

### *Calibration Requirements*

Verify that all critical instruments on the equipment have been logged into the calibration system, have calibration procedures in place, and are currently in calibration at the time of qualification testing. Record all information for the calibrated instruments used to control the mixer in Table 2.27.

### *Equipment Control Functions*

The objective of testing equipment control functions is to verify that the switches on the mixer operate per the manufacturer's specifications. The mixer will be

***Figure 2.8. OQ Elements of a Emulsifying Mixer***

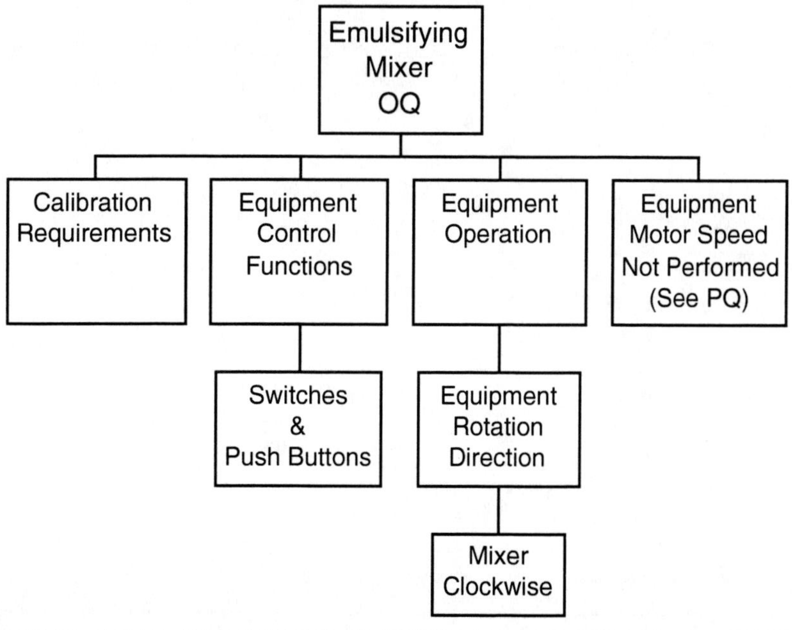

**Table 2.27. Calibrated and Noncalibrated Instruments**

| |
|---|
| *There were no calibrated or noncalibrated instruments on the blender.* |

operated with the impeller submerged in water. The two controls that need to be tested are the on switch and the off switch. The materials used in testing the switches are a mixing container and test fluid. The testing is performed as follows:

- Fill the mixing container with water to its maximum working volume of 103 L and record the amount used in Table 2.28. When operating the mixer, the head must be submerged in water to prevent damage to the mixer.

  Test fluid volume = 75% of 138 L = 103.5 L (rounded to 103 L)

- Press the on switch and verify that the mixer starts operating, then record the results in Table 2.29.
- Press the off switch and verify that the mixer stops operating, then record the results in Table 2.29.

### *Equipment Operation: Mixer Rotation Direction Test*

The objective of the mixer rotation direction test is to verify that the mixer motor rotates in the proper direction. The mixer will be operated with the impeller submerged in the same water as the Equipment Control Functions.

- Press the start key and observe the direction of rotation of the mixer motor as viewed from the top of the mixer; record the results in Table 2.30.

### *Equipment Operation: Mixer Speed Test*

The mixer speed test is not be performed during OQ because the mixer cannot be operated without the impeller being submerged in water or another representative fluid or powder. This test is performed in PQ.

**Table 2.28. Test Materials**

| Item | Specifications |
|---|---|
| Mixing container | *138 L* |
| Test fluid | *Water* |
| Test fluid volume | *103 L* |

**Table 2.29. *Control Function Test Results***

| Test Function | Expected Results | Acceptable (Yes/No) |
|---|---|---|
| Start switch operation | When the start switch is pressed, the mixer starts. | *Yes* |
| Stop switch operation | When the stop switch is pressed, the mixer stops. | *Yes* |

**Table 2.30. *Mixer Motor Rotation Direction Test Results***

| Item | Expected Results | Results | Acceptable (Yes/No) |
|---|---|---|---|
| Mixer motor rotation direction | Rotation should be clockwise as viewed from the top of the mixer. | *Clockwise rotation was observed.* | *Yes* |

## Performance Qualification

Once it has been established that the equipment is properly installed and functioning within specified operating parameters, it must be shown that the mixer can be operated reliably under routine, minimum, and maximum operating conditions (Figure 2.9).

### *Emulsifying Mixer Operation*

The emulsifying mixer operation test is designed to document the performance of the mixer using water and a colored dye. Water will be used for maximum loading conditions. This test is also intended to verify that the mixer can move the fluid about the container (i.e., mixing). The speed of the mixer will be measured and recorded.

To perform this test, a mixing container, a test fluid, a dye, and a tachometer is required. The following steps are performed.

- Fill the mixing container with water to its maximum working volume of 103 L and record the amount used in Table 2.31. When operating the mixer, the head must be submerged in water to prevent damage to the mixer.

  Test fluid volume = 75% of 138 L = 103.5 L (rounded to 103 L)

**Table 2.31. *Test Materials***

| Item | Specifications |
|---|---|
| Mixing container | *138 L* |
| Test fluid | *Water* |
| Dye | *FDC blue #1* |
| Test fluid volume | *103 L* |

**Figure 2.9. PQ Elements of an Emulsifying Mixer**

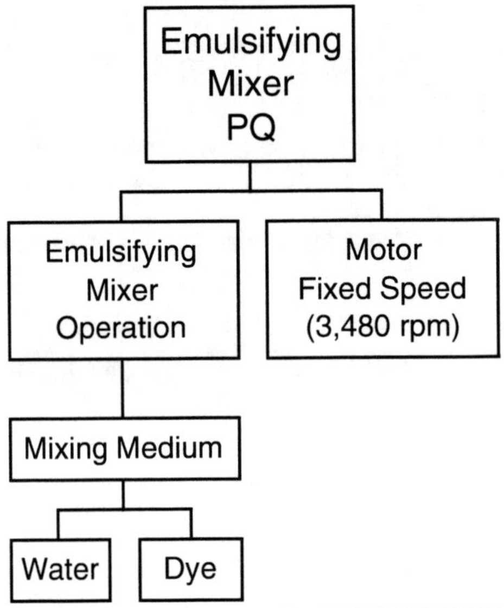


- Turn the mixer on and observe the motion of the fluid; record the results in Table 2.32. The results of this test are qualitative only and are based on observation.
- Add FDC blue #1 dye to the water, observe the mixing action of the mixer, and record the results in Table 2.32. Verify that the added dye is distributed uniformly throughout the container. Record the dye used in Table 2.31.
- Measure the speed of the mixer with a calibrated tachometer. Verify that the measured speed is within ±10 percent of the fixed speed of 3,480 rpm. Record the results in Table 2.33 and the instrument used to measure the speed in Table 2.34.

**Table 2.32. Mixer Performance Test Results**

| Item | Expected Results | Results* | Acceptable (Yes/No) |
|---|---|---|---|
| Fluid mixing capabilities | The mixer should move the fluid about the container. | *Fluid motion and vortex action was observed. There was increased motion with increased speed.* | *Yes* |
| Fluid mixing capabilities | The added dye should be distributed uniformly throughout the container. | *The dye was distributed uniformly throughout the container.* | *Yes* |

*The results of this test are qualitative only and are based on observation.

**Table 2.33. *Mixer Speed Test Results***

| Item | Specification (rpm) | Measured Speed (rpm) | Acceptable (Yes/No) |
|---|---|---|---|
| Mixer speed | 3,480 ± 10% | *3,521* | *Yes* |

**Table 2.34. *Instrument Used***

| Test Instrument | Identification Number | Calibration Due Date |
|---|---|---|
| *Tachometer* | *64020* | *06/21/97* |

Mixer speed specification = 3,480 rpm ± 10%

±10% of 3,480 = ±348

3,480 + 348 = 3,828

3,480 − 348 = 3,132

The measured rpm of 3,521 falls within ±10%.

## Shaker Mixer

The validation of a 55 L shaker mixer (Figure 2.10) will be discussed here. The ingredients are placed in a 55 L container that is secured in the shaker cage. The mixer is set for one of four fixed operating speeds. Changing the belt/pulley arrangement adjusts the mixer speeds. A switch lever manually controls the duration. The low-shear mixer is used to mix small batches of powders.

The validation protocol for a shaker mixer is described in the following sections. This protocol example can be used as a cGMP compliance guide, and it should be tailored to meet individual company requirements. Italicized words in the tables represent examples of information that could be recorded in a validation protocol for a shaker mixer.

## Installation Qualification

An IQ evaluation establishes confidence that the equipment is properly installed (Figure 2.11). The installation must meet the manufacturer's specified guidelines, along with design changes at installation. Also, the supporting electrical utilities must meet all electrical codes. The information required for an IQ evaluation is equipment identification, required documentation, equipment utility requirements, major component specifications, component material, lubricants, and equipment safety features.

**Figure 2.10. Turbula Shaker-Mixer (courtesy of Glen Mills, Inc.)**

### *Equipment Identification*

Record the equipment identification numbers in Table 2.35, along with the following information: equipment manufacturers' purchase order number, model number, serial number, company assigned equipment number, and the location of the equipment.

***Figure 2.11. IQ Elements of a Shaker Mixer***

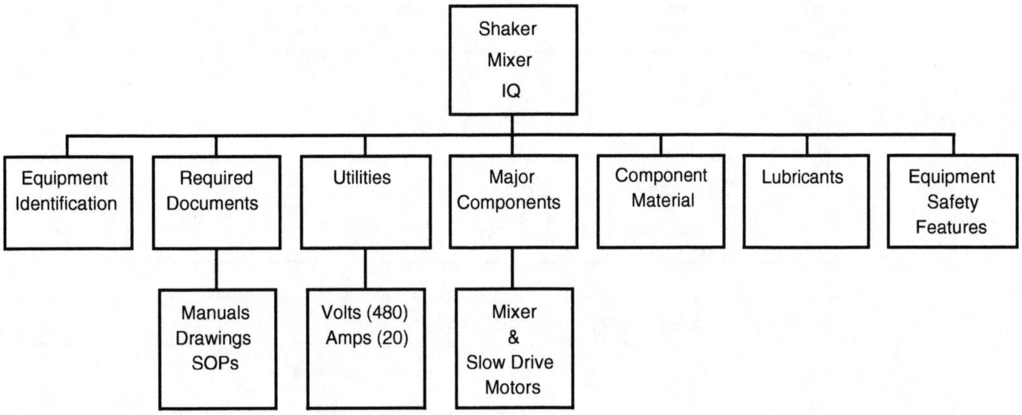


***Table 2.35. Equipment Identification***

| **Required Information** | **As-Found Conditions** |
|---|---|
| Manufacturer | *Any Mixer Co.* |
| Purchase order number | *007127* |
| Model number | *U 05 B* |
| Serial number | *412069* |
| Equipment number | *5133* |
| Location | *Room 659 (This equipment is portable.)* |

### *Required Documentation*

Record the equipment manufacturer's operation and maintenance manual and drawings in Table 2.36. Record the SOPs that cover the setup, operation, and cleaning of the mixer in Table 2.37.

### *Equipment Utility Requirements*

Compare the manufacturer's specified volts (V) and amps (A) requirements to their as-found conditions at the time of qualification testing and record the results in Table 2.38. Also record the location of the power supply source. Record the instrument used to measure the volts and amps in Table 2.39.

**Table 2.36. *Manufacturers' Manuals and Drawings***

| Number | Description | Date |
|---|---|---|
| *None* | *Shaker Mixer Operating Instructions for Model U 05 B* | *None* |
| *26800* | *Main Drive Drawing* | *04/26/95* |
| *12711* | *Basket Drawing* | *06/28/95* |
| *27800* | *Shaker Mixer Assembly Drawing* | *04/26/95* |
| *17800* | *Control System and Clutch Drawing* | *04/22/95* |

**Table 2.37. *Standard Operating Procedures***

| Number | Description | Release Date |
|---|---|---|
| *GRA021* | *Shaker Mixer Setup and Operation* | *10/23/92* |
| *GRA022* | *Shaker Mixer Cleaning Procedure* | *03/12/96* |

**Table 2.38. *Utilities***

| Utility | Specified | Measured Results | Acceptable (Yes/No) |
|---|---|---|---|
| Volts | Mixer motor = 480 ± 10% | *A-B 461; A-C 468 ; B-C 466* | *Yes* |
| Amps | Mixer motor = 3.55<br>Slow drive motor = 0.85 | 20 circuit rating | *Yes* |

Power supply source, breaker box *BB23B, wire numbers: 1, 3, 5*

**Table 2.39. *Instrument Used***

| Test Instrument | Identification Number | Calibration Due Date |
|---|---|---|
| *Multimeter* | *ME-025* | *04/19/97* |

***Volt calculation:***

Mixer motor volt specification = 480 V ± 10%

±10% of 480 = ±48

480 + 48 = 528

480 – 48 = 432

The measured voltage of 461/460/462 falls within ±10%.

***Amp calculation:***

Circuit rating = 20 A

Mixer motor = 3.55 A

Slow drive motor = 0.85 A

Equipment current draw = mixer motor A + slow drive motor A = 3.55 + 0.85 = 4.40 A

The circuit amp rating of 20 is greater than the maximum current draw of the equipment.

## *Major Component Specifications*

The component specifications of the protocol verifies that the mixer components purchased were delivered and installed. Record the major components in Table 2.40.

***Table 2.40.*** ***Major Components***

| Components | As-Found Conditions |
|---|---|
| Mixer motor | Manufacturer: *Any Motor Co.*<br>Model number: *A100LT8*<br>Serial number: *375211*<br>Volts: *480*<br>Amperes: *3.55*<br>Phases: *3*<br>Cycles: *60 Hz*<br>hp: *1*<br>rpm: *32 and 40* |
| Housing | Part number: *3*<br>Material: *Cast light alloy* |
| Protective cover | Part number: *102*<br>Material: *Plexiglas®* |
| Cage | Part number: *121*<br>Material: *Cast light alloy* |
| Removable collecting vessel | Part number: *104*<br>Material: *Cast light alloy* |
| Slow drive motor | Manufacturer: *Any Motor Co.*<br>Model number: *AM71KY40*<br>Serial number: *465157*<br>Volts: *480*<br>Amperes: *0.85*<br>Phases: *3*<br>Cycles: *60 Hz*<br>hp: *1/4*<br>rpm: *1,640* |

### *Component Material*

Record the material of each component that contacts the product in Table 2.41.

### *Lubricants*

Record the lubricants used to operate the mixer in Table 2.42 and state whether it makes contact with the product.

### *Equipment Safety Features*

The objective of testing equipment safety procedures is to verify that the safety features on the mixer function according to the manufacturer's specifications. This test will be performed with the mixer empty. Record all results in Table 2.43. The test procedure is as follows:

**Table 2.41. *Component Material***

| Component | Material |
|---|---|
| Mixing vessel | *316 stainless steel* |

**Table 2.42. *Lubricants***

| Where Used | Type | Manufacturer | Product Contact (Yes/No) |
|---|---|---|---|
| *Roller chains* | *Spara Flex 517 Grease* | *Any Oil Co.* | *No* |
| Is there a preventive maintenance procedure on file? (Yes/No) | | | *Yes* |

**Table 2.43. *Safety Features Test Results***

| Test Function | Expected Results | Acceptable (Yes/No) |
|---|---|---|
| Protective cover solenoid operation | When the mixer is operating, attempt to raise the hood: the hood cannot be raised. | *Yes* |
| | When the switch lever is put into position 3, the hood can be raised. | *Yes* |
| | When the hood is raised, attempt to start the mixer, the mixer: will not start. | *Yes* |
| Motor housing cover solenoid operation | When the mixer is switched on, with the motor housing removed, the mixer will not start. | *Yes* |
| | When the mixer is still on, replace the motor housing cover, and the mixer starts operating. | *Yes* |

- Lower the protective hood and switch on the mixer. Attempt to raise the hood while the mixer is operating and verify that the hood cannot be raised.
- Put the switch lever into position 3 and verify that the hood can be raised.
- Attempt to start the mixer when the hood is raised and verify that the mixer will not start.
- Remove the motor housing cover. Switch on the mixer and verify that the mixer will not start.
- Replace the motor housing cover while the mixer is still on and verify that the mixer starts operating.

## Operational Qualification

An OQ evaluation should establish that the equipment can operate within specified tolerances and limits (Figure 2.12). The mechanical ranges of the mixer are challenged, along with the basic mixer operations. The mixer will be validated for its operating ability, not how well it mixes powders. Information required for the OQ evaluation is calibration of the instruments used to control the mixer, equipment control functions (switches and push buttons), and equipment operation (mixer and slow drive motor rotation direction and mixer speed).

### *Calibration Requirements*

Verify that all critical instruments on the equipment have been logged into the calibration system, have calibration procedures in place, and are in calibration at the time of qualification testing. Record all information for the calibrated instruments used to control the mixer in Table 2.44.

### *Equipment Control Functions*

The objective of testing equipment control functions is to verify that the switches on the mixer operate per the manufacturer's specifications. The tests will be performed with the mixer empty. The controls that need testing are the power switch and switch lever. Use the following procedure to test the controls. Record all results in Table 2.45.

- Pull the power switch out until it locks in place. Move the switch lever to position 1 and verify that the mixer motor is running and the protection hood cannot be raised.
- Move the switch lever to position 0 and verify that the mixer is off and the protection hood cannot be raised.
- Move the switch lever to position 2 and verify that the slow drive motor is on, the basket moves into the charging position, and the protection hood cannot be raised.
- Move the switch lever to position 3 and verify that the machine is at a stand still with the basket set in position and the protection hood can be raised.

**Figure 2.12. OQ Elements of a Shaker Mixer**

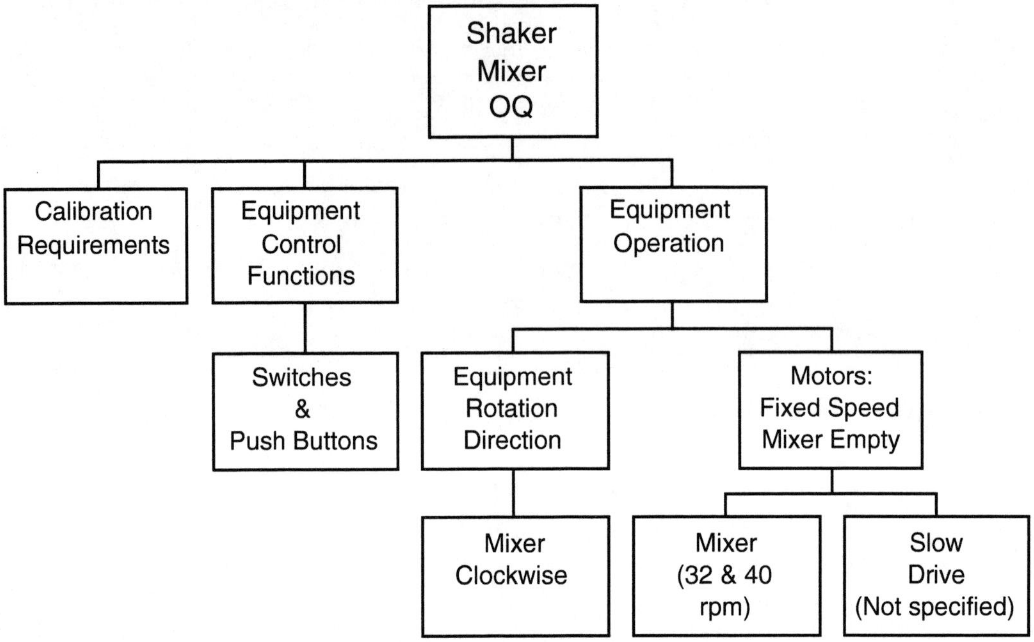


**Table 2.44. Calibrated and Noncalibrated Instruments**

*There were no calibrated or noncalibrated instruments on the blender.*

**Table 2.45. Control Function Test Results**

| Test Function | Expected Results | Acceptable (Yes/No) |
|---|---|---|
| Power switch and switch level position 1 operation | When the switch lever is in position 1 and the main motor is running, the protection hood cannot be raised. | *Yes* |
| Switch lever position 0 operation | When the switch lever is in position 0, the mixer is off, and the protection hood cannot be raised. | *Yes* |
| Switch lever position 2 operation | When the switch lever is in position 2 and the slow drive motor is on, the basket moves into the charging position, and the protection hood cannot be raised. | *Yes* |
| Switch lever position 3 operation | When the switch lever is in position 3 and the machine is at a stand still with the basket set in position, the protection hood can be raised. | *Yes* |

### *Equipment Operation: Mixer Rotation Direction Test*

The objective of the rotation direction test is to verify that the mixer rotates in the proper direction. The tests will be performed with the mixer empty. Set the mixer at speed I and observe the direction of rotation of the mixer as viewed from the front of the mixer. Record the results in Table 2.46.

### *Equipment Operation: Mixer Speed*

The objective of testing the mixer speed is to document the speed of the mixer with the mixer empty. A stopwatch will be required for the test.

The motion of the cage is not circular; therefore, use the knob on the cage as a spotting point and a line on the housing as a reference. Count the number of times that the knob passes the line to obtain the number of revolutions in 1 min. Record the number of revolutions below. Record the instrument used to measure the time in Table 2.47. Verify that the calculated speeds are within ±10 percent of the fixed speeds of 32 rpm minimum, setting I and 40 rpm maximum, setting IV. Record the results in Table 2.48.

***Mixer minimum speed calculation:***

Number of revolutions in 1 min = 34.5

Mixer minimum speed specification = 32 rpm ± 10%

$\pm 10\%$ of $32 = \pm 3.2$

$32 + 3.2 = 35.2$

**Table 2.46. *Mixer Motor Rotation Direction Test Results***

| Item | Expected Results | Results | Acceptable (Yes/No) |
|---|---|---|---|
| Mixer rotation direction | Rotation should be clockwise as viewed from the front of the mixer. | *Clockwise rotation was observed.* | *Yes* |

**Table 2.47. *Instrument Used***

| Test Instrument | Identification Number | Calibration Due Date |
|---|---|---|
| *Stopwatch* | *54039* | *02/16/97* |

**Table 2.48. *Mixer Speed Test Results***

| Item | Specification (rpm) | Measured Speed (rpm) | Acceptable (Yes/No) |
|---|---|---|---|
| Speed setting I | 32 ± 10% | *34.5* | *Yes* |
| Speed setting IV | 40 ± 10% | *40.5* | *Yes* |

32 – 3.2 = 28.8

The calculated rpm of 34.5 falls within ±10%.

***Mixer maximum speed calculation:***

Number of revolutions in 1 min = 40.5

Mixer maximum speed specification = 40 rpm ± 10%

±10% of 40 = ±4

40 + 4 = 44

40 – 4 = 36

The calculated rpm of 40.5 falls within ±10%.

## Performance Qualification

Once it has been established that the equipment is properly installed and functioning within specified operating parameters, it must be shown that the mixer can operate reliably under routine, minimum, and maximum operating conditions (Figure 2.13).

### *Shaker Mixer Operation*

The shaker mixer operation test is designed to document the performance of the mixer using a placebo and a colored dye. The placebo is used for maximum loading conditions. The two fixed speeds of the mixer will be calculated and recorded.

The materials and instruments required for this test are a mixing container, a placebo, a dye, a scale, and a stopwatch. Use the following procedure to test shaker mixer operation:

- Fill the mixer with placebo to its maximum working volume of 48 L and record the amount used in Table 2.49. Record the placebo blend used in Table 2.50.

  Total blender volume = 55 L

  Blending material volume = 85% of 55 = 47.75 (rounded to 48 L)

- Record the material, weight, and volume of the placebo used for each speed in Table 2.49. Record the instrument used to weigh the placebo in Table 2.52.
- Add FDC blue #1 dye powder and record the weight of the dye for each speed in Table 2.49. Record the instrument used to weigh the dye in Table 2.52.
- Blend the placebo for 5 min (at speed I and IV) and verify that the added dye is distributed uniformly throughout the container. Record the results in Table 2.51 Record the total mixing time in Table 2.50. Record the instruments used to measure the time in Table 2.52.
- The motion of the cage is not circular; therefore, use the knob on the cage as a spotting point and a line on the housing as a reference. Count the number

**Figure 2.13. *PQ Elements of a Shaker Mixer***

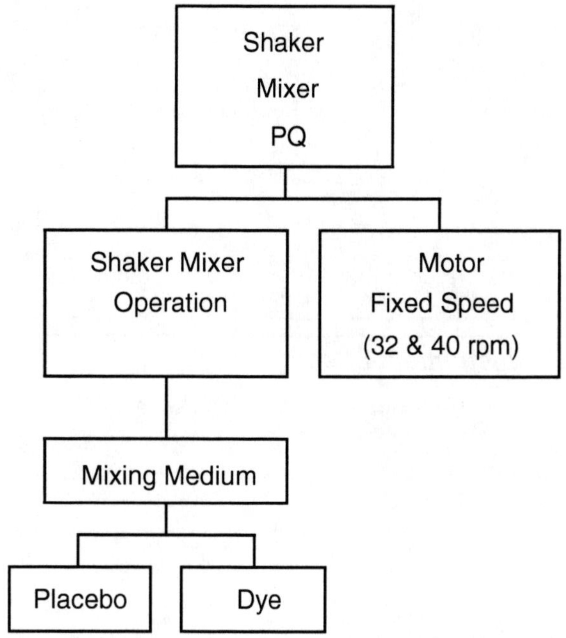


**Table 2.49. *Test Materials***

| Item | Specifications |
|---|---|
| Materials blended | *Lactose hydrous 316, microcrystalline cellulose* |
| Weight of placebo | *27.76 kg for speed I; 27 kg for speed IV* |
| Volume of material | *48 L* |
| Weight of dye | *0.01 kg of FDC Blue #1* |
| Mixing time for each test | *5 min* |

**Table 2.50. *Placebo Blend***

| Materials Used | Weight for Speed I | Weight for Speed IV |
|---|---|---|
| Lactose hydrous 316 | *26.2 kg* | *27.1 kg* |
| Microcrystalline cellulose 102 | *11.6 kg* | *10.4 kg* |
| FDC blue #1 dye | *0.01 kg* | *0.01 kg* |
| Mixing container | *8.5 kg* | *8.5 kg* |
| Total weight | *46.31 kg* | *46.01 kg* |

of times that the knob passes the line to obtain the number of revolutions in 1 min. Record the number of revolutions below. Verify that the calculated speeds are within ±10 percent of the fixed speeds of 32 rpm minimum,

**Table 2.51. *Mixer Performance Test Results***

| Item | Expected Results | Results* | Acceptable (Yes/No) |
|---|---|---|---|
| Powder mixing capabilities | The added dye should be distributed uniformly throughout the container. | *The dye was distributed uniformly throughout the container.* | *Yes* |
| Speed setting I | 32 rpm ± 10% | *33 rpm* | *Yes* |
| Speed setting IV | 40 rpm ± 10% | *39 rpm* | *Yes* |

*The results of this test are qualitative only and are based on observation.

**Table 2.52. *Instruments Used***

| Test Instrument | Identification Number | Calibration Due Date |
|---|---|---|
| *Scale* | *PE60* | *7/18/97* |
| *Stopwatch* | *54039* | *02/16/97* |

setting I and 40 rpm maximum, setting IV. Record the results in Table 2.51 Record the instrument used to measure the time in Table 2.52.

Mixer minimum speed calculation:

Number of revolutions = 33

Mixer minimum speed specification = 32 rpm ± 10%

±10% of 32 = ±3.2

32 + 3.2 = 35.2

32 – 3.2 = 28.8

The calculated rpm of 33 falls within ±10%.

Mixer maximum speed calculation:

Number of revolutions = 39

Mixer maximum speed specification = 40 rpm ± 10%

±10% of 40 = ±4

40 + 4 = 44

40 – 4 = 36

The calculated rpm of 39 falls within ±10%.

## MATRIX MIXER

The validation of a 150 L matrix mixer (Figure 2.14) will be discussed here. The high-shear mixer is used to mix or granulate pharmaceutical powders by adding a binder fluid to the powders while mixing. Powders are poured into the mixing bowl, and the impeller and granulator blades perform the mixing. The addition of a binder fluid performs the granulation of powders, while the impeller and granulator operate at specified slow and fast speeds. The mixer is motor driven, and a timer controls the mixing time.

The validation protocol for a matrix mixer is described in the following sections. This protocol example can be used as a cGMP compliance guide, and it should be tailored to meet individual company requirements. Italicized words in the tables represent examples of information that could be recorded in a validation protocol for a matrix mixer.

### Installation Qualification

An IQ evaluation establishes confidence that the equipment is properly installed (Figure 2.15). The installation must meet the manufacturer's specified guidelines,

***Figure 2.14. Matrix Mixer (courtesy of Niro Inc., Aeromatic-Fielder Division)***

**Figure 2.15. IQ Elements of a Matrix Mixer**

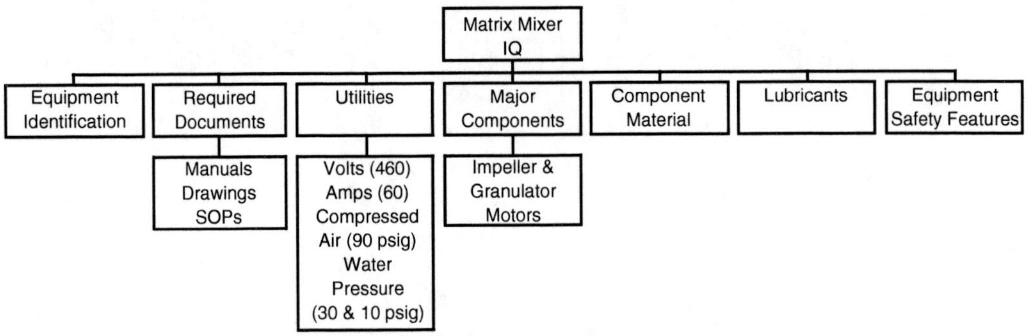


along with design changes at installation. Also, the supporting electrical utilities must meet all electrical codes. The information required for an IQ evaluation is equipment identification, required documentation, equipment utility requirements, major component specifications, component material, lubricants, and equipment safety features.

### *Equipment Identification*

Record the equipment identification numbers in Table 2.53, along with the following information: equipment manufacturer's purchase order number, model number, serial number, company assigned equipment number, and the location of the equipment.

### *Required Documentation*

Record the equipment manufacturer's operation and maintenance manual and drawings in Table 2.54. Record the SOPs that cover the setup, operation, and cleaning of the mixer in Table 2.55.

**Table 2.53. Equipment Identification**

| Required Information | As-Found Conditions |
|---|---|
| Manufacturer | *Any Mixer Co.* |
| Purchase order number | *0049436* |
| Model number | *MPA 015* |
| Serial number | *4318* |
| Equipment number | *0041* |
| Location | *Room 137* |

**Table 2.54. *Manufacturers' Manuals and Drawings***

| Number | Description | Date |
|---|---|---|
| *None* | *Installation, Operating, and Maintenance Instructions Manual* | *None* |
| *76709* | *System Wiring Diagram* | *08/19/92* |

**Table 2.55. *Standard Operating Procedures***

| Number | Description | Date |
|---|---|---|
| *GRA001* | *Matrix Mixer Setup and Operation* | *07/01/94* |
| *GRA002* | *Matrix Mixer Cleaning Procedure* | *08/11/94* |

### *Equipment Utility Requirements*

Compare the manufacturer's specified volts (V), amps (A), compressed air, and water pressure requirements to their as-found conditions at the time of qualification testing and record the results in Table 2.55. Also record the location of the power supply source. Record the instruments used to measure the volts, amps, compressed air, and water in Table 2.56.

**Table 2.56. *Utilities***

| Utility | Specified | Measured Results | Acceptable (Yes/No) |
|---|---|---|---|
| Volts | Impeller motor = 460 ± 10% | *A-B 461; A-C 468; B-C 466* | *Yes* |
| | Granulator motor = 460 ± 10% | *A-B 462; A-C 465; B-C 463* | *Yes* |
| Amps | Impeller motor = 16.6 | 60 circuit rating | *Yes* |
| | Granulator motor = 6.9 | | |
| Compressed air | ≥ 90 psig | *108 psig* | *Yes* |
| Water | Bowl jacket water pressure ″ 30 psig | *Bowl jacket is regulated to 30 psig* | *Yes* |
| | Washdown valve water pressure ″ 10 psig | *Washdown valves are regulated to 10 psig* | *Yes* |

Power supply source, breaker box *BB23B, wire numbers: 1, 3, 5 and 7, 9, 11*

**Table 2.57. Instruments Used**

| Test Instrument | Identification Number | Calibration Due Date |
|---|---|---|
| *Multimeter* | *ME-025* | *04/19/97* |
| *Air pressure gauge* | *PG-OPS-001* | *08/14/96* |
| *Water pressure gauge* | *PA-102* | *10/13/96* |

***Volt calculation:***

Impeller and granulator motors' volt specification = 460 V ± 10%

±10% of 460 = ±46

460 + 46 = 506

460 – 46 = 414

The measured impeller motor volts of 461/468/466 fall within ±10%.

The measured granulator motor volts of 462/465/463 fall within ±10%.

***Amp calculation:***

Circuit rating = 60 A

Impeller motor = 16.6 A

Granulator motor = 6.9 A

Equipment current draw = impeller motor A + granulator motor A = 16.6 + 6.9 = 23.5 A

The circuit amp rating of 60 is greater than the maximum current draw of the equipment.

***Compressed air calculation:***

Compressed air specification = ≥ 90 psig

The measured air pressure of 108 psig is greater than 90 psig.

***Water pressure calculation:***

Water supply specification: bowl jacket water pressure ″ 30 psig
Washdown valve water pressure ≥ 10 psig

Bowl jacket is regulated to 30 psig.

Washdown valves are regulated to 10 psig.

The measured water pressure is 62 psig (information only).

### *Major Component Specifications*

The component specifications section of the protocol verifies that the mixer components purchased were delivered and installed. Record the major components in Table 2.58.

### *Component Material*

Record the material of each component that contacts the product in Table 2.59.

### *Lubricants*

Record the lubricants used to operate the mixer in Table 2.60 and state whether it makes contact with the product.

**Table 2.58. Major Components**

| Components | As-Found Conditions |
|---|---|
| Mixing bowl | Volume: *150 L*<br>Material: *Stainless steel* |
| Impeller motor | Manufacturer: *Any Motor Co.*<br>Model number: *Not specified*<br>Serial number: *B365949*<br>Volts: *460*<br>Amperes: *16.6*<br>Phases: *3*<br>Cycles: *60 Hz*<br>hp: *15/10*<br>rpm: *130–260* |
| Solenoid valve, air supply | Part number: *M/43102/22*<br>Fail position: *Closed* |
| Discharge assembly | Material: *Stainless steel*<br>Size: *⊕4 in. diameter* |
| Granulator motor | Manufacturer: *Any Motor Co.*<br>Model number: *Not specified*<br>Serial number: *2659495*<br>Volts: *460*<br>Amperes: *6.9*<br>Phases: *3*<br>Cycles: *60 Hz*<br>hp: *5.5/4.2*<br>rpm: *1,740–3,460* |
| Liquid addition system (pressure pot) | Manufacture: *Any Pressure Pot Co.*<br>Model and serial number: *32932-001*<br>Material: *T316 L stainless steel*<br>Spray Nozzle: *4 ports*<br>Pressure rating: *140 psig*<br>Volume: *20 L* |

**Table 2.59. Component Material**

| Component | Material |
|---|---|
| Mixing bowl | *316 stainless steel* |
| Impeller | *Stainless steel* |
| Discharge assembly | *Stainless steel* |
| Granulator | *Stainless steel* |
| Pressure pot | *T316 L stainless steel* |

**Table 2.60. Lubricants**

| Where Used | Type | Manufacturer | Product Contact (Yes/No) |
|---|---|---|---|
| *Rotor bearings* | *Amoco Amolith #2* | *Any Oil Co.* | *No* |
| Is there a preventive maintenance procedure on file? (Yes/No) | | | *Yes* |

### *Equipment Safety Features*

The objective of testing equipment safety features is to verify that the safety features on the mixer function according to manufacturer specifications. Use the following procedure to test the equipment safety features. Record all results in Table 2.61.

**Table 2.61. Safety Features Test Results**

| Test Function | Expected Results | Acceptable (Yes/No) |
|---|---|---|
| Select speed II while the mixer is running in speed I. | The delay timer is activated | *Yes* |
| | The fast LED goes on. | *Yes* |
| | When the delay timer reaches zero, the mixer automatically changes to the fast speed | *Yes* |
| Select speed I while the mixer is running in speed II. | The motor is switched off. | *Yes* |
| | The delay timer is activated. | *Yes* |
| | The slow LED flashes. | *Yes* |
| | When the delay timer reaches zero, the mixer automatically changes to the slow speed | *Yes* |
| Turn airflow off and attempt to start the impeller. | The impeller will not start when the air pressure is below 58 psig. | *Yes* |
| Attempt to start the impeller at speed II. | The impeller cannot be started at speed II. | *Yes* |

- Select speed II (fast) while the mixer is running in speed I (slow) and verify that the delay timer is activated and the LED (light-emitting diode) goes on. When the delay timer reaches zero, the mixer automatically changes to the fast speed.
- Select speed I while the mixer is running in speed II and verify that the motor is switched off and the delay timer is activated. The slow LED flashes when the delay timer reaches zero, and the mixer automatically restarts in I.
- Attempt to start the impeller with the airflow off and verify that the impeller cannot be started.
- Attempt to start the impeller at the fast speed and verify that the impeller cannot be started.

## Operational Qualification

An OQ evaluation should establish that the equipment can operate within specified tolerances and limits (Figure 2.16). The mechanical ranges of the mixer are challenged, along with the basic mixer operations. The mixer will be validated for its operating ability, not how well it mixes powders. Information required for the OQ evaluation is calibration of the instruments used to control the mixer, equipment control functions (switches and push buttons), and equipment operation (mixer main and slow drive motor rotation direction and mixer speed).

### *Calibration Requirements*

Verify that all critical instruments on the equipment have been logged into the calibration system, have calibration procedures in place, and are in calibration at the time of qualification testing. Record all information for the calibrated instruments used to control the mixer in Table 2.62.

### *Equipment Control Functions*

The objective of testing the equipment control functions is to verify that the switches and push buttons on the mixer operate per the manufacturer's specifications. The tests will be performed with the mixer empty. Operate each control listed in Table 2.63 and verify their proper operation.

### *Equipment Operation: Impeller Blade and Granulator Blade Rotation Direction Test*

The objective of the blade rotation direction test is to verify that the mixer's impeller blade and granulator blade rotate in the proper direction. The tests will be performed with the mixer empty. Record the results in Table 2.64.

- With the mixer operating, verify that the impeller blades are rotating in a counterclockwise direction..
- With the mixer operating, verify that the granulator blades are rotating in a counterclockwise direction.

**Figure 2.16. OQ Elements of a Matrix Mixer**

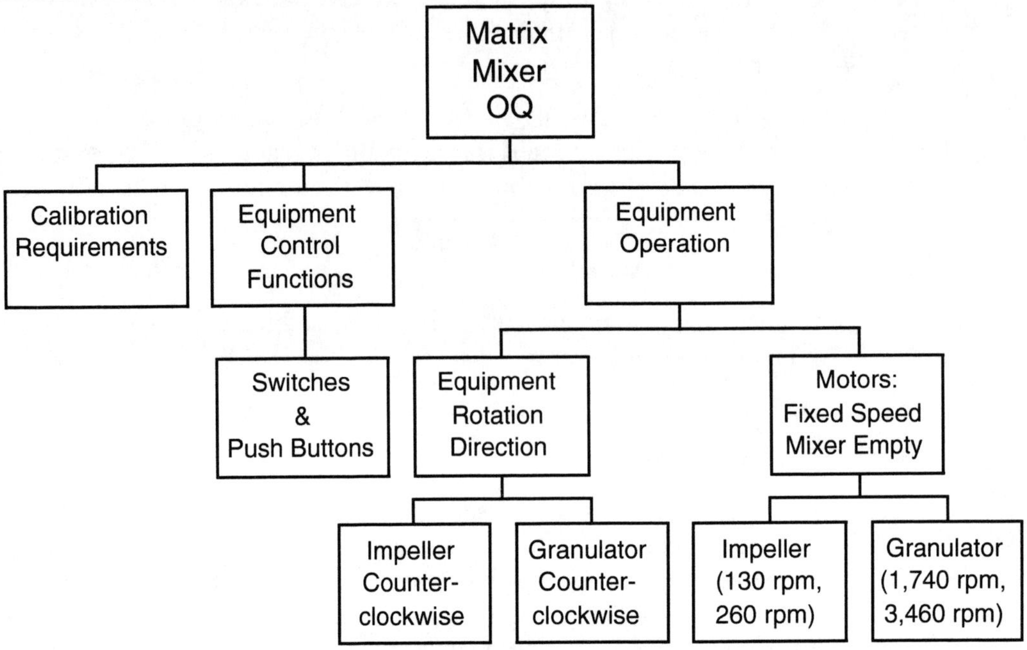


**Table 2.62. Calibrated and Noncalibrated Instruments**

| Calibrated Instrument | As-Found Conditions |
|---|---|
| Identification number | *PD0030* |
| Type | *Timer, analog* |
| Manufacturer | *Any Gauge Co.* |
| Model number | *54.032* |
| Serial number | *Not specified* |
| Range | *0–10 min* |
| Scale division | *0:12 sec* |
| Location | *Upper left side of the control panel* |
| Use | *Times and stops the mixer* |
| Calibration due date | *07/16/96* |
| Critical or not critical | *Not critical to the operation of the mixer* |

*continued on next page.*

*Table 2.62 continued*

| Calibrated Instrument | As-Found Conditions |
|---|---|
| Identification number | *PD0031* |
| Type | *Temperature gauge, analog* |
| Manufacturer | *Any Gauge Co.* |
| Model number | *JW72L* |
| Serial number | *Not specified* |
| Range | *0–100°C* |
| Scale division | *5°C* |
| Location | *Upper left side of the control panel* |
| Use | *Measures the temperature of the granulation* |
| Calibration due date | *07/16/96* |
| Critical or not critical | *Not critical to the operation of the mixer* |
| **Calibrated Instrument** | **As-Found Conditions** |
| Identification number | *PD237* |
| Type | *Air pressure gauge, analog* |
| Manufacturer | *Any Gauge Co.* |
| Model number | *Not specified* |
| Serial number | *Not specified* |
| Range | *0–23 lbf/in.*$^2$ |
| Scale division | *1 lbf/in.* |
| Location | *Right side of mixer* |
| Use | *Measures the impeller's seal pressure* |
| Calibration due date | *07/16/96* |
| Critical or not critical | *Critical to the operation of the mixer* |
| **Calibrated Instrument** | **As-Found Conditions** |
| Identification number | *PD238* |
| Type | *Air pressure gauge, analog* |
| Manufacturer | *Any Gauge Co.* |
| Model number | *Not specified* |
| Serial number | *Not specified* |
| Range | *0–23 lbf/in.*$^2$ |
| Scale division | *1 lbf/in.* |
| Location | *Right side of mixer* |
| Use | *Measures the impeller's seal pressure* |
| Calibration due date | *07/16/96* |
| Critical or not critical | *Critical to the operation of the mixer* |

*continued on next page.*

*Table 2.62 continued*

| Noncalibrated Instrument | As-Found Conditions |
|---|---|
| Identification number | *PD237* |
| Type | *Time recorder, mechanical* |
| Manufacturer | *Any Gauge Co.* |
| Model number | *Not specified* |
| Serial number | *Not specified* |
| Range | *0–99,999.99* |
| Scale division | *1 min* |
| Location | *On the outside of the main switch cabinet* |
| Use | *Records elapsed time of mixer use* |
| Calibration due date | *Calibration not required* |
| Critical or not critical | *Not critical to the operation of the mixer* |

**Table 2.63. Control Function Test Results**

| Test Function | Expected Results | Acceptable (Yes/No) |
|---|---|---|
| Process timer start/stop/reset push button | When the push button is pressed, the timer starts after a period has been set. | *Yes* |
| | When the push button is pressed two times, the timer restarts. | *Yes* |
| Mixer amp control dial operation | When the indicated current reaches the setpoint, the motor stops. | *Yes* |
| Discharge open button operation | When the button is pressed, the discharge door opens. | *Yes* |
| Discharge open indicator operation | When the door is opened, the indicator light goes on. | *Yes* |
| Discharge close button indicator operation | When the button is pressed, the discharge door closes. | *Yes* |
| Mixer slow button operation | When the button is pressed, the mixer operates at a slow speed. | *Yes* |
| Mixer fast button operation | When the button is pressed, the mixer operates at a fast speed. | *Yes* |
| Mixer stop button operation | When the button is pressed, the mixer stops. | *Yes* |
| Emergency stop button operation | When the button is pressed, the mixer stops. | *Yes* |
| Granulator slow/off/switch operation | When the switch is set to slow, the mixer runs at a slow speed. | *Yes* |
| | When the switch is set to off, the mixer stops. | *Yes* |
| | When the switch is set to fast, the mixer runs at a fast speed. | *Yes* |
| Washdown valve operation | When the valves are open, water enters the mixer. | *Yes* |

**Table 2.64. Impeller Blade and Granulator Blade Rotation Direction Test Results**

| Item | Expected Results | Results | Acceptable (Yes/No) |
|---|---|---|---|
| Impeller blade rotation direction | Rotation should be counter-clockwise as viewed through the inspection port. | *Counterclockwise rotation was observed.* | *Yes* |
| Granulator blade rotation direction | Rotation should be counter-clockwise as viewed through the inspection port. | *Counterclockwise rotation was observed.* | *Yes* |

### *Equipment Operation: Impeller Blade and Granulator Blade Speed Test*

The objective of the speed test is to document the speed of the impeller and granulator blades with the mixer empty. A tachometer will be required for this test. Record the results in Table 2.65. Record the instrument used to measure the speed in Table 2.66.

- Operate the mixer with the impeller and granulator set at the slow speed setting. Measure the speed of the impeller blades with a calibrated tachometer and verify that the measured speed is within ±10 percent of the fixed speed of 130 rpm.

  Impeller blade speed specification = 130 rpm ± 10%

  ±10% of 130 = ±13

  130 + 13 = 143

  130 –13 = 117

  The measured rpm of 134.9 falls within ±10%.

- Measure the speed of the granulator blades with a calibrated tachometer and verify that the measured speed is within ±10 percent of the fixed speed of 1,740 rpm.

**Table 2.65. Mixer Speed Test Results**

| Impeller Blade | | Granulator Blade | |
|---|---|---|---|
| Speed Setting | Measured Results (rpm) | Speed Setting | Measured Results (rpm) |
| I | 134.9 | I | 1,799.4 |
| I | 134.5 | II | 3,584.0 |
| II | 269.9 | I | 1,799.7 |
| II | 269.8 | II | 3,599.1 |

**Table 2.66.** ***Instrument Used***

| Test Instrument | Identification Number | Calibration Due Date |
|---|---|---|
| *Tachometer* | *64020* | *06/21/97* |

Granulator blade speed specification = 1,740 rpm ± 10%

±10% of 1,740 = ±174

1,740 + 174 = 1,914

1,740 – 174 = 1,566

The measured rpm of 1,799.4 falls within ±10%.

- Operate the mixer with the impeller set at speed I and the granulator set at speed II. Measure the speed of the impeller blades with a calibrated tachometer and verify that the measured speed is within ±10 percent of the fixed speed of 130 rpm.

Impeller blade speed specification = 130 rpm ± 10%

±10% of 130 = ±13

130 + 13 = 143

130 – 13 = 117

The measured rpm of 134.5 falls within ±10%.

- Measure the speed of the granulator blades with a calibrated tachometer and verify that the measured speed is within ±10 percent of the fixed speed of 3,460 rpm.

Granulator blade speed specification = 3,460 rpm ± 10%

±10% of 3,460 = ±346

3,460 + 346 = 3,806

3,460 – 346 = 3,114

The measured rpm of 3,584.0 falls within ±10%.

- Operate the mixer with the impeller set at speed II and the granulator set at speed I. Measure the speed of the impeller blades with a calibrated tachometer and verify that the measured speed is within ±10 percent of the fixed speed of 260 rpm.

Impeller blade speed specification = 260 rpm ± 10%

±10% of 260 = ±26

260 + 26 = 286

260 – 26 = 234

The measured rpm of 269.9 falls within ±10%.

- Measure the speed of the granulator blades with a calibrated tachometer and verify that the measured speed is within ±10 percent of the fixed speed of 1,740 rpm.

  Granulator blade speed specification = 1,740 rpm ± 10%

  ±10% of 1,740 = ±174

  1,740 + 174 = 1,914

  1,740 – 174 = 1,566

  The measured rpm of 1,799.7 falls within ±10%.

- Operate the mixer with the impeller and granulator set at speed II. Measure the speed of the impeller blades with a calibrated tachometer and verify that the measured speed is within ±10 percent of the fixed speed of 260 rpm.

  Impeller blade speed specification = 260 rpm ± 10%

  ±10% of 260 = ±26

  260 + 26 = 286

  260 – 26 = 234

  The measured rpm of 269.8 falls within ±10%.

- Measure the speed of the granulator blades with a calibrated tachometer and verify that the measured speed is within ±10 percent of the fixed speed of 3,460 rpm.

  Granulator blade speed specification = 3,460 rpm ± 10%

  ±10% of 3,460 = ±346

  3,460 + 346 = 3,806

  3,460 – 346 = 3,114

  The measured rpm of 3,599.1 falls within ±10%.

## Performance Qualification

Once it has been established that the equipment is properly installed and functioning within specified operating parameters, it must be shown that the mixer can perform reliably under routine, minimum, and maximum operating conditions (Figure 2.17).

### *Matrix Mixer Operation*

The objective of testing the operation of the matrix mixer is to document its performance with a placebo. The placebo is used for maximum loading conditions.

**Figure 2.17. PQ Elements of a Matrix Mixer**

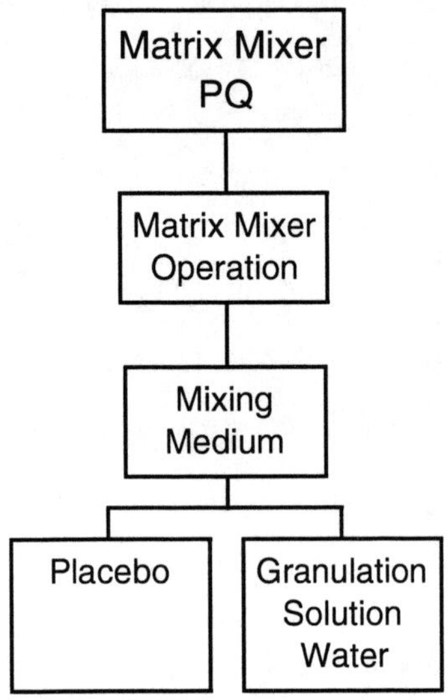


Four tests will be performed at various speed settings. The speed test cannot be performed because the impeller and granulator blades are covered by placebo.

The materials and instruments required for this test are a placebo, a pressure pot, a granulation solution, and a scale. Four separate tests will be performed according to the following procedures:

***Test 1:***

- Fill the mixer with placebo to its maximum working volume of 98 L and record the placebo used in Table 2.67. Record the weight of the placebo in Table 2.68. Record the instrument used to measure the weight in Table 2.69.

  Total mixer volume = 150 L

  Mixing material volume = 65% of 150 = 97.5 (rounded to 98 L)

- Set the impeller and granulator to speed I.
- Start mixing the placebo and add water until it is granulated. Record the results and mixing time in Table 2.68.

***Test 2:***

- Fill the mixer with placebo to its maximum working volume of 98 L and record the placebo used in Table 2.67. Record the weight of the placebo in Table 2.68. Record the instrument used to measure the weight in Table 2.69.
- Set the impeller to speed I and the granulator to speed II.

**Table 2.67. *Test Materials***

| Item | Specifications |
|---|---|
| Test placebo | *Microcrystalline cellulose and lactose* |
| Test fluid volume | *98 L* |
| Liquid addition system (pressure pot) | *20 L* |
| Granulation solution | *Water* |
| Graduated cylinder | *4,000 mL* |

**Table 2.68. *Mixer Performance Test Materials***

| Item | Test 1 | Test 2 | Test 3 | Test 4 |
|---|---|---|---|---|
| Impeller speed setting | I | I | II | II |
| Granulator speed setting | I | II | I | II |
| Placebo weight | *55 kg dry* | *55 kg dry* | *53.8 kg dry* | *53.8 kg dry* |
| Granulation solution volume | *23 L* | *23 L* | *23 L* | *23 L* |
| Granulation lumpy? | *No* | *No* | *No* | *No* |
| Mixing time from mixer's timer | *7 min* | *7 min* | *7 min* | *7 min* |

**Table 2.69. *Instrument Used***

| Test Instrument | Identification Number | Calibration Due Date |
|---|---|---|
| *Scale* | *PER600* | *07/18/97* |

- Start mixing the placebo and add water until it is granulated. Record the results in Table 2.68.

***Test 3:***

- Fill the mixer with placebo to its maximum working volume of 98 L and record the placebo in Table 2.67. Record the weight of the placebo in Table 2.68. Record the instrument used to measure the weight in Table 2.69.
- Set the impeller to speed II and the granulator to speed I.
- Start mixing the placebo and add water until it is granulated. Record the results and mixing time in Table 2.68.

***Test 4:***

- Fill the mixer with placebo to its maximum working volume of 98 L and record the placebo in Table 2.67. Record the weight of the placebo in Table 2.68. Record the instrument used to measure the weight in Table 2.69.

- Set the impeller to speed II and the granulator to the speed II.
- Start mixing the placebo and add water until it is granulated. Record the results and mixing time in Table 2.68.

## CHILLER

The validation of a chiller (Figure 2.18) that is used to cool the jacket water of a 150 L matrix mixer will be described here. Warm jacket water enters an evaporator in a direction counter to the flow of refrigerant. As the refrigerant boils, it picks up its heat of evaporation from the jacket water, which in turn drops in temperature and is supplied to cool the process. Vapor enters the compressor and adds heat of compression to the refrigerant before delivering the hot compressed gas to the

**Figure 2.18. Accuchiller (courtesy of Thermal Care)**

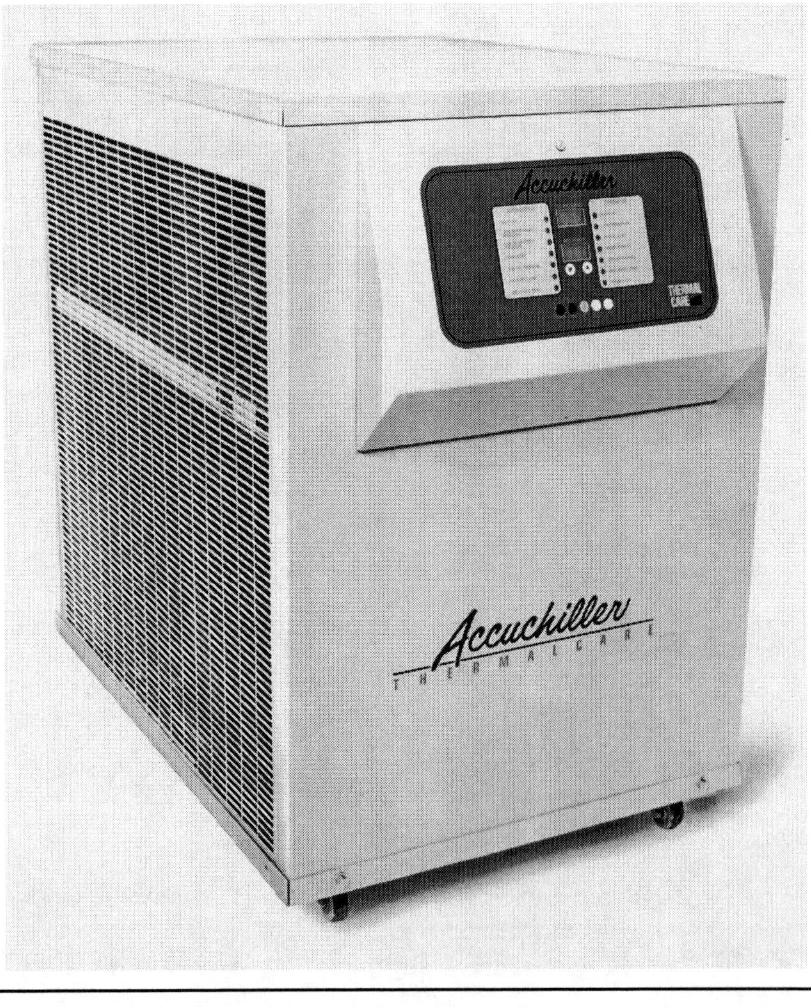

condenser. The refrigerant cools and liquefies in the condenser. After passing through the filter, an expansion valve allows the proper amount of refrigerant to enter the evaporator, where it picks up the process heat from the chiller water circuit again.

The validation protocol for a chiller is described in the following sections. This protocol example can be used as a cGMP compliance guide, and it should be tailored to meet individual company requirements. Italicized words in the tables represent examples of information that could be recorded in a validation protocol for a chiller.

## Installation Qualification

An IQ evaluation establishes confidence that the equipment is properly installed (Figure 2.19). The installation must meet the manufacturer's specified guidelines, along with design changes at installation. Also, the supporting electrical utilities must meet all electrical codes. The information required for an IQ evaluation is equipment identification, required documentation, equipment utility requirements, major component specifications, component material, lubricants, and equipment safety features.

### *Equipment Identification*

Record the equipment identification numbers in Table 2.70, along with the following information: equipment manufacturer's purchase order number, model number, serial number, company assigned equipment number, and the location of the equipment.

***Figure 2.19 IQ Elements of a Chiller***

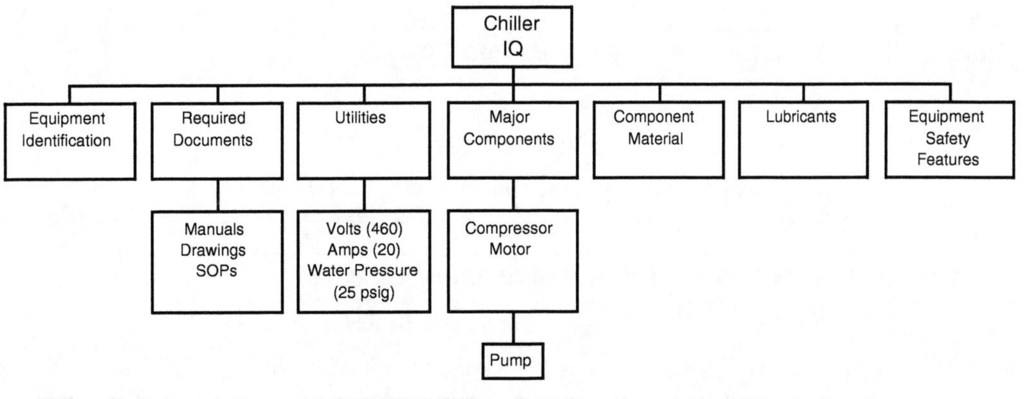

**Table 2.70. Equipment Identification**

| Required Information | As-Found Conditions |
|---|---|
| Manufacturer | *Any Chiller Co.* |
| Purchase order number | *008650* |
| Model number | *AQ0A0403* |
| Serial number | *CH-14795* |
| Equipment number | *6610* |
| Location | *Room 361 (This equipment is portable.)* |

### *Required Documentation*

Record the equipment manufacturer's operation and maintenance manual and drawings in Table 2.71. Record the SOPs that cover the setup, operation, and cleaning of the chiller in Table 2.72.

### *Equipment Utility Requirements*

Compare the manufacturer's specified volts (V), amps (A), and water pressure requirements to their as-found conditions at the time of qualification testing and record the results in Table 2.73. Also record the location of the power supply source. Record the instruments used to measure the volts, amps, and water pressure in Table 2.74.

***Volt calculation:***

Motor volt specification = 460 V ± 10%

±10% of 460 = ±46

460 + 46 = 482

**Table 2.71. Manufacturers' Manuals and Drawings**

| Number | Description | Date |
|---|---|---|
| *1020* | *Chiller Operation and Installation Manual* | *10/94* |
| *SPC-6110* | *Typical Piping Schematic, 1/2–3 Tons, Drawing* | *10/91* |

**Table 2.72. Standard Operating Procedures**

| Number | Description | Date |
|---|---|---|
| *GRA021* | *Chiller Setup and Operation* | *04/23/96* |

**Table 2.73. *Utilities***

| Utility | Specified | Measured Results | Acceptable (Yes/No) |
|---|---|---|---|
| Volts | 460 ± 10% | *A-B 461; AC 460; B-C 462* | *Yes* |
| Amps | 9 | 20 circuit rating | *Yes* |
| Water | Water pressure ≥ 25 psig | *65 psig* | *Yes* |
| Power supply source, breaker box *BB10, wire numbers: 3, 5, 7* | | | |

**Table 2.74. *Instruments Used***

| Test Instrument | Identification Number | Calibration Due Date |
|---|---|---|
| *Multimeter* | *ME-025* | *04/19/97* |
| *Water pressure gauge* | *P-102* | *02/10/97* |

460 – 46 = 438

The measured volts of 461/460/462 fall within ±10%.

***Amp calculation:***

Circuit rating = 20 A

Equipment current draw = 9 A

The circuit rating of 20 amps is greater than the maximum current draw of the equipment.

***Water pressure calculation:***

Water supply specification ≥ 25 psig

The measured water pressure of 65 psig is greater than 25 psig.

### *Major Component Specifications*

The component specifications of the protocol is designed to verify that the chiller components purchased were delivered and installed. Record the major components in Table 2.75.

### *Component Material*

Record the material of each component that contacts the product in Table 2.76.

**Table 2.75. Major Components**

| Components | As-Found Conditions |
|---|---|
| Compressor motor | Manufacturer: *Any Motor Co.*<br>Model number: *E07020*<br>Serial number: *65443*<br>Volts: *460*<br>Amperes: *9*<br>Phases: *3*<br>Cycles: *60 Hz*<br>hp: *3*<br>rpm: *1,210* |
| Pump | Manufacturer: *Any Pump Co.*<br>Model number: *644 H*<br>Serial number: *025-78B* |
| Condenser refrigerant | Material: *R-22, 121 lb* |
| Evaporator | *Not specified, sprayed on insulation covered the unit* |
| Reservoir | *Not specified, sprayed on insulation covered the unit* |

**Table 2.76. Component Material**

| Component | Material |
|---|---|
| *None of the components make contact with the product.* | |

### *Lubricants*

Record the lubricants used to operate the chiller in Table 2.77 and indicate if they make contact with the product.

### *Equipment Safety Features*

The operating thermostat senses the process supply water temperature and displays "to process" or the setpoint temperature. The control activates the hot gas bypass

**Table 2.77. Lubricants**

| Where Used | Type | Manufacturer | Product Contact (Yes/No) |
|---|---|---|---|
| *Reservoir coolant* | *R-22, 121 lb* | *Any Coolant Co.* | *No* |
| *Pump motor* | *APG 90 Oil* | *Any Oil Co.* | *No* |
| Is there a preventive maintenance procedure on file? (Yes/No) | | | *Yes* |

solenoid valve. This valve enables the metering of hot gas into the evaporator after the expansion valve, thereby maintaining chilled water temperature. The valve enables continuous compressor operation during partial loads as well as 36°F (34°F) accuracy of chilled water temperature. The control is also designed to cycle the compressor when the required capacity is below the handling capacity of the hot gas bypass. The sight glass and moisture indicator provide a warning signal in the event moisture has entered the system. When moisture enters the system, the green dot turns yellow.

## Operational Qualification

An OQ evaluation should establish that the equipment can operate within specified tolerances and limits (Figure 2.20). The mechanical ranges of the chiller are challenged, along with the basic chiller operations that the operator will use. The chiller will be validated for its operating ability. Information required for the OQ evaluation is calibration of the instruments used to control the chiller, equipment control functions (switches and push buttons), and equipment operation (compressor motor rotation direction).

### *Calibration Requirements*

Verify that all critical instruments on the equipment have been logged into the calibration system, have calibration procedures in place, and are in calibration at the

***Figure 2.20. OQ Elements of a Chiller***

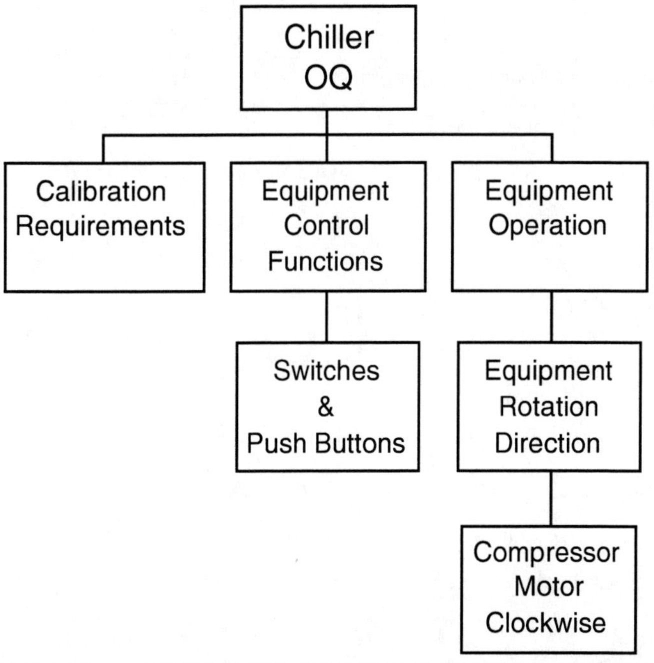

time of qualification testing. Record all information for the calibrated instruments used to control the chiller in Table 2.78.

### *Equipment Control Functions*

The objective of testing equipment control functions is to verify that the switches and push buttons on the chiller operate per the manufacturer's specifications. Operate each control listed in Table 2.79 and verify their proper operation.

### *Equipment Operation: Compressor Motor Rotation Direction Test*

The objective of the rotation direction test is to verify that the compressor motor rotates in the proper direction. Press the compressor on/off switch and observe the direction of rotation of the compressor motor as viewed from behind the motor. Record the results in Table 2.80.

## Performance Qualification

Once it has been established that the equipment is properly installed and functioning within specified operating parameters, it must be shown that the chiller can perform reliably under routine, minimum, and maximum operating conditions (Figure 2.21).

**Table 2.78. Calibrated and Noncalibrated Instruments**

*There were no calibrated instruments on the blender.*

| Noncalibrated Instrument | As-Found Conditions |
|---|---|
| Identification number | *6567* |
| Type | *Water pressure gauge, analog* |
| Manufacturer | *Any Gauge Co.* |
| Model number | *Not specified* |
| Serial number | *Not specified* |
| Range | *0–200 psig* |
| Scale division | *5 psig* |
| Location | *Mounted on the front of the unit* |
| Use | *Monitors the water pressure to the process* |
| Calibration due date | *Calibration not required* |
| Critical or not critical | *Not critical to the operation of the blender* |

*continued on next page.*

*Table 2.79 continued*

| Noncalibrated Instrument | As-Found Conditions |
|---|---|
| Identification number | *6768* |
| Type | *Temperature gauge, digital* |
| Manufacturer | *Any Gauge Co.* |
| Model number | *Not specified* |
| Serial number | *Not specified* |
| Range | *-20 to +120°F* |
| Scale division | *2°F* |
| Location | *Mounted on the front of the unit* |
| Use | *Monitors the water temperature to the process* |
| Calibration due date | *Calibration not required* |
| Critical or not critical | *Not critical to the operation of the mixer* |
| **Noncalibrated Instrument** | **As-Found Conditions** |
| Identification number | *6769* |
| Type | *Temperature gauge, analog* |
| Manufacturer | *Any Gauge Co.* |
| Model number | *Not specified* |
| Serial number | *Not specified* |
| Range | *0–200°F* |
| Scale division | *2°F* |
| Location | *Mounted on the front of the unit* |
| Use | *Monitors the water temperature from the process* |
| Calibration due date | *Limited, reference calibration 01/26/96* |
| Critical or not critical | *Not critical to the operation of the mixer* |

### *Chiller Operation*

The objective of testing the chiller's operation is to show that the chiller performs according to the manufacturer's specifications and process requirements. The temperature to the process and from the process will be measured and recorded. The following procedure is used to test the operation of a chiller:

- Set the water temperature to the process at 50°F with the "to the process" temperature gauge. When the water temperature to the process reaches the setpoint, verify that the water temperature from the process is within ±10 percent of the water temperature to the process by reading the gauge installed on the outlet line. Record the results in Table 2.81.

**Table 2.79. *Control Function Test Results***

| Test Function | Expected Results | Acceptable (Yes/No) |
|---|---|---|
| Power on/off switch operation | When the power on/off switch is pressed, the chiller goes on or off. | *Yes* |
| Compressor on/off switch operation | When the compressor on/off switch is pressed, the compressor goes on or off. | *Yes* |
| Pump on switch operation | When the pump on switch is pressed, the pump goes on. | *Yes* |
| * symbol operation | When the symbol * is pressed, the setpoint is displayed. | *Yes* |
| *▲/*▼ symbol operation | When the symbol *▲/*▼ is pressed, the setpoint increases or decreases. | *Yes* |
| Shut off valve "to process" operation | When the shut off valve "to process" is rotated, the flow of water to the process is stopped. | *Yes* |
| Shut off valve "from process" operation | When the shut off valve "from process" is rotated, the flow of water from the process is stopped. | *Yes* |

**Table 2.80. *Compressor Motor Rotation Direction Test Results***

| Item | Expected Results | Results | Acceptable (Yes/No) |
|---|---|---|---|
| Compressor motor rotation direction | Rotation should be clockwise as viewed from behind the motor. | *Clockwise rotation was observed.* | *Yes* |

**Figure 2.21. *PQ Elements of a Chiller***

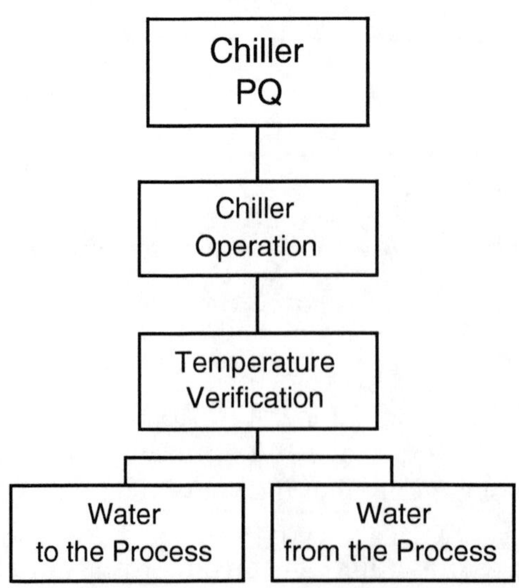

**Table 2.81. *Chiller Performance Test Results***

| Setpoint Temperature | Water Temperature to Process, Gauge Reading | Water Temperature from Process, Gauge Reading | Acceptable (Yes/No) |
|---|---|---|---|
| 50°F | *50°F* | *51°F* | *Yes* |
| 60°F | *60°F* | *59°F* | *Yes* |

Setpoint temperature specification = 50°F ± 10%

±10% of 50 = ±5

50 + 5 = 55

50 – 5 = 45

The temperature from the process of 51°F falls within the ±10% of the temperature to the process.

- Set the water temperature to the process at 60°F using the "to the process" temperature gauge. When the water temperature to the process reaches the setpoint, verify that the water temperature from the process is within ±10 percent of the water temperature to the process by reading the gauge installed on the outlet line. Record the results in Table 2.81.

Setpoint temperature specification = 60°F ± 10%

±10% of 60 = ±6

60 + 6 = 66

60 – 6 = 54

The temperature from the process of 59°F falls within the ±10% of the temperature to the process.

## OVEN

The validation of an oven (Figure 2.22) that is used to dry wet granulation and intermediate excipients will be described here. Product is placed on drying trays that are stacked on racks. The racks and trays are rolled into the oven, and the granulation or intermediates are dried at specified temperatures. Sampling the granulation or intermediates determines the drying endpoint; then perform a loss on drying (LOD) analysis. The unit consists of an oven with an air handling system with horizontal airflow, a heating coil with controls to achieve specified temperatures, and a 24 h chart recorder to monitor the temperature.

The validation protocol for an oven is described in the following sections. This protocol example can be used as a cGMP compliance guide, and it should be

***Figure 2.22. Gruenberg Oven (Courtesy of Gruenberg)***

tailored it to meet individual company requirements. Italicized words in the tables represent examples of information that could be recorded in a validation protocol for an oven.

## Installation Qualification

An IQ evaluation establishes confidence that the equipment is properly installed (Figure 2.23). The installation must meet the manufacturer's specified guidelines, along with design changes at installation. Also, the supporting electrical utilities must meet all electrical codes. The information required for an IQ evaluation is equipment identification, required documentation, equipment utility requirements, major component specifications, component material, lubricants, and equipment safety features.

### *Equipment Identification*

Record the equipment identification numbers in Table 2.82, along with the following information: equipment manufacturer's purchase order number, model number, serial number, company assigned equipment number, and the location of the equipment.

### *Required Documentation*

Record the equipment manufacturer's operation and maintenance manual and drawings in Table 2.83. Record the SOPs that cover the setup, operation, and cleaning of the oven in Table 2.84.

***Figure 2.23. IQ Elements of an Oven***

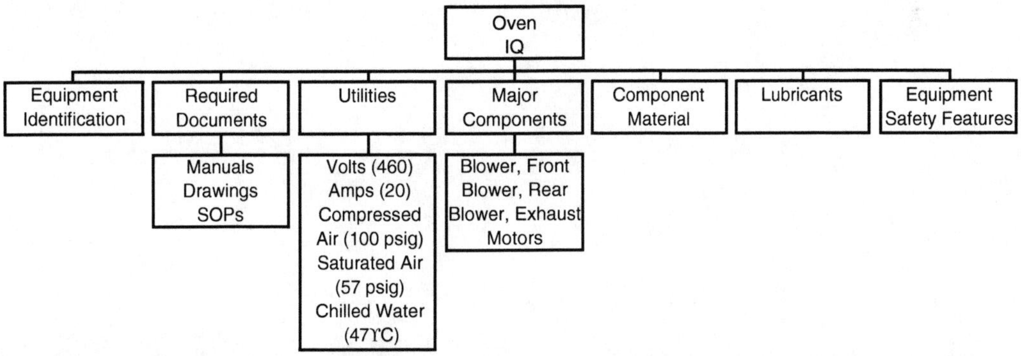


***Table 2.82. Equipment Identification***

| Required Information | As-Found Conditions |
|---|---|
| Manufacturer | *Any Oven Co.* |
| Purchase order number | *008750* |
| Model number | *SS11.142SH71T* |
| Serial number | *CH-14795* |
| Equipment number | *9182* |
| Location | *Room 981* |

**Table 2.83. Manufacturers' Manuals and Drawings**

| Number | Description | Date |
|---|---|---|
| *None* | *Oven Instructions Manual* | *None* |
| *1001D* | *Electrical Schematic* | *07/18/94* |
| *1002D* | *P&ID* | *07/25/94* |
| *100D* | *Class B Oven Configuration* | *01/13/95* |
| *300D* | *Room Layout* | *08/15/94* |
| *400D* | *Control Console* | *08/15/94* |

**Table 2.84. Standard Operating Procedures**

| Number | Description | Date |
|---|---|---|
| *GRA038* | *Oven Setup and Operation* | *01/96* |
| *GRA039* | *Oven Cleaning Procedure* | *01/96* |

### *Equipment Utility Requirements*

Compare the manufacturer's specified volts (V), amps (A), compressed air pressure, saturated air pressure, and chilled water temperature requirements to their as-found conditions at the time of qualification testing and record the results in Table 2.85. Also record the location of the power supply source. Record the instruments used to measure the volts, amps, compressed air pressure, saturated air pressure, and chilled water temperature in Table 2.86.

***Volt calculation:***

Motor supplied volt specification = 460 V ± 10%

±10% of 460 = ± 46

460 + 46 = 506

460 – 46 = 414

The measured volts of 459/463/461 fall within ±10%.

***Amp calculation:***

Circuit rating = 20 A

Blower motor inlet, front = 4.5 A

Blower motor inlet, rear = 4.5 A

Blower motor, exhaust = 4.5 A

Equipment current draw = blower motor inlet, front A + blower motor inlet, rear A + blower motor, exhaust A = 4.5 + 4.5 + 4.5 = 13.5 A

**Table 2.85. Utilities**

| Utility | Specified | Measured Results | Acceptable (Yes/No) |
|---|---|---|---|
| Volts | Motor = 460 ± 10% | *A-B 459; A-C 463; B-C 461* | *Yes* |
| Amps | Blower motor inlet, front = 4.5<br>Blower motor inlet, rear = 4.5<br>Blower motor, exhaust = 4.5 | 20 circuit rating | *Yes* |
| Compressed air | 100–115 psig | *110 psig* | *Yes* |
| Saturated air | 57 psig ± 10% | *52 psig* | *Yes* |
| Chilled water | 47°F ± 10% | *45.5°F* | *Yes* |

Power supply source, breaker box *BB21C, wire numbers: 1, 3, 5*

**Table 2.86. Instruments Used**

| Test Instrument | Identification Number | Calibration Due Date |
|---|---|---|
| *Multimeter* | *ME-025* | *04/19/97* |
| *Air pressure gauge* | *PG-001* | *08/14/97* |
| *Water pressure gauge* | *P-102* | *02/10/97* |

The circuit amp rating of 20 is greater than the maximum current draw of the equipment.

***Compressed air calculation:***

Compressed air specification = 100 to 115 psig

The measured air pressure of 110 psig falls within the range.

***Saturated air calculation:***

Saturated air specification = 57 psig ± 10%

±10% of 57 = ±5.7

57 + 5.7 = 62.7

57 – 5.7 = 51.3

The measured saturated air of 52 psig falls within ±10%.

***Chilled water calculation:***

Chilled water specification = 47°F ± 10%

±10% of 47 = ±4.7

$47 + 4.7 = 51.7$

$47 - 4.7 = 42.3$

The measured saturated air of 45.5°F falls within ±10%.

### *Major Component Specifications*

The component specifications section of the protocol is designed to verify that the oven components purchased were delivered and installed. Record the major components in Table 2.87.

### *Component Material*

Record the material of each component that contacts the product in Table 2.88.

**Table 2.87. Major Components**

| Components | As-Found Conditions |
|---|---|
| Oven cabinet | Material: 304 *stainless steel* |
| Blower motor, front | Manufacturer: *Any Motor Co.*<br>Model number: *EM366IT*<br>Serial number: *W5-19*<br>Volts: *460*<br>Amperes: *4.5*<br>Phases: *3*<br>Cycles: *60 Hz*<br>hp: *3*<br>rpm: *1,725* |
| Blower motor, rear | Manufacturer: *Any Motor Co.*<br>Model number: *EM366IT*<br>Serial number: *W5-20*<br>Volts: *460*<br>Amperes: *4.5*<br>Phases: *3*<br>Cycles: *60 Hz*<br>hp: *3*<br>rpm: *1,725* |
| Blower motor, exhaust | Manufacturer: *Any Motor Co.*<br>Model number: *EM366IT*<br>Serial number: *W5-21*<br>Volts: *460*<br>Amperes: *4.5*<br>Phases: *3*<br>Cycles: *60 Hz*<br>hp: *3*<br>rpm: *1,725* |

**Table 2.88. Component Material**

| Component | Material |
|---|---|
| *None of the oven components make contact with the product.* | |

### *Lubricants*

Record the lubricants used to operate the oven in Table 2.89 and indicate if they make contact with the product.

### *Equipment Safety Features*

The door can be pushed open from the inside, if shut inside. The exhaust motor, intake motor, and heat circuit will not operate when the door is open.

## Operational Qualification

An OQ evaluation should establish that the equipment can operate within specified tolerances and limits (Figure 2.24). The mechanical ranges of the oven are challenged, along with the basic oven operations that the operator will use. The oven will be validated for its operating ability, not how well it dries the granulation. Information required for the OQ evaluation is calibration of the instruments used to control the oven, equipment control functions (switches and push buttons) and equipment operation (cycle sequencing, open door leak, differential pressure, timing operation, circulation airflow, exhaust airflow, high temperature limit, and oven door open).

### *Calibration Requirements*

Verify that all critical instruments on the equipment have been logged into the calibration system, have calibration procedures in place, and are in calibration at the

**Table 2.89. Lubricants**

| Where Used | Type | Manufacturer | Product Contact (Yes/No) |
|---|---|---|---|
| *Motors* | *SAE-02 Oil* | *Any Oil Co.* | *No* |
| *Pump motor* | *Lithrom #2 Grease* | *Any Oil Co.* | *No* |
| Is there a preventive maintenance procedure on file? (Yes/No) | | | *Yes* |

**Figure 2.24. OQ Elements of an Oven**

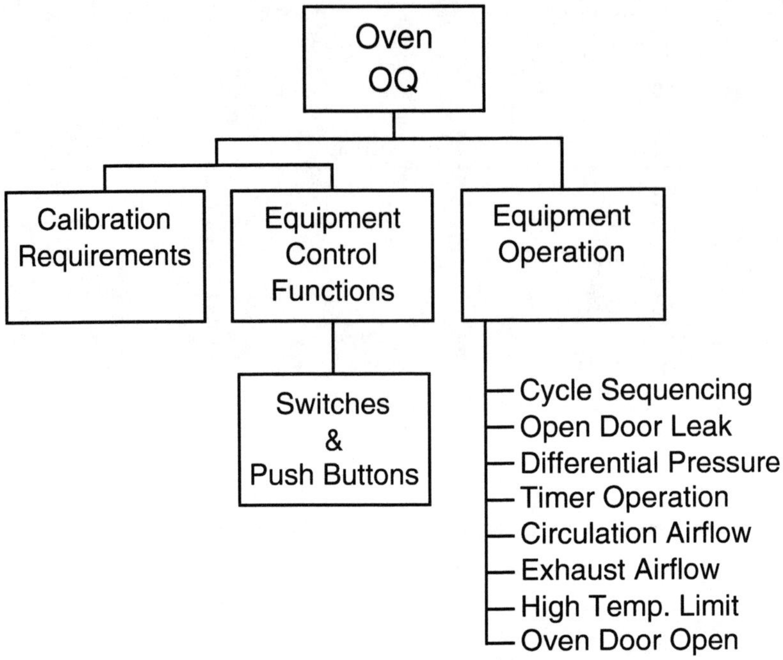


time of qualification testing. Record all information for the calibrated instruments used to control the oven in Table 2.90.

### *Equipment Control Functions*

The objective of testing equipment control functions is to verify that the push buttons, switches, and indicator lights on the oven operate per the manufacturer's specifications. Operate each control listed in Table 2.91 and 2.92. Then verify their proper operation.

### *Equipment Operation: Cycle Sequencing Verification Test*

The objective of the cycle sequencing verification test is to verify that the oven can perform a preprogrammed sequence of operations: heat, hold temperature, and cool. Execute a drying cycle and verify that sequencing occurs as programmed. Then record the results in Table 2.93.

### *Equipment Operation: Oven Door Leak Test*

The objective of the open door leak test is to verify that the oven is sealed when the door is closed. A smoke stick will be required for this test. Execute a drying cycle and trace the oven door seal with a smoke stick to verify that the smoke is not blown away from or drawn into the oven. Record the results in Table 2.94.

**Table 2.90. *Calibrated and Noncalibrated Instruments***

| Calibrated Instrument | As-Found Conditions |
|---|---|
| Identification number | *TCR-100* |
| Type | *Chart recorder, mechanical* |
| Manufacturer | *Any Gauge Co.* |
| Model number | *CRM0007* |
| Serial number | *86915149* |
| Range | *24 h, 90–130°F calibrated* |
| Scale division | *15 min, 2°F chart, 1°F digital* |
| Location | *On top of the control panel* |
| Use | *Records the temperature of the process* |
| Calibration due date | *10/03/96* |
| Critical or not critical | *Critical to the operation of the oven* |
| **Calibrated Instrument** | **As-Found Conditions** |
| Identification number | *ITD-100* |
| Type | *High limit sensor* |
| Manufacturer | *Any Gauge Co.* |
| Model number | *CPA 0003 Type J* |
| Serial number | *Not specified* |
| Range | *32–170°F* |
| Scale division | *1°F* |
| Location | *On the left side of the control panel* |
| Use | *Monitor and alarm for the process temperature high limit* |
| Calibration due date | *09/29/96* |
| Critical or not critical | *Critical to the operation of the oven* |
| **Calibrated Instrument** | **As-Found Conditions** |
| Identification number | *ITD-200* |
| Type | *Temperature gauge, digital* |
| Manufacturer | *Any Gauge Co.* |
| Model number | *CIM 0611* |
| Serial number | *19225149* |
| Range | *32–210°F calibrated* |
| Scale division | *1°F* |
| Location | *In the center of the control panel* |
| Use | *Records the preheated steam temperature* |
| Calibration due date | *10/01/96* |
| Critical or not critical | *Critical to the operation of the oven* |

*continued on next page.*

*Table 2.90 continued*

| Calibrated Instrument | As-Found Conditions |
|---|---|
| Identification number | *ITD-300* |
| Type | *Temperature gauge, digital* |
| Manufacturer | *Any Gauge Co.* |
| Model number | *CIM 0611* |
| Serial number | *8B2251499* |
| Range | *32–210°F calibrated* |
| Scale division | *1°F* |
| Location | *On the right side of the control panel* |
| Use | *Records the chilled water temperature* |
| Calibration due date | *10/03/96* |
| Critical or not critical | *Critical to the operation of the oven* |
| **Noncalibrated Instrument** | **As-Found Conditions** |
| Identification number | *GP-100* |
| Type | *Air pressure gauge, digital* |
| Manufacturer | *Any Gauge Co.* |
| Model number | *Not specified* |
| Serial number | *Not specified* |
| Range | *0–60 psig* |
| Scale division | *2 psig* |
| Location | *Inside the control panel* |
| Use | *Positioning valves* |
| Calibration due date | *Calibration not required* |
| Critical or not critical | *Not critical to the operation of the oven* |
| **Calibrated Instrument** | **As-Found Conditions** |
| Identification number | *GPD-100* |
| Type | *Differential air pressure gauge, digital* |
| Manufacturer | *Any Gauge Co.* |
| Model number | *Series 0002* |
| Serial number | *Not specified* |
| Range | *0–0.50 in. w.c.* |
| Scale division | *0.05 in. w.c.* |
| Location | *On the front of the oven at the top* |
| Use | *Indicates the differential air pressure* |
| Calibration due date | *10/03/96* |
| Critical or not critical | *Critical to the operation of the oven* |

*continued on next page.*

*Table 2.90 continued*

| Calibrated Instrument | As-Found Conditions |
|---|---|
| Identification number | *MIT-100* |
| Type | *Timer, digital* |
| Manufacturer | *Any Gauge Co.* |
| Model number | *6A202CX* |
| Serial number | *44463* |
| Range | *0.1 sec to 59 min and 59 sec* |
| Scale division | *0.1 sec* |
| Location | *On the front of the control panel* |
| Use | *Times the process* |
| Calibration due date | *11/20/96* |
| Critical or not critical | *Critical to the operation of the oven* |

**Table 2.91. *Equipment Control Function Test Results***

| Test Function | Expected Results | Acceptable (Yes/No) |
|---|---|---|
| Power disconnect switch operation | When the switch is pressed, the oven goes on and off and the control panel door can be opened. | *Yes* |
| On switch operation | When the on switch is pressed, the drying cycle starts. | *Yes* |
| Off switch operation | When the off switch is pressed, the drying cycle stops. | *Yes* |
| Door switch operation | When the door is open, the oven maintains normal operation. | *Yes* |
| Bell silence selector switch operation | When the switch is pressed, the end-of-the-cycle bell is disabled. | *Yes* |
| Exhaust airflow switch operation | When the switch is pressed, the heat is de-energized, an audible alarm sounds, and the exhaust air failure light goes on. | *Yes* |
| Circulation airflow switch operation | When the switch is pressed, the circulation blower is de-energized, an audible alarm sounds, and the exhaust failure light goes on. | *Yes* |
| High limit reset switch operation | When the switch is pressed, after a high limit failure, the high limit controller is reset. | *Yes* |
| High limit temperature controller | When oven temperature exceeds a preset limit, the heaters are de-energized and an alarm sounds. | *Yes* |

**Table 2.92. *Indicator Lights Test Results***

| Test Function | Expected Results | Acceptable (Yes/No) |
|---|---|---|
| Oven in operation | When the oven is in operation, verify that the indicator light goes on. | *Yes* |
| Oven on | When the oven is on, verify that the indicator light goes on. | *Yes* |
| Exhaust air failure | When there is an exhaust failure, verify that the indicator light goes on. | *Yes* |
| Cycle complete | When a cycle is complete, verify that the indicator light goes on. | *Yes* |
| Circulation air failure #1 | When there is a circulation air failure #1, verify that the indicator light goes on. | *Yes* |
| Circulation air failure #2 | When there is a circulation air failure #2, verify that the indicator light goes on. | *Yes* |
| High temperature alarm | When there is a high temperature alarm, verify that the indicator light goes on. | *Yes* |

**Table 2.93. *Cycle Sequencing Verification Test Results***

| Item | Results |
|---|---|
| Setpoint = 110°F ± 10°F | *Setpoint = 110°F with high limit at 120°F* |
| Dry time = 30 min | *Dry time was set for 30 min.* |
| Control points around setpoint (timer start or heat-up) | *Timer began to elapse when the temperature reached 100°F* |

| Expected Results | Acceptable (Yes/No) |
|---|---|
| When the on switch was pressed, the oven began to heat up. | *Yes* |
| When the displayed temperature reached 100°F, the oven timer started. | *Yes* |
| The oven timer operated continuously through the drying period, with no alarms activated. | *Yes* |
| When the over timer deactivated at the end of the drying period, the heater shut down and the over started to cool. | *Yes* |

**Table 2.94. *Open Door Leak Test Results***

| Test Function | Results |
|---|---|
| Top of the oven door | *The smoke was not blown away from or drawn into the oven.* |
| Left side of the oven door | *The smoke was not blown away from or drawn into the oven.* |
| Bottom of the oven door | *The smoke was not blown away from or drawn into the oven.* |
| Right side of the oven door | *The smoke was not blown away from or drawn into the oven.* |

### *Equipment Operation: Oven Differential Pressure Test*

The objective of the differential pressure test is to verify that the air pressure in the oven during operation is negative with respect to the area outside the oven. A differential pressure gauge will be required. Execute a drying cycle and measure the differential pressure between the inside and the outside the oven, using a calibrated differential pressure gauge. Record the results in Table 2.95. Record the instrument used to measure the differential pressure in Table 2.96.

### *Equipment Operation: Timer*

The objective of testing the timer is to verify that the timer can maintain time control of the oven while the oven is operating. Execute a drying cycle for 15 min and record the test information on a chart recorder. Verify the normal oven operation during the drying cycle and record the results in Table 2.97. Record the instrument used to log the information in Table 2.98.

### *Equipment Operation: Circulation Airflow Failure Test*

The objective of testing the circulation airflow is to verify that the timer reacts properly in a failure condition where the circulation airflow is stopped. The following procedure is used:

- Execute a drying cycle for 15 min and record the test information on a chart recorder. Create a circulation airflow failure by disconnecting the differential pressure gauge tubes. Verify the suspended oven time and record the results in Table 2.99. Record the instrument used to log the information in Table 2.98.
- Reconnect the tubes after several minutes and verify that the drying cycle restarts. Record the results in Table 2.99.

**Table 2.95. Differential Pressure Test Results**

| Test Function | Results | Acceptable (Yes/No) |
|---|---|---|
| Differential pressure | *The oven differential pressure was negative with respect to the area outside the oven.* | *Yes* |
| Average differential pressure in w.c. | *25 in. w.c.* | *Yes* |

**Table 2.96. Instrument Used**

| Test Instrument | Identification Number | Calibration Due Date |
|---|---|---|
| *Differential pressure gauge* | *100GPD9182* | *04/22/096* |

**Table 2.97. Timer Operation Test Results**

| Test Function | Results | Acceptable (Yes/No) |
|---|---|---|
| Oven in operation | *Oven operation light was on during the cycle.* | *Yes* |
| Cycle complete | *Cycle complete light was on at the end of the cycle.* | *Yes* |
| Audible alarm on | *The audible alarm was on at the end of the cycle.* | *Yes* |
| Heat off | *When the temperature recorder dropped 10°F, the heat went off.* | *Yes* |
| Timer stops | *The timer stopped at the end of the cycle.* | *Yes* |

**Table 2.98. Instrument Used**

| Test Instrument | Identification Number | Calibration Due Date |
|---|---|---|
| *Scale* | *PER600* | *07/18/97* |
| *Data recorder* | *DL-066* | *03/26/97* |

**Table 2.99. Circulation Airflow Test Results**

| Test Function | Results | Acceptable (Yes/No) |
|---|---|---|
| Timer operation | *After the tubes were disconnected, the timer stopped.* | *Yes* |
| Alarm on | *After the tubes were disconnected, the alarm was audible.* | *Yes* |
| Circulation airflow failure light | *After the tubes were disconnected, the circulation airflow light went on.* | *Yes* |
| Heat off | *When the temperature recorder dropped more than 10°F, the heat went off.* | *Yes* |
| Oven and timer operation | *After the tubes were reconnected, the oven and timer started operating.* | *Yes* |

### *Equipment Operation: Exhaust Airflow Failure Test*

The objective of the exhaust airflow test is to verify that the timer reacts properly in a failure condition where the exhaust airflow is stopped. The following procedure is used:

- Execute a drying cycle for 15 min and record the test information on a chart recorder. Create an exhaust airflow failure by turning the solvent/aqueous key to the vertical position. Verify the suspended oven time and record the results in Table 2.100. Record the instrument used to log the information in Table 2.98.
- Return the solvent/aqueous key to the horizontal position after several minutes and verify that the drying cycle restarts. Record the results in Table 2.100.

**Table 2.100. Exhaust Airflow Test Results**

| Test Function | Results | Acceptable (Yes/No) |
|---|---|---|
| Timer operation | *After the key was turned to a vertical position, the timer stopped.* | *Yes* |
| Alarm on | *After the key was turned to a vertical position, the alarm was audible.* | *Yes* |
| Exhaust airflow failure light | *After the key was turned to a vertical position, the exhaust airflow light went on.* | *Yes* |
| Heat off | *When the temperature recorder dropped more than 10°F, the heat went off.* | *Yes* |
| Oven and timer operation | *After the key was turned to a horizontal position, the oven and timer started operating.* | *Yes* |

### *Equipment Operation: High Temperature Limit Failure Test*

The objective of verifying the temperature limits is to determine if the timer reacts properly during failure conditions where the process temperature exceeds the high temperature limit. Use the following procedure to test high temperature limits:

- Execute a drying cycle for 15 min and record the test information on a chart recorder. Set the high temperature to 130°F and the process temperature to 135°F. Verify the suspended oven time and record the results in Table 2.101.
- Reset the temperature controllers to their default settings and verify that the drying cycle restarts. Record the results in Table 2.101.

**Table 2.101. High Temperature Limit Failure Test Results**

| Test Function | Results | Acceptable (Yes/No) |
|---|---|---|
| Timer operation | *After the high temperature limit was passed, the timer stopped.* | *Yes* |
| Alarm on | *After the high temperature limit was passed, the alarm was audible.* | *Yes* |
| Exhaust airflow failure light | *After the high temperature limit was passed, the exhaust airflow light went on.* | *Yes* |
| Heat off | *When the temperature recorder dropped more than 10°F, the heat went off.* | *Yes* |
| Oven and timer operation | *After the high temperature limit was passed, the oven and timer started operating.* | *Yes* |

### *Equipment Operation: Oven Door Open Test*

The objective of the oven door open test is to verify that the timer does not suspend the drying operation when the door is open. Execute a drying cycle for 5 min and record the test information on a chart recorder. Open the door while the oven is operating and verify that it keeps operating. Record the results in Table 2.102. Record the instrument used to log the information in Table 2.98.

## Performance Qualification

Once it has been established that the equipment is properly installed and functioning within specified operating parameters, it must be shown that the oven can perform reliably under routine, minimum, and maximum operating conditions (Figure 2.25).

### *Oven Operation*

The operation of the oven is tested to show that the oven performs according to the manufacturer's specifications and process requirements. The oven will be tested

**Table 2.102. Open Door Test Results**

| Test Function | Results | Acceptable (Yes/No) |
|---|---|---|
| Timer operation | *When the door is open, the timer does not pause.* | *Yes* |
| Oven on light | *When the door is open, the oven in operation light remains on.* | *Yes* |
| Heat off | *When the door is open, the temperature does not fall.* | *Yes* |
| Oven and timer | *When the door is open, the oven and timer keep operating.* | *Yes* |

**Figure 2.25. PQ Elements of an Oven**

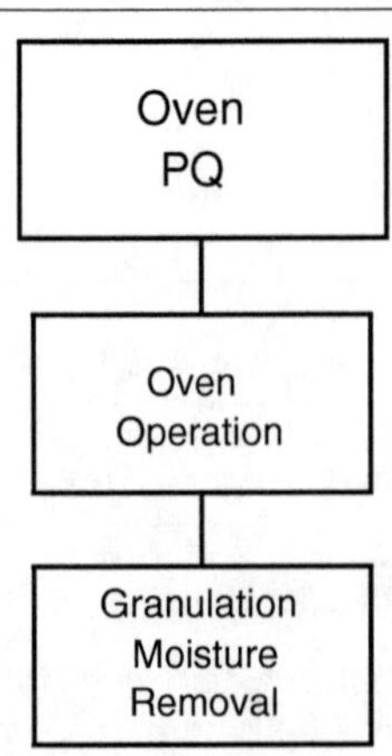

with racks and trays in place. The oven will be operated through a complete drying cycle.

Monitor the oven for 1 h at 160°F after a 25 minute stabilization period. Verify that the oven provided stable temperatures of 160°F ± 10°F during the test. Produce a mapping of the temperature and record the results in Table 2.103.

***Summary of Temperature Mapping (only a partial record is shown):***

***Date: 09/14/97***

| Time | TC1 | TC2 | TC3 | TC4 | TC5 | TC6 | TC7 | TC8 | TC9 |
|---|---|---|---|---|---|---|---|---|---|
| 18:25:59 | 160.2 | 160.8 | 160.2 | 160.9 | 161.0 | 159.6 | 159.3 | 158.8 | 158.0 |
| . | . | . | . | . | . | . | . | . | . |
| . | . | . | . | . | . | . | . | . | . |
| . | . | . | . | . | . | . | . | . | . |
| 19:25:37 | 160.9 | 161.1 | 160.0 | 161.7 | 161.1 | 159.2 | 160.0 | 159.0 | 157.9 |

### *Moisture Removal Test*

The objective of the moisture removal test is to verify that the oven removes moisture from wet granulation as evidenced by tray weight loss and LOD results. The moisture removal test is performed as follows:

- Put wet granulation on the top, middle, and bottom trays of a rack. Weigh each tray with a calibrated scale and record the results in Table 2.104. Record the granulation that was use in Table 2.105. Record the instrument used to weigh the granulation in Table 2.98.
- Place the racks in the oven. Set the heat controller to 130°F. Set the heat variation to 10°F. Dry the wet granulation for 8 h and record the information on the chart recorder. Weigh the trays every hour and record the results in Table 2.104.
- At the end of the cycle, weigh the trays and pull final LOD samples.

***Table 2.103. Oven Performance Test Results***

| Minimum Temperature | Maximum Temperature | Average Temperature | Acceptable (Yes/No) |
|---|---|---|---|
| *156.9°F* | *161.8°F* | *159.6°F* | *Yes* |

**Table 2.104. Moisture Removal Test Results**

| Time (h) | LOD | Top Tray Weight (kg) | Middle Tray Weight (kg) | Bottom Tray Weight (kg) |
|---|---|---|---|---|
| | *Initial 12.3%* | *Initial 11.6* | *9.4* | *11.4* |
| 1 | | *11.0* | *9.0* | *10.8* |
| 2 | | *10.4* | *8.4* | *10.4* |
| 3 | | *10.2* | *8.2* | *10.2* |
| 4 | | *10.0* | *8.0* | *10.0* |
| 5 | | *10.0* | *7.8* | *9.8* |
| 6 | | *9.8* | *7.8* | *9.6* |
| 7 | | *9.8* | *7.8* | *9.6* |
| 8 | | *9.8* | *7.8* | *9.6* |
| | *Final 2.1%* | *Final 9.8* | *7.8* | *9.6* |

Acceptance criteria met? (Yes/No) *Yes*

**Table 2.105. Test Materials and Conditions**

| Item | Specifications |
|---|---|
| Test placebo | *Dical phosphate, 32.4 kg* |

## Tray and Rack Washer

The validation of a tray and rack washer (Figure 2.26) used to wash granulation drying trays and racks will be described here. The unit is designed to wash and dry up to 4 racks using preprogrammed cycles of 1 to 40 min. A programmable controller operates the washer. A wash cycle consists of wash, rinse, and dry phases. During the wash and initial rinse phases, the reservoir in the washer is filled with water that is passed through a steam heat exchanger before entering the washer. Detergent is added through a separate line during the wash phases. A steam coil in the reservoir keeps the water at the selected setpoint temperature while the wash solution or rinse water is recirculated during the wash or rinse phase. The final rinse water makes one pass through the washer, then it is drained. During the drying phase, the dryer fan and exhaust fan are activated to circulate air through the washer to dry the trays and racks.

The validation protocol for a tray and rack washer is described in the following sections: This protocol example can be used as a cGMP compliance guide, and it should be tailored to meet individual company requirements. Italicized words in the tables represent examples of information that could be recorded in a validation protocol for a tray and rack washer.

***Figure 2.26. Cabinet Washer (courtesy of Girton)***

## Installation Qualification

An IQ evaluation establishes confidence that the equipment is properly installed (Figure 2.27). The installation must meet the manufacturer's specified guidelines, along with design changes at installation. Also, the supporting electrical utilities must meet all electrical codes. The information required for an IQ evaluation is equipment identification, required documentation, equipment utility requirements, major component specifications, component material, lubricants, and equipment safety features.

### *Equipment Identification*

Record the equipment identification numbers in Table 2.106, along with the following information: equipment manufacturer's purchase order number, model number, serial number, company assigned equipment number, and the location of the equipment.

**Figure 2.27. IQ Elements of a Tray and Rack Washer**

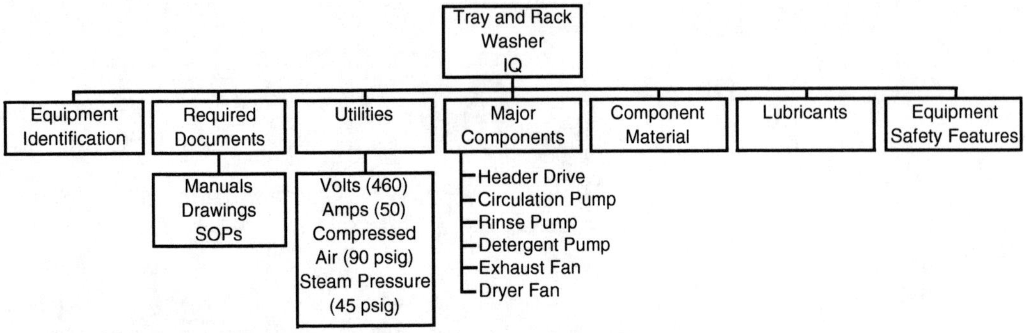


**Table 2.106. Equipment Identification**

| Required Information | As-Found Conditions |
|---|---|
| Manufacturer | *Any Washer Co.* |
| Purchase order number | *34451* |
| Model number | 08-O |
| Serial number | *10918049* |
| Equipment number | *3082* |
| Location | *Room 581* |

## *Required Documentation*

Record the equipment manufacturer's operation and maintenance manual and drawings in Table 2.107. Record the SOPs that cover the setup, operation, and cleaning of the washer in Table 2.108.

**Table 2.107. Manufacturers' Manuals and Drawings**

| Number | Description | Date |
|---|---|---|
| *None* | *Cabinet Washer Manual* | *None* |
| *111EZ* | *08-O Cabinet Washer Drawing PID* | *09/01/94* |
| *None* | *Piping and Instrumentation Diagram* | *None* |
| *None* | *Electrical Enclosure Front Panel Layout* | *None* |

**Table 2.108. Standard Operating Procedures**

| Number | Description | Date |
|---|---|---|
| *GRA056* | *Cabinet Washer Setup, Operation, and Cleaning* | *11/30/95* |

### *Equipment Utility Requirements*

Compare the manufacturer's specified volts (V), amps (A), compressed air pressure, and steam pressure requirements to their as-found conditions at the time of qualification testing and record the results in Table 2.109. Also record the location of the power supply source. Record the instruments used to measure the volts, amps, compressed air pressure, and steam pressure in Table 2.110.

***Volt calculation:***

Motor volt specification = 460 V ± 10%

±10% of 460 = ±46

460 + 46 = 482

460 – 46 = 438

The measured volts of 465/459/463 fall within ±10%.

***Table 2.109.* *Utilities***

| Utility | Specified | Measured Results | Acceptable (Yes/No) |
|---|---|---|---|
| Volts | Motor = 460 ± 10% | *A-B 465; A-C 459;*<br>*B-C 463* | *Yes* |
| Amps | Header drive motor = 1.4<br>Circulation pump motor = 6<br>Rinse pump motor = 6<br>Detergent pump motor = 6<br>Exhaust fan motor = 1.8<br>Dryer fan motor = 9 | 50 circuit rating | *Yes* |
| Compressed air | Air pressure ≥ 90 psig | *110 psig* | *Yes* |
| Steam | Maximum pressure ″ 45 psig | *44 psig* | *Yes* |

Power supply source, breaker box *BB21, wire numbers: 25, 27, 29*

***Table 2.110.* *Instruments Used***

| Test Instrument | Identification Number | Calibration Due Date |
|---|---|---|
| *Multimeter* | *ME-025* | *04/19/97* |
| *Air pressure gauge* | *P-102* | *02/10/97* |
| *Steam pressure gauge* | *P-104* | *09/22/97* |

***Amp calculation:***

Circuit rating = 50 A

Header drive motor = 1.4 A

Circulation pump motor = 6 A

Rinse pump motor = 6 A

Detergent pump motor = 6 A

Exhaust fan motor = 1.8 A

Dryer fan motor = 9 A

Equipment current draw = 30.2 A (6 motors)

The circuit amp rating of 50 amps is greater than the maximum current draw of the equipment.

***Compressed air pressure calculation:***

Compressed air specification ≥ 90 psig

The measured compressed air pressure of 110 psig is greater than 90 psig.

***Steam pressure calculation:***

Steam pressure specification ″ 45 psig

The measured steam pressure of 44 psig is less than 45 psig.

### *Major Component Specifications*

The component specifications of the protocol verifies that the washer components purchased were delivered and installed. Record the major components in Table 2.111.

### *Component Material*

Record the material of each component that contacts the product in Table 2.112.

### *Lubricants*

Record the lubricants used to operate the washer in Table 2.113 and indicate if they make contact with the product.

### *Equipment Safety Features*

The washer has a normally open drain valve that allows the wash tank to empty when the valve is de-energized after an alarm. Restarting the washer may cause the pump to run dry and become damaged. Therefore, it is recommended that all programs be aborted after an alarm and restarted from the beginning.

**Table 2.111. Major Components**

| Components | As-Found Conditions |
|---|---|
| Washer cabinet | Manufacturer: *Any Cabinet Co.*<br>Material: 304 *stainless steel*<br>Size: *39 in. × 83 in. × 89 in.* |
| Header drive motor | Manufacturer: *Any Motor Co.*<br>Model number: *A771E*<br>Serial number: *GH597Z*<br>Volts: *460*<br>Amperes: *1.4*<br>Phases: *3*<br>Cycles: *60 Hz*<br>hp: 1/*3*<br>rpm: *30* |
| Circulation pump motor | Manufacturer: *Any Motor Co.*<br>Model number: *MDC-B-SN23011*<br>Serial number: *33670*<br>Volts: *460*<br>Amperes: *6*<br>Phases: *3*<br>Cycles: *60 Hz*<br>hp: *20*<br>rpm: *3,525* |
| Rinse pump motor | Manufacturer: *Any Motor Co.*<br>Model number: *AH-72525*<br>Serial number: *227543*<br>Volts: *460*<br>Amperes: *6*<br>Phases: *3*<br>Cycles: *60 Hz*<br>hp: *2*<br>rpm: *3,450* |
| Detergent pump motor | Manufacturer: *Any Motor Co.*<br>Model number: *PX-0319-PK*<br>Serial number: *16739*<br>Volts: *460*<br>Amperes: *6*<br>Phases: *3*<br>Cycles: *60 Hz*<br>hp: *Not specified*<br>rpm: *3,450* |

*continued on next page.*

*Table 2.111 continued*

| Components | As-Found Conditions |
|---|---|
| Exhaust fan motor | Manufacturer: *Any Motor Co.*<br>Model number: *02DN2100210*<br>Serial number: *33670*<br>Volts: *460*<br>Amperes: *1.8*<br>Phases: *3*<br>Cycles: *60 Hz*<br>hp: *1/2*<br>rpm: *1,740* |
| Dryer pump motor | Manufacturer: *Any Motor Co.*<br>Model number: *AH-72525*<br>Serial number: *334921*<br>Volts: *460*<br>Amperes: *9*<br>Phases: *3*<br>Cycles: *60 Hz*<br>hp: *3*<br>rpm: *2,857* |

**Table 2.112. Component Material**

| Component | Material |
|---|---|
| *None of the oven components make contact with the product.* | |

**Table 2.113. Lubricants**

| Where Used | Type | Manufacturer | Product Contact (Yes/No) |
|---|---|---|---|
| *Motors* | *SAE-02 Oil* | *Any Oil Co.* | *No* |
| Is there a preventive maintenance procedure on file? (Yes/No) | | | *Yes* |

## Operational Qualification

An OQ evaluation should establish that the equipment can operate within specified tolerances and limits (Figure 2.28). The mechanical ranges of the washer are challenged, along with the basic washer operations that the operator will use. The washer will be validated for its operating ability. Black box software testing is performed to evaluate the microprocessor system against the documented functional requirements or intended uses of the equipment. Information required for the OQ evaluation is calibration of the instruments used to control the washer and equipment control functions (switches and push buttons).

***Figure 2.28.* *OQ Elements of a Tray and Rack Washer***

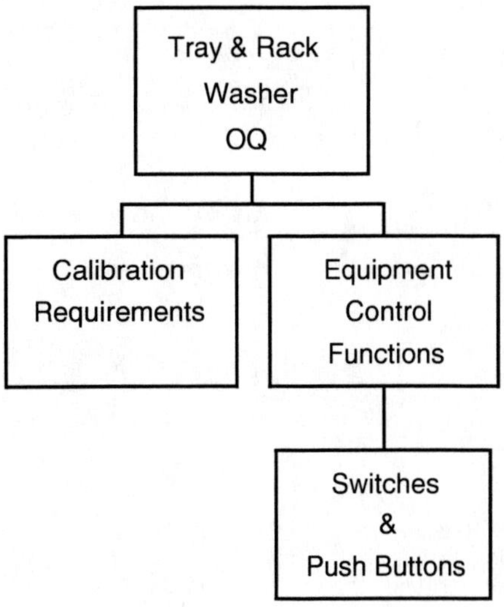


### *Calibration Requirements*

Verify that all critical instruments on the equipment have been logged into the calibration system, have calibration procedures in place, and are in calibration at the time of qualification testing. Record all information for the calibrated instruments used to control the washer in Table 2.114.

### *Equipment Control Functions*

The objective of testing equipment control functions is to verify that the switches, push buttons, and alarms on the washer operate per the manufacturer's specifications. Operate each control listed in Tables 2.115, 2.116, and 2.117 and verify their proper operation. Record the results in Tables 2.115, 2.116, and 2.117.

## Performance Qualification

Once it has been established that the equipment is properly installed and functioning within specified operating parameters, it must be shown that the washer can operate reliably under routine, minimum, and maximum operating conditions (Figure 2.29).

### *Tray and Rack Washer Operation: Timer Test*

The objective of the timer test is to ensure that the rinse timers operate according to the manufacturer's specifications. A stopwatch will be needed for this test. Operate the washer with the wash and rinse timers set at normal operation. Measure the times of the wash and rinse cycles and verify that the measured times are within

**Table 2.114. Calibrated and Noncalibrated Instruments**

| Calibrated Instrument | As-Found Conditions |
|---|---|
| Identification number | *100-GP/3082* |
| Type | *Water pressure gauge, analog* |
| Manufacturer | *Any Gauge Co.* |
| Model number | *Not specified* |
| Serial number | *Not specified* |
| Range | *0–60 psig* |
| Scale division | *5 psig* |
| Location | *Mounted below the control panel* |
| Use | *Indicates the wash pump pressure* |
| Calibration due date | *04/16/97* |
| Critical or not critical | *Not critical to the operation of the washer* |
| **Calibrated Instrument** | **As-Found Conditions** |
| Identification number | *200-GP/3028* |
| Type | *Steam pressure gauge, analog* |
| Manufacturer | *Any Gauge Co.* |
| Model number | *Not specified* |
| Serial number | *Not specified* |
| Range | *0–100 psig* |
| Scale division | *1 psig* |
| Location | *Mounted on steam supply line at heat exchanger* |
| Use | *Monitors the steam pressure in the supply line* |
| Calibration due date | *04/16/97* |
| Critical or not critical | *Not critical to the operation of the washer* |
| **Calibrated Instrument** | **As-Found Conditions** |
| Identification number | *100-ITD/3082* |
| Type | *Temperature gauge, digital* |
| Manufacturer | *Any Gauge Co.* |
| Model number | *Not specified* |
| Serial number | *Not specified* |
| Range | *0–999°F* |
| Scale division | *2°F* |
| Location | *Mounted on the control panel* |
| Use | *Indicates the wash cycle water temperature* |
| Calibration due date | *04/15/97* |
| Critical or not critical | *Critical to the operation of the washer* |

*continued on next page.*

*Table 2.115 continued.*

| Calibrated Instrument | As-Found Conditions |
|---|---|
| Identification number | *100-MIT/3082* |
| Type | *Timer, digital* |
| Manufacturer | *Any Gauge Co.* |
| Model number | *Not specified* |
| Serial number | *Not specified* |
| Range | *00:00 to 99:99 (hours:minutes)* |
| Scale division | *1 min* |
| Location | *Mounted on the control panel* |
| Use | *Indicates and controls the cycle time* |
| Calibration due date | *04/16/97* |
| Critical or not critical | *Critical to the operation of the washer* |
| **Noncalibrated Instrument** | **As-Found Conditions** |
| Identification number | *GP-200* |
| Type | *Air pressure gauge, analog* |
| Manufacturer | *Any Gauge Co.* |
| Model number | *Not specified* |
| Serial number | *Not specified* |
| Range | *0–160 psig* |
| Scale division | *2 psig* |
| Location | *Mounted on the control panel* |
| Use | *Indicates regulated air pressure* |
| Calibration due date | *Calibration not required* |
| Critical or not critical | *Not critical to the operation of the washer* |

**Table 2.115. Equipment Control Function Test Results**

| Test Function | Expected Results | Acceptable (Yes/No) |
|---|---|---|
| Control on/off switch operation | When the control on/off switch is pressed, power is supplied to the control system. | *Yes* |
| Start/reset push button | When the start push button is pressed, with the control in standby mode, the machine starts operating. | *Yes* |
| Stop push button | When the stop push button is pressed, the machine stops operating. | *Yes* |
| Wash time selector dial operation | When the dial is set to 10 min, the machine washes for 10 min. | *Yes* |
| Rinse time selector dial operation | When the dial is set to 5 min, the machine rinses for 5 min. | *Yes* |
| Final rinse time selector dial operation | When the dial is set to 1 min, the machine rinses for 1 min. | *Yes* |

**Table 2.116. Internal Switches**

| Switch Number/ Identification | Switch Position | Switch Setting | Acceptable (Yes/No) |
|---|---|---|---|
| 1. Prewash time | *4* | *1 min* | *Yes* |
| 2. Drain time | *7* | *1 min 45 sec* | *Yes* |
| 3. Prewash and wash temperature | *0* | *130°F* | *Yes* |
| 4. Detergent time | *5* | *25 sec* | *Yes* |
| 5. Flush delay time | *0* | *0 sec* | *Yes* |
| 6. Dryer temperature | *D* | *195°F* | *Yes* |
| 7. Not used | — | — | — |
| 8. Pause time | *0* | *0 sec* | *Yes* |
| 9. Heat | *3* | *" 130°F* | *Yes* |
| 10. Dryer time | *2* | *10 min* | *Yes* |

**Table 2.117. Alarms**

| Test Function | Expected Results | Acceptable (Yes/No) |
|---|---|---|
| Washer door operation | When the washer door is opened while a program is running, the alarm will sound and the message "door" is displayed. | *Yes* |
| | When the washer door is closed, the message "OFF4" is displayed. | *Yes* |
| Stop button | When the stop button is pressed, the alarm sounds and the message "OFF1" is displayed. | *Yes* |

**Figure 2.29. PQ Elements of a Tray and Rack Washer**

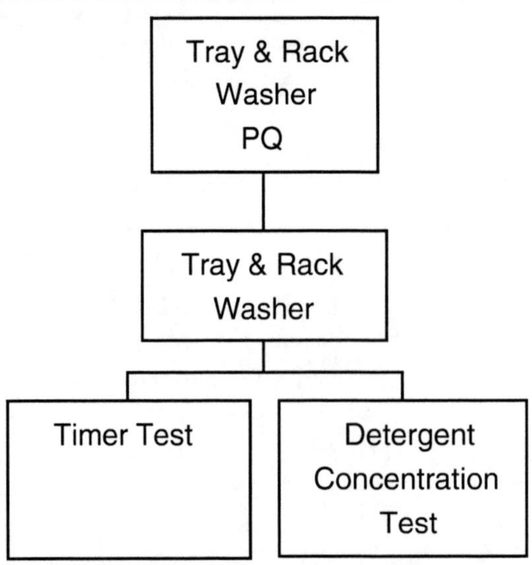

±10 percent of the setpoint times. Record the results in Table 2.118. Record the instrument used to measure the time in Table 2.119.

### *Tray and Rack Washer Operation: Detergent Concentration Test*

The objective of the detergent concentration test is to verify the concentration of the detergent used in the washer. This test is for informational purposes only. A stopwatch will be required. Use the following procedure during testing:

- Remove the hose connected to the discharge side of the detergent pump. Place a container under the discharge port to collect detergent. Operate the pump by holding the prime button to collect detergent. Then measure the time with a calibrated stopwatch. Record the results in Table 2.120 and the instrument used to measure the time in Table 2.119.

  1,150 mL ÷ 1,000 = 1.15 L/min

- Start the washer using the same cycle used in a normal operation. Use a calibrated stopwatch to measure the time that the detergent pump operates during the wash phase. (Do this operation three times.) Calculate the volume pumped, then record the results in Table 2.121 and the instrument used to measure the time in Table 2.119.

  1.15 L/min ÷ 60 = 0.019 L/sec

  57 sec(0.019) = 1.080 L

  55 sec(0.019) = 1.050 L

  55 sec(0.019) = 1.050 L

**Table 2.118. Timer Test**

| Washer Phase | Dial Setting (min:sec) | Results (min:sec) | Acceptable (Yes/No) |
|---|---|---|---|
| Wash | 13:57 | *13:57* | *Yes* |
| Rinse | 9:59 | *9:57* | *Yes* |

**Table 2.119. Instrument Used**

| Test Instrument | Identification Number | Calibration Due Date |
|---|---|---|
| *Stopwatch* | *54039* | *02/16/97* |

**Table 2.120. Pumping Capacity**

| Test Function | Expected Results | Acceptable (Yes/No) |
|---|---|---|
| Detergent pump time | *1 min* | *Yes* |
| Volume pumped | *1,150 L* | *Yes* |
| Pumping capacity in L/min | *1.15 L/min* | *Yes* |

**Table 2.121. *Volume Pumped***

| Test Function | Expected Results | Acceptable (Yes/No) |
|---|---|---|
| Operation 1, time pumped | *57 sec* | *Yes* |
| Calculated volume pumped | *1.080 L* | *Yes* |
| Operation 2, time pumped | *55 sec* | *Yes* |
| Calculated volume pumped | *1.050 L* | *Yes* |
| Operation 3, time pumped | *55 sec* | *Yes* |
| Calculated volume pumped | *1.050 L* | *Yes* |

- At the end of the wash cycle, stop the washer and open the door. Collect a sample of the wash water and submit it to the lab for potassium hydroxide concentration testing. (Perform this operation three times.) Record the results in Table 2.122. This data was obtained from the lab test results.

## Comminuting Mill

The validation of a comminuting mill (Figure 2.30) that is motor driven and manually operated will be described here. Control of the unit consists of, on/off push buttons and speed adjustment for the mill and auger feed. A dial on the control box adjusts the speed of the auger feed. The mill blade is set at one of three speeds by repositioning belts on a multistep flywheel and motor pulley. The unit is used to reduce the size of granules. The feed auger to the mill conveys product from the feed hopper at a controlled rate. The rotating blades of the mill reduce the size of the granules. Size reduction is determined by the type of screen in the mill. Granules that are small enough to pass through the screen drop through the discharge chute to a collection container.

The validation protocol for a comminuting mill is described in the following sections. This protocol example can be used as a cGMP compliance guide, and it should be tailored to meet individual company requirements. Italicized words in

**Table 2.122. *Wash Water Concentration***

| Test Function | Expected Results (mL/L $H_2O$) | Acceptable (Yes/No) |
|---|---|---|
| Sample #1 wash water concentration | *0.0095* | *Yes* |
| Sample #2 wash water concentration | *0.0125* | *Yes* |
| Sample #3 wash water concentration | *0.0087* | *Yes* |
| Average results | *0.0101* | *Yes* |

the tables represent examples of information that could be recorded in a validation protocol for a comminuting mill.

## Installation Qualification

An IQ evaluation establishes confidence that the equipment is properly installed (Figure 2.31). The installation must meet the manufacturer's specified guidelines,

***Figure 2.30. FitzMill Comminuting Machine (courtesy of Fitzpatrick Co.)***

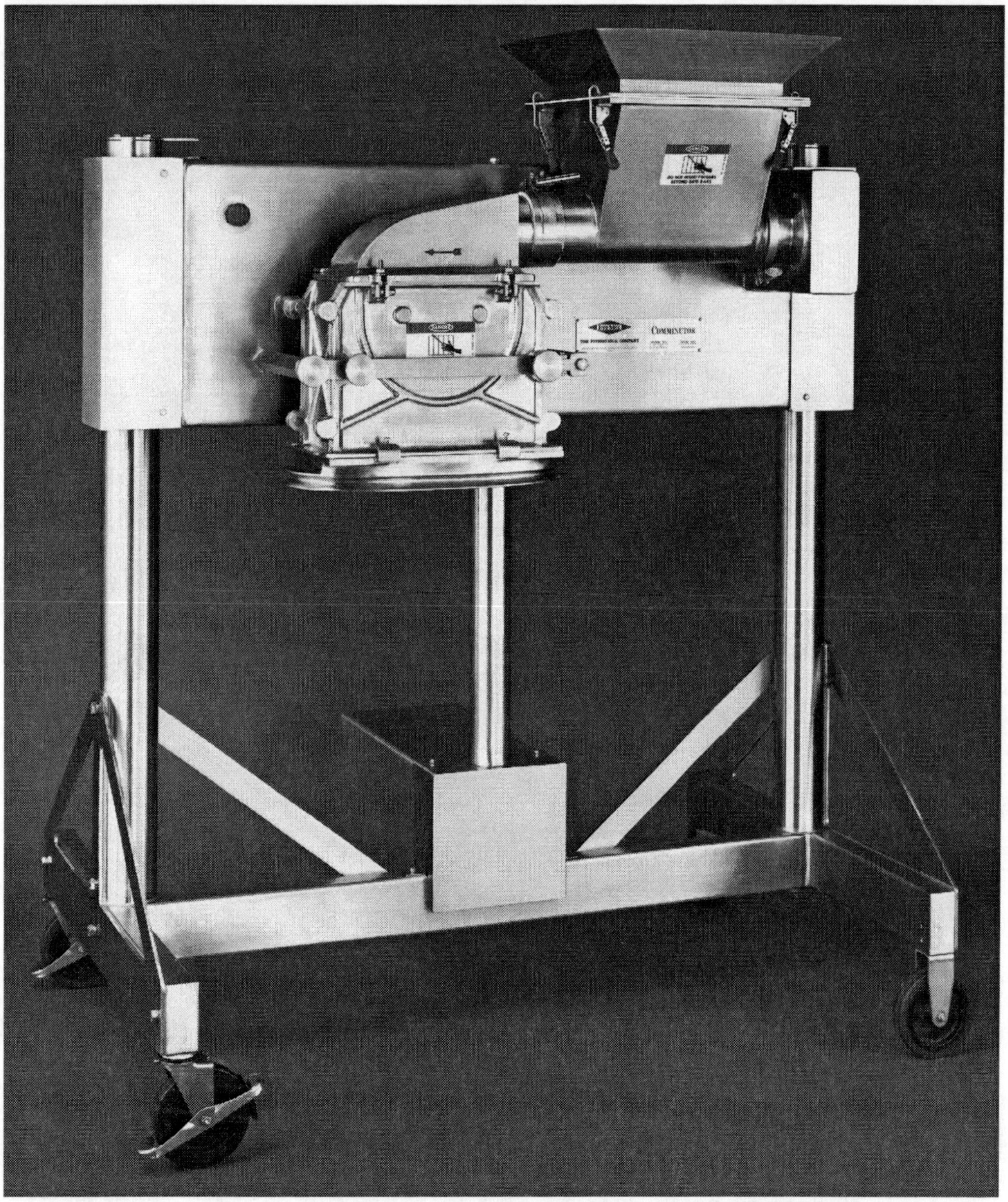

along with design changes at installation. Also, the supporting electrical utilities must meet all electrical codes. The information required for an IQ evaluation is equipment identification, required documentation, equipment utility requirements, major component specifications, component material, lubricants, and equipment safety features.

### *Equipment Identification*

Record the equipment identification numbers in Table 2.123, along with the following information: equipment manufacturer's purchase order number, model number, serial number, company assigned equipment number, and the location of the equipment.

### *Required Documentation*

Record the equipment manufacturer's operation and maintenance manual and drawings in Table 2.124. Record the SOPs that cover the setup, operation, and cleaning of the mill in Table 2.125.

**Figure 2.31. IQ Elements of a Comminuting Mill**

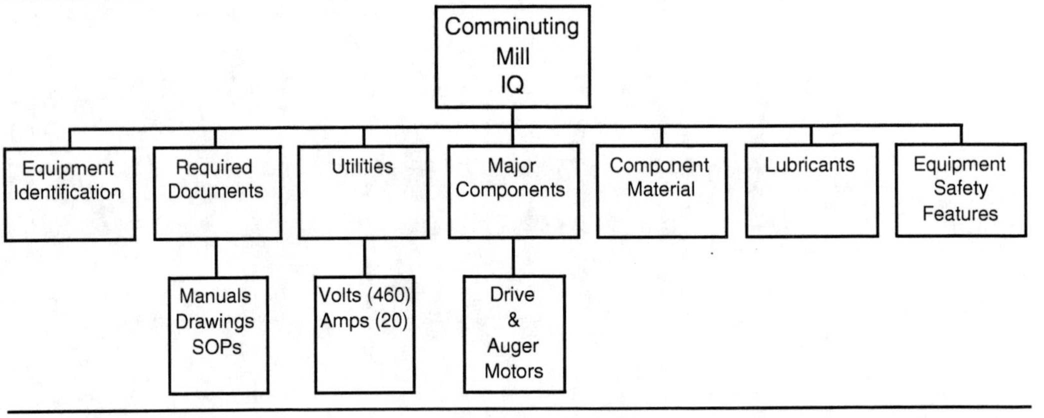


**Table 2.123. Equipment Identification**

| Required Information | As-Found Conditions |
|---|---|
| Manufacturer | *Any Mill Co.* |
| Purchase order number | *009235* |
| Model number | DAS60 |
| Serial number | *78801* |
| Equipment number | *6492* |
| Location | *Room 270 (This equipment is portable.)* |

**Table 2.124. *Manufacturers' Manuals and Drawings***

| Number | Description | Date |
|---|---|---|
| *M5095* | *Comminuting Machine Operating and Instructions Manual* | *None* |
| *3001-7400* | *Typical SFV Comminutor Drawing* | *02/28/78* |
| *3001-9153* | *SFV-SAD60-BSS Drawing* | *06/23/95* |
| *3001-6730* | *NMEA 4X 460 V AC Control Panel Layout Drawing* | *05/04/95* |

**Table 2.125. *Standard Operating Procedures***

| Number | Description | Date |
|---|---|---|
| *GRA005* | *Mill Auger Feed Setup, Operation, and Cleaning* | *02/16/96* |
| *GRA006* | *Mill Auger Feed Cleaning Procedure* | *02/23/96* |

### *Equipment Utility Requirements*

Compare the manufacturer's specified volts (V) and amps (A) requirements to their as-found conditions at the time of qualification testing and record the results in Table 2.126. Also record the location of the power supply source. Record the instrument used to measure the volts and amps in Table 2.127.

***Volt calculation:***

Supplied volt specification = 460 V ± 10%

±10% of 460 = ±46

460 + 46 = 506

**Table 2.126. *Utilities***

| Utility | Specified | Measured Results | Acceptable (Yes/No) |
|---|---|---|---|
| Volts | Drive motor = 460 ± 10% | *A-B 455; A-C 457; B-C 456* | *Yes* |
| Amps | Drive motor = 11<br>Auger motor = 1.17 | 20 circuit rating | *Yes* |

Power supply source, breaker box *BB37, wire numbers: 7, 9, 11*

**Table 2.127. *Instrument Used***

| Test Instrument | Identification Number | Calibration Due Date |
|---|---|---|
| *Multimeter* | *ME-025* | *04/19/97* |

460 – 46 = 414

The measured volts of 455/457/456 fall within ±10%.

***Amp calculation:***

Circuit rating = 20 A

Drive motor = 11 A

Auger motor = 1.17 A

Equipment current draw = drive motor A + auger motor A = 11 + 1.17 = 12.17 A

The circuit amp rating of 20 is greater than the maximum current draw of the equipment.

### *Major Component Specifications*

The component specifications of the protocol verifies that the mill components purchased were delivered and installed. Record the major components in Table 2.128.

### *Component Material*

Record the material of each component that contacts the product in Table 2.129.

### *Lubricants*

Record the lubricants used to operate the mill in Table 2.130 and indicate if they make contact with the product.

### *Equipment Safety Features*

The objective of testing equipment safety features is to verify that the safety features on the mill function according to manufacturer specifications. Use the following procedure and record the results in Table 2.131.

- Press the mill and auger feed push buttons and verify that the mill and auger motors will not start with the throat removed.
- Press the mill and auger feed push buttons and verify that the mill and auger motors will not start with the safety grid removed.

## Operational Qualification

An OQ evaluation should establish that the equipment can operate within specified tolerances and limits (Figure 2.32). The mechanical ranges of the mill are challenged, along with the basic mill operations that the operator will use. The mill will be validated for its operating ability, not how well it mills granulation. Information required for the OQ evaluation is calibration of the instruments used to control the

**Table 2.128. Major Components**

| Components | As-Found Conditions |
|---|---|
| Drive motor | Manufacturer: *Any Motor Co.*<br>Model number: *A519A*<br>Serial number: *F304K053X*<br>Volts: *460*<br>Amperes: *11*<br>Phases: *3*<br>Cycles: *60 Hz*<br>hp: *3*<br>rpm: *1,000, 2,450, 4,600* |
| Rotor blades | Material: *Stainless steel*<br>Type: *Knife/impact blades* |
| Drive motor | Manufacturer: *Any Motor Co.*<br>Model number: *A381A*<br>Serial number: *F611R223X*<br>Volts: *460*<br>Amperes: *1.17*<br>Phases: *3*<br>Cycles: *60 Hz*<br>hp: *1/4*<br>rpm: *21–208* |
| Auger feed | Material: *R-22, 121 lb*<br>Size: *2-7/8 in. diameter* |
| Screen | Part number: *A5600-1351*<br>Perforation size: *0.033 in.*<br>Shape: *Round*<br>Material: *316 stainless steel* |

**Table 2.129. Component Material**

| Component | Material |
|---|---|
| Mill chamber | *316 stainless steel* |
| Rotor blades | *316 stainless steel* |
| Auger feed | *Stainless steel* |
| Screen | *316 stainless steel* |

**Table 2.130. Lubricants**

| Where Used | Type | Manufacturer | Product Contact (Yes/No) |
|---|---|---|---|
| *Gearbox* | *Amolith #2 Oil* | *Any Oil Co.* | *No* |
| *Pillow block bearings and casters* | *Lubriplate 630-2 grease* | *Any Oil CO.* | *No* |
| Is there a preventive maintenance procedure on file? (Yes/No) | | | *Yes* |

**Table 2.131. Safety Features Test Results**

| Test Function | Expected Results | Acceptable (Yes/No) |
|---|---|---|
| Safety interlock switch operation | When the mill and auger feed push buttons are pressed, the mill and auger motors will not start with the throat removed. | *Yes* |
| | When the mill and auger feed push buttons are pressed, the mill and auger motors will not start with the safety grid removed. | *Yes* |

**Figure 2.32. OQ Elements of a Comminuting Mill**

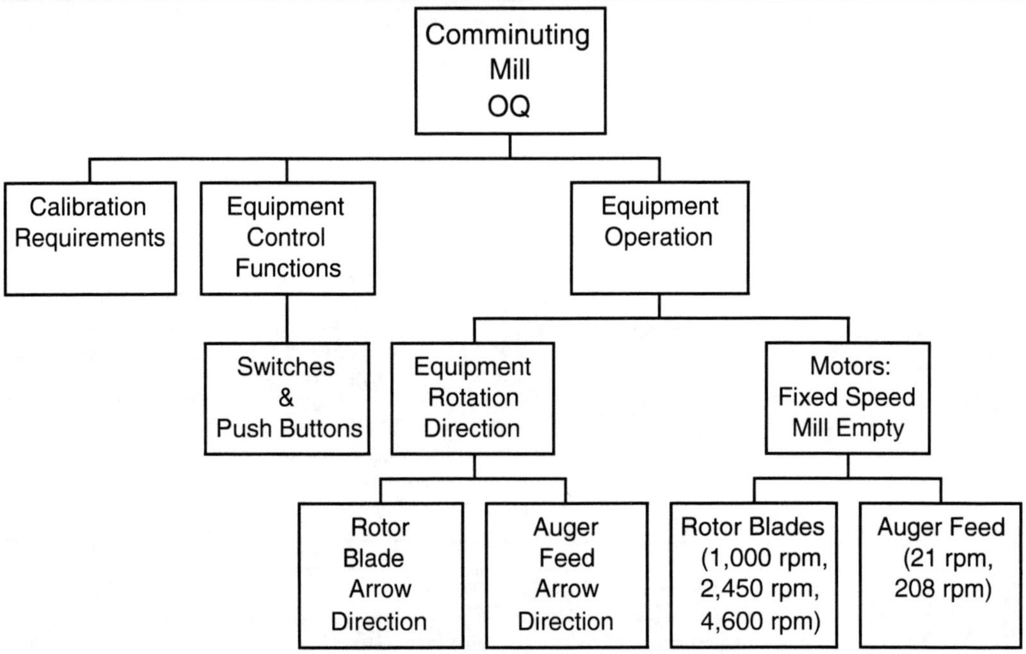


mill, equipment control functions (switches and push buttons), and equipment operation (mill and auger motor rotation direction, rotor blade and auger feed speed tests).

### *Calibration Requirements*

Verify that all critical instruments on the equipment have been logged into the calibration system, have calibration procedures in place, and are in calibration at the time of qualification testing. Record all information for the calibrated instruments used to control the mill in Table 2.132.

**Table 2.132. *Calibrated and Noncalibrated Instruments***

*There were no calibrated instruments on the blender.*

| Noncalibrated Instrument | As-Found Conditions |
|---|---|
| Identification number | *AM-1462* |
| Type | *Ammeter* |
| Manufacturer | *Any Gauge Co.* |
| Model number | *5449* |
| Serial number | *Not specified* |
| Range | *0–20 A* |
| Scale division | *0.5 A* |
| Location | *Mounted on the control panel* |
| Use | *Monitors the drive motor amperage* |
| Calibration due date | *Calibration not required* |
| Critical or not critical | *Not critical to the operation of the mill* |

### *Equipment Control Functions*

The objective of testing equipment control functions is to verify that the switches and push buttons on the mill operate per the manufacturer's specifications. Operate each control listed in Table 2.133 and verify its proper operation. Record the results in Table 2.133.

### *Equipment Operation: Rotor Blade and Auger Feed Rotation Direction Test*

The objective of the rotation direction test is to verify that the rotor blades and the auger feed rotate in the proper direction. The following tests will be performed with the mill empty. Record the results in Table 2.134.

- Jog the drive motor and verify that the rotor blades are rotating in the same direction as the incoming product from the auger feed and the mill direction arrow.
- Jog the drive motor and verify that the auger feed is rotating in the same direction as the mill direction arrow.

### *Equipment Operation: Rotor Blade and Auger Feed Speed Test*

The objective of the speed test is to document the speed of the rotor blades and the auger feed with the mill empty. Repositioning the belts on a multistep flywheel and the motor pulley produces multiple speeds of the rotor blades. The three settings provide speeds of approximately 1,000, 2,450, and 4,600 rpm. The auger feed speed control dial controls the auger feed. The 0 to 6 speed settings are gear ratios, and the actual speed range is approximately 21 to 208 rpm.

**Table 2.133. Equipment Control Function Test Results**

| Test Function | Expected Results | Acceptable (Yes/No) |
|---|---|---|
| Mill start push button operation | When the mill start push button is pressed, the mill motor starts and the green light goes on. | *Yes* |
| Mill stop push button operation | When the mill stop push button is pressed, the mill motor stops and the green light goes off. | *Yes* |
| Auger feed start push button operation | When the auger feed start push button is pressed with the mill motor running, the auger feed motor starts and the green light goes on. | *Yes* |
| Auger feed stop push button operation | When the auger feed stop push button is pressed with the mill motor running, the auger feed motor stops and the green light goes off. | *Yes* |
| Auger feed control dial operation | When the auger feed speed control dial is rotated from 0 to 6 and 6 to 0, the auger feed's speed increases and decreases. | *Yes* |
| Disconnect handle operation | When the mill start push button is pressed and the disconnect handle is turned on, the mill starts. | *Yes* |
| | When the mill start push button is pressed, and the disconnect handle is turned off, the mill does not start. | *Yes* |

**Table 2.134. Rotor Blade and Auger Feed Direction Test Results**

| Item | Expected Results | Results | Acceptable (Yes/No) |
|---|---|---|---|
| Rotor blade rotation direction | The rotor blade should be rotating in the same direction as the in-coming product from the auger feed and the mill direction arrow. | *The rotor blades rotated in the same direction as the mill direction arrow.* | *Yes* |
| Auger feed rotation direction | The auger feed should be rotating in the same direction as the mill direction arrows. | *The auger feed rotated in the same direction as the mill direction arrow.* | *Yes* |

A tachometer is required for the speed test. The following procedure is to be used:

- Measure the speed of the rotor blades with a calibrated tachometer. Verify that the measured speed is within ± 10 percent of the fixed speeds of 1,000, 2,450, and 4,600 rpm and record the results in Table 2.135. Record the instrument used to measure the speed in Table 2.137.

  Rotor blade speed specification = 1,000 rpm ± 10%

  ±10% of 1,000 = ±100

  1,000 + 100 = 1,100

  1000 – 100 = 900

  The measured rpm of 1,028 falls within ±10%.

**Table 2.135. *Rotor Blade Test Results***

| Specification (rpm ± 10%) | Rotor Flywheel Pulley Setting | Drive Motor Pulley Setting | Measured Speed (rpm) | Acceptable (Yes/No) |
|---|---|---|---|---|
| 1,000 | Large | Small | *1,028* | *Yes* |
| 2,450 | Small | Medium | *2,580* | *Yes* |
| 4,600 | Small | Large | *4,747* | *Yes* |

Rotor blade speed specification = 2,450 rpm ± 10%

±10% of 2,450 = ±245

2,450 + 245 = 2,695

2,450 – 245 = 2,205

The measured rpm of 2,580 falls within ±10%.

Rotor blade speed specification = 4,600 rpm ±10%

±10% of 4,600 = ±460

4,600 + 460 = 5,060

4,600 – 460 = 4,140

The measured rpm of 4,747 falls within ±10%.

- Measure the speed of the auger feed with a calibrated tachometer. Verify that the measured speed is within ±10 percent of the fixed speed of 21 rpm at speed setting 0 and 208 rpm at speed setting 6. Record the results in Table 2.136. Record the instrument used to measure the speed in Table 2.137.

**Table 2.136. *Auger Feed Speed Test Results***

| Auger Feed Speed Control Dial Setting | Speed Specification (rpm ± 10%) | Measured Speed (rpm) | Acceptable (Yes/No) |
|---|---|---|---|
| 0 | 21 | *21* | *Yes* |
| 6 | 208 | *211* | *Yes* |

**Table 2.137. *Instrument Used***

| Test Instrument | Identification Number | Calibration Due Date |
|---|---|---|
| *Tachometer* | *64020* | *06/21/97* |

Auger feed speed specification = 21 rpm ± 10%

±10% of 21 = ±2.1

21 + 2.1 = 23.1

21 – 2.1 = 18.9

The measured rpm of 21 falls within ±10%.

Auger feed speed specification = 208 rpm ± 10%

±10% of 208 = ±20.8

208 + 20.8 = 228.8

208 – 20.8 = 187.2

The measured rpm of 211 falls within ±10%.

## Performance Qualification

Once it has been established that the equipment is properly installed and functioning within specified operating parameters, it must be shown that the mill can perform reliably under routine, minimum, and maximum operating conditions (Figure 2.33).

***Figure 2.33. PQ Elements of a Comminuting Mill***

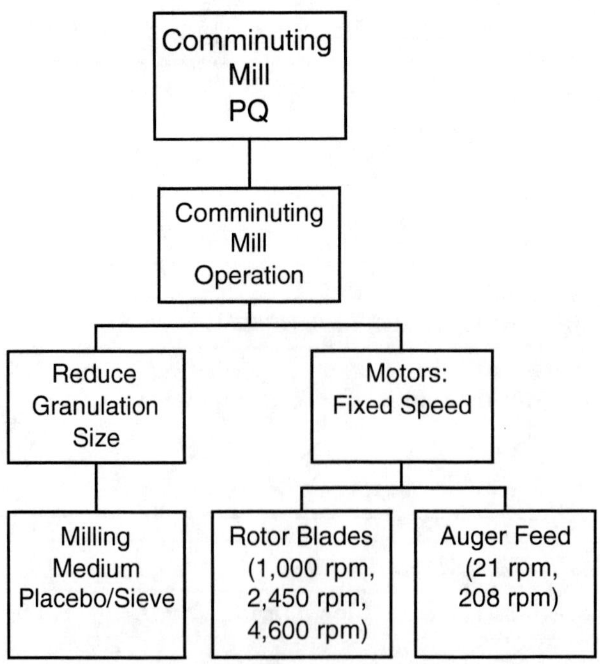

### *Mill Operation*

The mill operation test is designed to document the performance of the mill using placebo granulation. The placebo is used for maximum loading conditions. This test is also designed to verify that the mill can reduce the size of granules.

The materials and instruments required for this test are a placebo, a sieve, a pan, a scale, and a tachometer. Weigh the granulation in the sieve and pan before and after milling to verify the reduction in particle size. Use the following procedure to test mill operation:

- Pour approximately 1 kg of placebo through a sieve of mesh size sufficient to retain some granulation and catch the tailings in a pan. Record the tared weight of the granulation tailings from the sieve and pan in Table 2.138. Record the sieve used in Table 2.139.
- Pour the granulation tailings from the sieve and pan into a container. Record the tared weight of the sieved granulation to verify any loss from the original sample in Table 2.138.
- Setup the mill with a screen that has the same perforations as the sieve that is recorded in Table 2.139. Pour the granulation into the mill with the mill running at a low speed and then at a high speed and record the speeds in Table 2.140. Catch the milled granulation in a container.
- Pour the milled granulation through the original sieve and catch the tailings in the original pan used above. Record the tared weight of the granulation tailings from the sieve and the pan in Table 2.138.
- Verify that the milling operation reduces the overall particle size distribution by decreasing granulation retained by the sieve by weight.
- Measure the speed of the rotor blades with a calibrated tachometer. Verify that the measured speed is within ±10 percent of the fixed speed of 1,000, 2,450, and 4,600 rpm and record the results in Table 2.140. Record the instrument used to measure the speed in Table 2.142.

***Table 2.138. Mill Performance Test Results***

| Equipment | Granulation Weight Before Milling | Granulation Weight After Milling | Acceptable (Yes/No) |
|---|---|---|---|
| Sieve | *262 g* | *0 g* | *Yes* |
| Pan | *738 g* | *1,000 g* | *Yes* |
| Was the granulation size reduced? (Yes/No) | | | *Yes* |

***Table 2.139. Test Materials and Conditions***

| Item | Specifications |
|---|---|
| Lactose hydrous | *1 kg* |
| Sieve | *80 mesh size* |

***Table 2.140. Rotor Blade Test Results***

| Specification (rpm ± 10%) | Rotor Flywheel Pulley Setting | Drive Motor Pulley Setting | Measured Speed (rpm) | Acceptable (Yes/No) |
|---|---|---|---|---|
| 1,000 | Large | Small | *1,010* | *Yes* |
| 2,450 | Small | Medium | *2,461* | *Yes* |
| 4,600 | Small | Large | *4,652* | *Yes* |

Rotor blade speed specification = 1,000 rpm ± 10%

±10% of 1,000 = ±100

1,000 + 100 = 1,100

1,000 – 100 = 900

The measured rpm of 1,010 falls within ±10%.

Rotor blade speed specification = 2,450 rpm ± 10%

±10% of 2,450 = ±245

2,450 + 245 = 2,695

2,450 – 245 = 2,205

The measured rpm of 2,461 falls within ±10%.

Rotor blade speed specification = 4,600 rpm ± 10%

±10% of 4,600 = ±460

4,600 + 460 = 5,060

4,600 – 460 = 4,140

The measured rpm of 4,652 falls within ±10%.

- Measure the speed of the auger feed with a calibrated tachometer. Verify that the measured speed is within ±10 percent of the fixed speed of 21 rpm at speed setting 0 and 208 rpm at speed setting 6. Record the results in Table 2.141. Record the instrument used to measure the speed in Table 2.142.

Auger feed speed specification = 21 rpm ± 10%

±10% of 21 = ±2.1

21 + 2.1 = 23.1

21 – 2.1 = 18.9

The measured rpm of 20 falls within ±10%.

**Table 2.141. *Auger Feed Speed Test Results***

| Auger Feed Speed Control Dial Setting | Speed Specification (rpm ± 10%) | Measured Speed (rpm) | Acceptable (Yes/No) |
|---|---|---|---|
| 0 | 21 | *20* | *Yes* |
| 6 | 208 | *209* | *Yes* |

**Table 2.142. *Instruments Used***

| Test Instrument | Identification Number | Calibration Due Date |
|---|---|---|
| *Scale* | *PE600* | *07/18/97* |
| *Stopwatch* | *54039* | *02/16/97* |
| *Tachometer* | *64020* | *06/21/97* |

Auger feed speed specification = 208 rpm ± 10%

±10% of 208 = ±20.8

208 + 20.8 = 228.8

208 – 20.8 = 187.2

The measured rpm of 209 falls within ±10%.

## Sieve Shaker

The validation of a sieve shaker (Figure 2.34) that is motor driven, operates at a fixed speed, and is manually operated is described here. The sieve shaker consists of a sieving head assembly that vibrates via power provided by a 1 hp motor. The sieving head consists of the main hopper rim, screen rim frame, mesh support base, and discharge funnel. Screens may be changed to control the size of the particles resulting from the sieving operation. Eight screens range in size from 14 to 80 mesh. The vibrasonic control unit consists of a probe attached to the middle of a screen, an electrical wire, and a control panel that controls the probe. The sieve shaker is used for delumping powder or granulation to a desired particle size distribution.

The validation protocol for a sieve shaker is described in the following sections. This protocol example can be used as a cGMP compliance guide and it should be tailored to meet individual company requirements. Italicized words in the tables represent examples of information that could be recorded in a validation protocol for a sieve shaker.

***Figure 2.34. Sieve Shaker (courtesy of Russell Finex, Inc.)***

## Installation Qualification

An IQ evaluation establishes confidence that the equipment is properly installed (Figure 2.35). The installation must meet the manufacturer's specified guidelines, along with design changes at installation. Also, the supporting electrical utilities must meet all electrical codes. The information required for an IQ evaluation is equipment identification, required documentation, equipment utility requirements, major component specifications, component material, lubricants, and equipment safety features.

***Figure 2.35. IQ Elements of a Sieve Shaker***

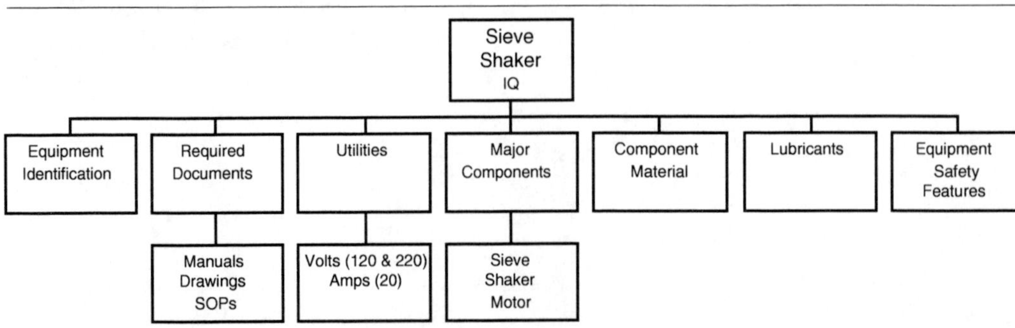

### *Equipment Identification*

Record the equipment identification numbers in Table 2.143, along with the following information: equipment manufacturer's purchase order number, model number, serial number, company assigned equipment number, and the location of the equipment.

### *Required Documentation*

Record the equipment manufacturer's operation and maintenance manual and drawings in Table 2.144. Record the SOPs that cover the setup, operation, and cleaning of the sieve shaker in Table 2.145.

### *Equipment Utility Requirements*

Compare the manufacturer's specified volts (V) and amps (A) requirements to their as-found conditions at the time of qualification testing and record the results in Table 2.146. Also record the location of the power supply source. Record the instrument used to measure the volts and amps in Table 2.147.

**Table 2.143. Equipment Identification**

| Required Information | As-Found Conditions |
|---|---|
| Manufacturer | *Any Sieve Co.* |
| Purchase order number | *007734* |
| Model number | 51052 |
| Serial number | *3010* |
| Equipment number | *2362* |
| Location | *PRD lab (This equipment is portable.)* |

**Table 2.144. Manufacturers' Manuals and Drawings**

| Number | Description | Date |
|---|---|---|
| *400-FIN* | *Sieve Shaker Operating and Instructions Manual* | *02/28/93* |
| *1500-FBR* | *Main Sieve Rim and Parts Drawing* | *01/22/93* |

**Table 2.145. Standard Operating Procedures**

| Number | Description | Date |
|---|---|---|
| *GRA023* | *Sieve Shaker Setup, Operation, and Cleaning* | *03/14/95* |
| *GRA024* | *Sieve Shaker Cleaning Procedure* | *09/22/95* |

**Table 2.146. Utilities**

| Utility | Specified | Measured Results | Acceptable (Yes/No) |
|---|---|---|---|
| Volts | Control unit = 120 ± 10% | *124* | *Yes* |
| | Sieve shaker = 220 ± 10% | *A-B 221; A-C 219; B-C 223* | *Yes* |
| Amps | Sieve shaker motor = 3 | 20 circuit rating | *Yes* |

Power supply source, breaker box *BB28, wire numbers: 1 and 13, 15, 17*

**Table 2.147. Instrument Used**

| Test Instrument | Identification Number | Calibration Due Date |
|---|---|---|
| *Multimeter* | *ME-025* | *04/19/97* |

***Volt calculation:***

Control unit volt specification = 120 V ± 10%

+10% of 120 = ±12

120 + 12 = 132

120 – 12 = 108

The measured volts of 124 falls within ±10%.

Sieve shaker supplied volt specification = 220 V ± 10%

±10% of 220 = ±22

220 + 22 = 242

220 – 22 = 198

The measured volts of 221/219/223 fall within ±10%.

***Amp calculation:***

Circuit rating = 20 A

Equipment current draw = 3 A

The circuit amp rating of 20 is greater than the maximum current draw of the equipment.

### *Major Component Specifications*

The component specifications section of the protocol verifies that the sieve shaker components purchased were delivered and installed. Record the major components in Table 2.148.

**Table 2.148. Major Components**

| Components | As-Found Conditions |
|---|---|
| Sieve shaker motor | Manufacturer: *Any Motor Co.*<br>Model number: *51-FBR*<br>Serial number: *63251*<br>Volts: *220*<br>Amperes: *3*<br>Phases: *3*<br>Cycles: *60 Hz*<br>hp: *1.0*<br>rpm: *3,450* |
| Screens | Material: *316 stainless steel*<br>Mesh sizes: *14, 20, 24, 30, 40, 50, 60, 80*<br>Type: *Tensile bolting cloth* |
| Main rim | Part number: *A45015*<br>Material: *316 stainless steel*<br>Size: *2 in. deep* |

### *Component Material*

Record the material of each component that contacts the product in Table 2.149.

### *Lubricants*

Record the lubricants used to operate the sieve shaker in Table 2.150 and indicate if they make contact with the product.

**Table 2.149. Component Material**

| Component | Material |
|---|---|
| Screens | *316 stainless steel* |
| Shallow funnel | *316 stainless steel* |
| Main rim | *316 stainless steel* |
| Vibrasonic probe | *Stainless steel* |

**Table 2.150. Lubricants**

| Where Used | Type | Manufacturer | Product Contact (Yes/No) |
|---|---|---|---|
| *Flywheel bearings* | *Ore-Lube K2* | *Any Oil Co.* | *No* |
| Is there a preventive maintenance procedure on file? (Yes/No) | | | *Yes* |

### *Equipment Safety Features*

There are no safety features on this equipment. Nevertheless, there are some general safety precautions that should be followed when operating a sieve shaker. Do not make any adjustments when the equipment is operating. Always ensure that when a electrical lead is disconnected, the vibrasonic generator/control box is disconnected from the power supply.

## Operational Qualification

An OQ evaluation should establish that the equipment can operate within specified tolerances and limits (Figure 2.36). The mechanical ranges of the sieve shaker are challenged, along with the basic operations. The sieve shaker will be validated for its operating ability, not how well it sieves granulation. Information required for the OQ evaluation is calibration of the instruments used to control the sieve shaker, equipment control functions (switches and push buttons) and equipment operation (sieve shaker motor rotation direction and motor speed).

### *Calibration Requirements*

Verify that all critical instruments on the equipment have been logged into the calibration system, have calibration procedures in place, and are in calibration at the time of qualification testing. Record all information for the calibrated instruments used to control the sieve shaker in Table 2.151.

***Figure 2.36. OQ Elements of a Sieve Shaker***

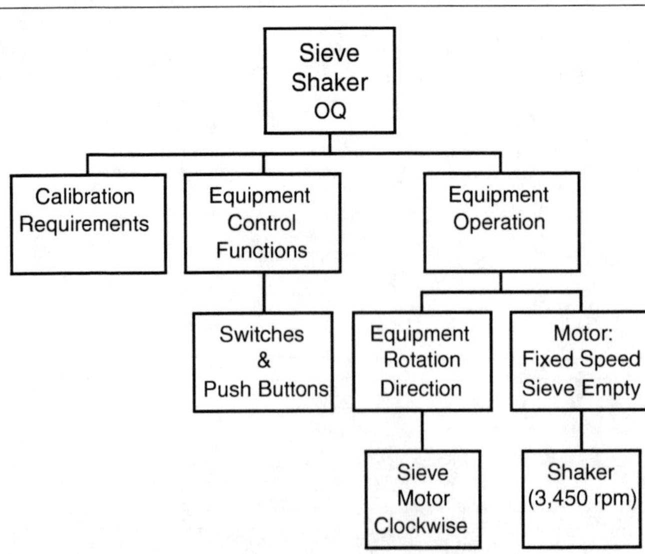

**Table 2.151. Calibrated and Noncalibrated Instruments**

*There were no calibrated or noncalibrated instruments on the blender.*

### *Equipment Control Functions*

The objective of testing equipment control functions is to verify that the switches, push buttons, and indicators on the sieve shaker operate per the manufacturer's specifications. Operate each control listed in Tables 2.152 and 2.153 and verify its proper operation. Record the results in Tables 2.152 and 2.153.

### *Equipment Operation: Sieve Shaker Motor Rotation Direction Test*

The objective of the rotation direction test is to verify that the sieve shaker motor rotates in the proper direction. The test is performed with the sieve shaker empty. Jog the motor and verify that the motor rotates in a clockwise direction as viewed from the top of the motor. Record the results in Table 2.154.

**Table 2.152. Equipment Control Function Test Results**

| Test Function | Expected Results | Acceptable (Yes/No) |
|---|---|---|
| ON/OFF switch for sieve shaker | When the ON/OFF switched is turned on and off, the sieve shaker motor goes on and off. | *Yes* |
| ON/OFF switch for vibrasonic probe | When the vibrasonic probe switch is turned on and off, the vibrasonic probe goes on and off. | *Yes* |

**Table 2.153. Equipment Indicator Function Test Results**

| Test Function | Expected Results | Acceptable (Yes/No) |
|---|---|---|
| Normal running operation | When the equipment is in normal operation, LED 1 goes on. | *Yes* |
| Undervoltage at output operation | When the output voltage of the generator is too low, LED 2 goes on. | *Yes* |
| Overvoltage at output operation | When the output voltage of the generator is too high, LED 3 goes on. | *Yes* |
| Open circuit in output operation | When there is an open circuit in the input lead/connections from the unit, LEDs 2 and 4 go on. | *Yes* |
| Short circuit in output operation | When there is a short circuit in the output of the equipment, LEDs 2 and 5 go on. | *Yes* |

**Table 2.154. Rotor Blade and Auger Feed Rotation Direction Test Results**

| Item | Expected Results | Results | Acceptable (Yes/No) |
|---|---|---|---|
| Sieve shaker motor rotation direction | Rotation should be in a clockwise direction as viewed from the top of the motor. | *The motor rotated in a clockwise direction.* | *Yes* |

### *Equipment Operation: Sieve Shaker Motor Speed Test*

The objective of the motor speed test is to document the speed of the sieve shaker motor with the sieve shaker empty. The motor operates at a fixed speed of 3,450 rpm. Measure the speed of the motor with a calibrated tachometer and verify that the measured speed is within ±10 percent of the fixed speed of 3,450. Record the results in Table 2.155. Record the instrument used to measure the speed in Table 2.156.

Motor speed specification = 3,450 rpm ± 10%

±10% of 3,450 = ±345

3,450 + 345 = 3,795

3,450 – 345 = 3,105

The measured rpm of 3,455 falls within ±10%.

## Performance Qualification

Once it has been established that the equipment is properly installed and functioning within specified operating parameters, it must be shown that the sieve shaker can operate reliably under routine, minimum, and maximum operating conditions (Figure 2.37).

**Table 2.155. Sieve Shaker Motor Speed Test Results**

| Speed Specification (rpm ± 10%) | Measured Speed (rpm) | Acceptable (Yes/No) |
|---|---|---|
| 3,450 | *3,455* | *Yes* |

**Table 2.156. Instrument Used**

| Test Instrument | Identification Number | Calibration Due Date |
|---|---|---|
| *Tachometer* | *64020* | *06/21/97* |

***Figure 2.37. PQ Elements of a Sieve Shaker***

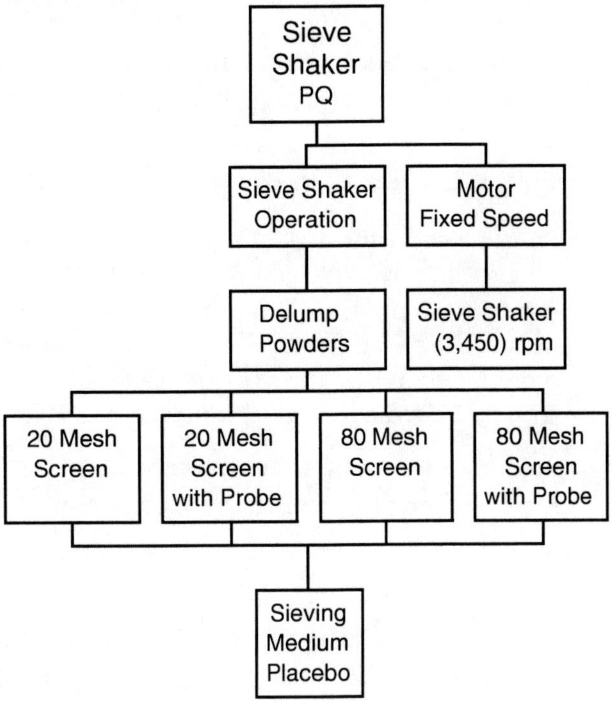


### *Sieve Shaker Operation*

The sieve shaker operation test is designed to document the performance of the sieve shaker using placebo granulation. The placebo is used for maximum loading conditions. This test is also designed to verify that the sieve shaker can delump powder or granulation to a desired particle size distribution.

The materials and instruments required for this test are a placebo, screens, a scale, and a tachometer. The following procedure is used:

- Pour placebo through the sieve shaker with a 20 mesh screen in place. Record the results in Table 2.158. Pour the sieve-characterized placebo through a reference 20 mesh screen. Record the results in Table 2.158. Verify that at least 95 percent of the placebo passes through the reference 20 mesh screen. Record the placebo and the screen used in Table 2.157. Record the instrument used to weigh the placebo in Table 2.160.

  Reference 20 mesh screen placebo weight ÷ 20 mesh screen placebo weight = percent of placebo that passed through the reference screen.

  486.3 g ÷ 527.8 g = 97%

- Pour placebo through the sieve shaker with a 20 mesh screen and a vibrasonic probe in place and running. Record the results in Table 2.158. Pour the sieve-characterized placebo through a reference 20 mesh screen. Record the results in Table 2.158. Verify that at least 95 percent of the placebo passes through the reference 20 mesh screen. Record the placebo and the screen used in Table 2.157. Record the instrument used to weigh the placebo in Table 2.160.

Reference 20 mesh screen placebo weight with probe ÷ 20 mesh screen placebo weight = percent of placebo that passed through the reference screen.

387.8 g ÷ 395.8 g = 98%

- Pour placebo through the sieve shaker with an 80 mesh screen in place. Record the results in Table 2.158. Pour the sieve-characterized placebo through a reference 80 mesh screen. Record the results in Table 2.158. Verify that at least 95 percent of the placebo passes through the reference 80 mesh screen. Record the placebo and the screen used in Table 2.157. Record the instrument used to weigh the placebo in Table 2.160.

  Reference 80 mesh screen placebo weight ÷ 80 mesh screen placebo weight = percent of placebo that passed through the reference screen.

  45.7 g ÷ 47.6 g = 97%

- Pour placebo through the sieve shaker with an 80 mesh screen and a vibrasonic probe in place and running. Record the results in Table 2.158. Pour the sieve-characterized placebo through a reference 80 mesh screen. Record the results in Table 2.158. Verify that at least 95 percent of the placebo passes through the reference 80 mesh screen. Record the placebo and the screen used in Table 2.157. Record the instrument used to weigh the placebo in Table 2.160.

**Table 2.157. Test Materials**

| Item | Specifications |
|---|---|
| Placebo blend | *1 kg* |
| Screens (mesh size) | *20, 20 with prove, 80, 80 with probe* |

**Table 2.158. Sieve Shaker Performance Test Results**

| Screen Mesh Size | Placebo Weight Sieve | Characterized Placebo Weight Reference | Percentage | Acceptable (Yes/No) |
|---|---|---|---|---|
| 20 | *527.8 g* | *486.3 g* | *97* | *Yes* |
| 20 with probe | *395.8 g* | *387.8 g* | *98* | *Yes* |
| 80 | *47.6 g* | *45.7 g* | *97* | *Yes* |
| 80 with probe | *26.6 g* | *26.1 g* | *97* | *Yes* |
| Did at least 95% of the placebo pass through the reference screen? (Yes/No) | | | | *Yes* |

Reference 80 mesh screen placebo weight with probe ÷ 80 mesh screen placebo weight = percent of placebo that passed through the reference screen.

26.1 g ÷ 26.6 g = 97%

- Measure the speed of the sieve shaker mixer motor with a calibrated tachometer and verify that the measured speed is within ±10 percent of the fixed speed of 3,450. Record the results in Table 2.159. Record the instrument used to measure the speed in Table 2.160.

  Motor speed specification = 3,450 rpm ± 10%

  ±10% of 3,450 = ±345

  3,450 + 345 = 3,795

  3,450 – 345 = 3,105

  The measured rpm of 3,449 falls within ±10%.

**Table 2.159. Sieve Shaker Motor Speed Test Results**

| Speed Specification (rpm ± 10%) | Measured Speed (rpm) | Acceptable (Yes/No) |
|---|---|---|
| 3,450 | *3,449* | *Yes* |

**Table 2.160. Instruments Used**

| Test Instrument | Identification Number | Calibration Due Date |
|---|---|---|
| *Scale* | *PE600* | *07/18/97* |
| *Tachometer* | *64020* | *06/21/97* |

# Chapter 3

# Validation of Tableting Equipment

This chapter will cover the qualification testing of equipment used to perform operations in the tableting department (Figure 3.1).

## Tablet Press

The validation of a 45-station tablet press (Figure 3.2) will be described here. The tablet press is an automatic, high-speed rotary press. A motor drives the press at speeds that vary from 410 to 1,630 tablets per minute (tpm). The material being tableted is fed from a hopper by gravity through the feed frame into dies. Regulating the weight adjusting cam controls the weight of material in each tablet

***Figure 3.1. Tableting Equipment***

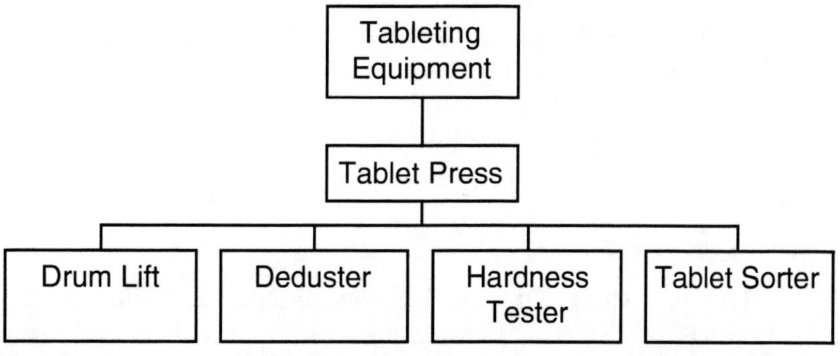

**Figure 3.2. Fette Press (courtesy of Fette America, Inc.)**

volumetrically. The tablets are made when material is compressed between the punches as they pass between the pressure rolls. For ejection, the upper punches are raised out of the die and clear the feed frame; the lower ejection cam lifts the lower punches to push the tablets out of the dies. Takeoff bars on the feed frame remove tablets from the press.

The validation protocol for a tablet press is described in the following sections. This protocol example can be used as a cGMP compliance guide, and it should be tailored to meet individual company requirements. Italicized words in the tables represent examples of information that could be recorded in a validation protocol for a tablet press.

## Installation Qualification

An IQ evaluation establishes confidence that the equipment is properly installed (Figure 3.3). The installation must meet the manufacturer's specified guidelines,

***Figure 3.3. IQ Elements of a Tablet Press***

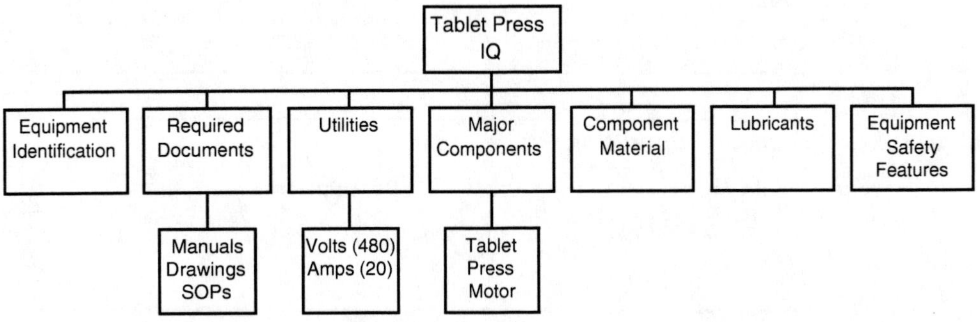


along with design changes at installation. Also, the supporting electrical utilities must meet all electrical codes. The information required for an IQ evaluation is equipment identification, required documentation, equipment utility requirements, major component specifications, component material, lubricants, and equipment safety features.

### *Equipment Identification*

Record the equipment identification numbers in Table 3.1, along with the following information: equipment manufacturer's purchase order number, model number, serial number, company assigned equipment number, and the location of the equipment.

### *Required Documentation*

Record the equipment manufacturer's operation and maintenance manual and drawings in Table 3.2. Record the SOPs that cover the setup, operation, and cleaning of the tablet press in Table 3.3.

***Table 3.1. Equipment Identification***

| **Required Information** | **As-Found Conditions** |
|---|---|
| Manufacturer | *Any Tablet Press Co.* |
| Purchase order number | *045839* |
| Model number | *45 station* |
| Serial number | *852966TS* |
| Equipment number | *4000* |
| Location | *Room 516* |

**Table 3.2.** ***Manufacturers' Manuals and Drawings***

| Number | Description | Date |
|---|---|---|
| *None* | *Tablet Press Installation, Operation, and Maintenance Manual* | *10/25/94* |

**Table 3.3.** ***Standard Operating Procedures***

| Number | Description | Release Date |
|---|---|---|
| *TAB021* | *45-Station Tablet Press Setup and Operation* | *10/01/94* |
| *TAB022* | *45-Station Tablet Press Cleaning Procedure* | *09/23/94* |

### *Equipment Utility Requirements*

Compare the manufacturer's specified volts (V) and amps (A) requirements to their as-found conditions at the time of qualification testing and record the results in Table 3.4. Also record the location of the power supply source. Record the instrument used to measure the volts and amps in Table 3.5.

***Volt calculation:***

Tablet press motor volt specification = 480 V ± 10%

±10% of 480 = ±48

480 + 48 = 528

480 – 48 = 432

The measured volts of 480/489/479 fall within ±10%.

***Amp calculation:***

Circuit rating = 20 A

Equipment current draw = 8.5 A

The circuit amp rating of 20 is greater than the maximum current draw of the equipment.

### *Major Component Specifications*

The component specifications section of the protocol verifies that the tablet press components purchased were delivered and installed. Record the major components in Table 3.6.

### *Component Material*

Record the material of each component that contacts the product in Table 3.7.

**Table 3.4. *Utilities***

| Utility | Specified | Measured Results | Acceptable (Yes/No) |
|---|---|---|---|
| Volts | 480 ± 10% | *A-B 480; A-C 489; B-C 479* | *Yes* |
| Amps | Motor = 8.5 | 20 circuit rating | *Yes* |

Power supply source, breaker box *BB15, wire numbers: 7, 9, 11*

**Table 3.5. *Instrument Used***

| Test Instrument | Identification Number | Calibration Due Date |
|---|---|---|
| *Multimeter* | *ME-025* | *04/19/97* |

**Table 3.6. *Major Components***

| Components | As-Found Conditions |
|---|---|
| Tablet press motor | Manufacturer: *Any Motor Co.*<br>Model number: *T1163M*<br>Serial number: *67451*<br>Volts: *480*<br>Amperes: *8.5*<br>Phases: *3*<br>Cycles: *60 Hz*<br>hp: *5*<br>rpm: *1,725* |
| Clutch | Manufacturer: *Any Clutch Co.*<br>Part number: *C4170*<br>Serial number: *57219* |

**Table 3.7. *Component Material***

| Component | Material |
|---|---|
| Rotor head | *Mild steel* |
| Hopper | *Stainless steel* |
| Punches | *Mild steel* |
| Dies | *Mild steel* |

### *Lubricants*

Record the lubricants used to operate the tablet press in Table 3.8 and indicate if they make contact with the product.

**Table 3.8. Lubricants**

| Where Used | Type | Manufacturer | Product Contact (Yes/No) |
|---|---|---|---|
| *Motor* | *APG 90 Oil* | *Any Oil Co.* | *No* |
| *Rotor head gears* | *Lubriplate NLG1 No. 1 Grease* | *Any Oil Co.* | *No* |
| *Punches and dies* | *Lubriplate Super FML-2* | *Any Oil Co.* | *Yes* |
| Is there a preventive maintenance procedure on file? (Yes/No) | | | *Yes* |

### *Equipment Safety Features*

The objective of testing equipment safety features is to verify that the safety features on the tablet press function according to the manufacturer's specifications. This test is performed with the tablet press empty. Verify that all of the guards are present and record the results in Table 3.9.

## Operational Qualification

An OQ evaluation should establish that the equipment can operate within specified tolerances and limits (Figure 3.4). The mechanical ranges of the tablet press are challenged, along with basic tablet press operations. The tablet press will be validated for its operating ability, not how well it makes tablets. Information required for the OQ evaluation is calibration of the instruments used to control the tablet press, equipment control functions (switches and push buttons) and equipment operation (cam tracks, upper punches, lower punches, feed frames, takeoff bars, rotor head rotation direction, tablet press speed).

**Table 3.9. Safety Features Test Results**

| Test Function | Expected Results | Acceptable (Yes/No) |
|---|---|---|
| Gear guard | The gear guard was present and functional. | *Yes* |
| Belt guard | The belt guard was present and functional. | *Yes* |
| Upper punch guard | The upper punch guard was present and functional. | *Yes* |
| Rear guard apron | The rear guard apron was present and functional. | *Yes* |
| Clutch guard | The clutch guard was present and functional. | *Yes* |
| Splash guard | The splash guard was present and functional. | Yes |

**Figure 3.4. OQ Elements of a Tablet Press**

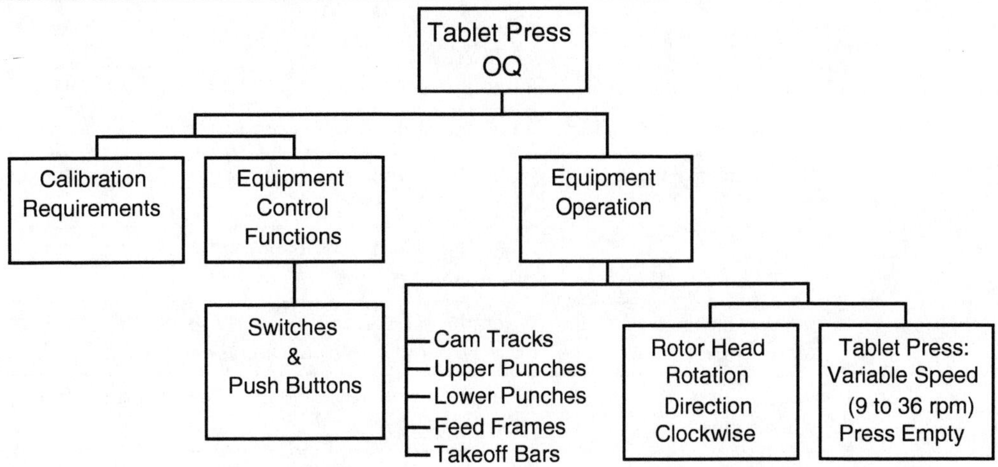


## *Calibration Requirements*

Verify that all critical instruments on the equipment have been logged into the calibration system, have calibration procedures in place, and are in calibration at the time of qualification testing. Record all information for the calibrated instruments used to control the tablet press in Table 3.10.

## *Equipment Control Functions*

The objective of testing equipment control functions is to verify that the switches and push buttons on the tablet press operate per the manufacturer's specifications. Record the results in Table 3.11. The tests will be performed with the tablet press empty. Operate each control listed in Table 3.11 and verify its proper operation.

## *Equipment Operation: Cam Tracks Test*

The objective of the cam tracks test is to verify that the upper and lower cam tracks make contact with the upper punches according to the manufacturer's specification. Use the following procedure and record the results in Table 3.12:

- Install the punches and verify that the cams are contacting the punch head angles on both sides of the double-sided cams.
- Verify that the punches are contacting one side of the single-sided cam through a full cam track, upper and lower.

**Table 3.10. Calibrated and Noncalibrated Instruments**

| |
|---|
| *There were no calibrated or noncalibrated instruments on the blender.* |

**Table 3.11. Equipment Control Function Test Results**

| Test Function | Expected Results | Acceptable (Yes/No) |
|---|---|---|
| Start switch operation | When the start switch is pressed, the motor starts. | *Yes* |
| Stop switch operation | When the stop switch is pressed, the motor stops. | *Yes* |
| Clutch lever operation | When the clutch lever is pushed clockwise, the clutch is engaged. | *Yes* |
| | When the clutch lever is pushed counterclockwise, the clutch is disengaged. | *Yes* |
| Safety-adjusting wheel operation | When the safety-adjusting wheel is turned, the pointer moves up and down indicating compression force. | *Yes* |
| Pressure-adjusting wheel operation | When the pressure-adjusting wheel is turned, the pressure is increased. | *Yes* |
| Weight-adjusting wheel operation | When the weight-adjusting wheel is turned, the lower punches are lifted. | *Yes* |
| Handwheel operation | When the handwheel is turned, the rotor head rotates clockwise. | *Yes* |
| Motor-adjusting wheel operation | When the motor-adjusting wheel is turned clockwise, the press speed increases. | *Yes* |
| | When the motor-adjusting wheel is turned counter-clockwise, the press speed decreases. | *Yes* |

**Table 3.12. Cam Tracks Test Results**

| Item | Expected Results | Results | Acceptable (Yes/No) |
|---|---|---|---|
| Cam tracks | Cam tracks should touch the head angles on both sides of the cams. | *Cam tracks touched the heads on both sides.* | *Yes* |
| | Punches should contact one side of a single-sided cam through a full cam track, upper and lower. | *Punches contacted one side through a full cam track, upper and lower.* | *Yes* |

### *Equipment Operation: Upper Punch Test*

The objective of the upper punch test is to verify that the upper punch penetration is according to the manufacturer's specification. A Vernier caliper is required for this test, which is performed as follows:

- Attach a piece of tape to mark the depth of penetration of an upper punch when it is set to a standard depth.
- Remove the upper punch and use a calibrated Vernier caliper to measure the depth of penetration into the die. Record the results in Table 3.13 and record the instrument used to measure the depth in Table 3.14.

**Table 3.13. *Upper Punch Test Results***

| Item | Expected Results | Results | Acceptable (Yes/No) |
|---|---|---|---|
| Upper punch penetration depth | 0.250 in. ± 0.125 in. | *0.237 in.* | *Yes* |

**Table 3.14. *Test Instrument Used***

| Test Instrument | Identification Number | Calibration Due Date |
|---|---|---|
| *Vernier caliper* | *20989* | *06/06/97* |

Depth of penetration specification = 0.250 in. ± 0.125 in.

0.250 + 0.125 = 0.375

0.250 – 0.125 = 0.125

The measured depth of 0.237 in. falls within the specification.

### *Equipment Operation: Lower Punch Test*

The objective of the lower punch test is to verify that the lower punch height is set according to the manufacturer's specification. A dial indicator is required for this test. Measure the height of the lower punch above the die with a dial indicator and record the results in Table 3.15. Record the instrument used to measure the height in Table 3.16.

Height of lower punch specification = 0.003 in. to 0.005 in.

The measured depth of 0.005 in. falls within the specification.

**Table 3.15. *Lower Punch Test Results***

| Item | Expected Results | Results | Acceptable (Yes/No) |
|---|---|---|---|
| Lower punch height | 0.003 in. ± 0.005 in. | *0.005 in.* | *Yes* |

**Table 3.16. *Test Instrument Used***

| Test Instrument | Identification Number | Calibration Due Date |
|---|---|---|
| *Dial indicator* | *7556A1* | *06/07/97* |

### *Equipment Operation: Feed Frame Test*

The objective of the feed frame test is to verify that the feed frame distance above the rotor head is according to the manufacturer's specification. A feeler gauge is required for this test. Measure the clearance between the feed frame and the rotor head with a feeler gauge and record the results in Table 3.17. Record the instrument used to measure the clearance in Table 3.18.

Feed frame clearance specification = 0.003 in. to 0.005 in.

The measured clearance of 0.004 in. falls within the specification.

### *Equipment Operation: Takeoff Bar Test*

The objective of the takeoff bar test is to verify that the takeoff bars do not make contact with the lower punches. Turn the tablet press by hand and verify that the takeoff bars do not make contact with the lower punches. Record the results in Table 3.19.

### *Equipment Operation: Tablet Press Rotation Direction*

The objective of the rotation direction test is to verify that the rotor head rotates in the proper direction. The tests will be performed with the tablet press empty. Press the start switch and observe the direction of rotation of the rotor head as viewed from in front of the press and record the results in Table 3.20.

**Table 3.17. Feed Frame Test Results**

| Item | Expected Results | Results | Acceptable (Yes/No) |
|---|---|---|---|
| Feed frame clearance | 0.003 in. ± 0.005 in. | *0.004 in.* | *Yes* |

**Table 3.18. Test Instrument Used**

| Test Instrument | Identification Number | Calibration Due Date |
|---|---|---|
| *Feeler gauge* | *10975* | *06/07/97* |

**Table 3.19. Takeoff Bar Test Results**

| Item | Expected Results | Results | Acceptable (Yes/No) |
|---|---|---|---|
| Takeoff bars | When turning the tablet press by hand, verify that takeoff bars do not make contact with the lower punches. | *Takeoff bars did not make contact with the lower punches.* | *Yes* |

**Table 3.20. *Rotor Head Rotation Direction Test Results***

| Item | Expected Results | Results | Acceptable (Yes/No) |
|---|---|---|---|
| Rotor head rotation direction | Rotation should be clockwise as viewed from in front of the tablet press. | *Clockwise rotation was observed.* | *Yes* |

### *Equipment Operation: Tablet Press Speed*

The objective of the speed test is to verify that the measured speeds are within ± 10 percent of the manufacturer's specification of a minimum of 9 rpm and a maximum of 36 rpm. This test will be performed with the press empty. A stopwatch is required for this test. Measure the speed of the rotor head with a calibrated stopwatch. Verify that the measured speeds are within ±10 percent of the manufacturer's specification and record the results in Table 3.21. Record the instrument used to measure the speed in Table 3.22.

Tablet press minimum speed specification = 9 rpm ± 10%

±10% of 9 = ±0.9

9 + 0.9 = 9.9

9 – 0.9 = 8.1

The measured rpm of 9.3 falls within ±10%.

Tablet press maximum speed specification = 36 rpm ± 10%

±10% of 36 = ±3.6

36 + 3.6 = 39.6

36 – 3.6 = 32.4

The measured rpm of 37 falls within ±10%.

**Table 3.21. *Tablet Press Speed Test Results***

| Item | Specification (rpm) | Measured Speed (rpm) | Acceptable (Yes/No) |
|---|---|---|---|
| Minimum | 9 ± 10% | *9.3* | *Yes* |
| Maximum | 36 ± 10% | *37* | *Yes* |

**Table 3.22. *Instrument Used***

| Test Instrument | Identification Number | Calibration Due Date |
|---|---|---|
| *Stopwatch* | *54039* | *02/16/97* |

## Performance Qualification

Once it has been established that the equipment is properly installed and functioning within specified operating parameters, it must be shown that the tablet press can operate reliably under routine, minimum, and maximum operating conditions (Figure 3.5).

### *Tablet Press Operation: Tablet Weight and Hardness*

The objective of the weight and hardness test is to verify that tablet weight and hardness can be maintained consistently throughout the entire weight and hardness setting range.

The materials and instruments required for this test are a placebo and a weight, hardness, and thickness gauge. Compress tablets using placebo granulation. Record the placebo used in Table 3.23. Obtain the average weight and hardness of 5 tablets at start-up, 10, 20, and 30 min, and record the results in Tables 3.24 and 3.25. Record the instrument used to measure the weight and hardness in Table 3.26.

***Figure 3.5. PQ Elements of a Tablet Press***

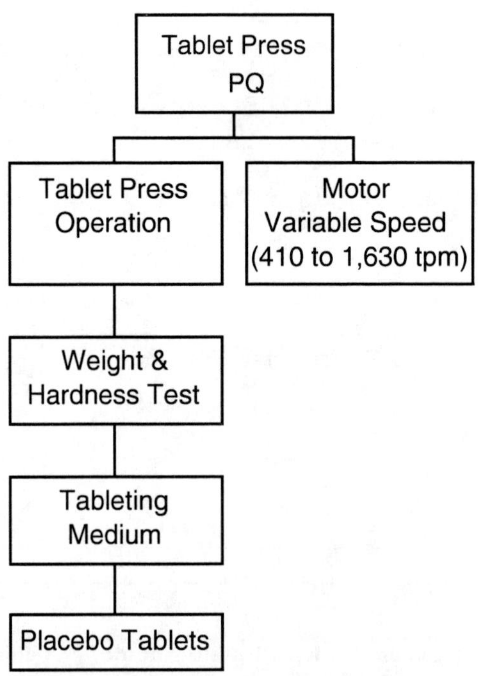

**Table 3.23. Test Material**

| Item | Substance |
|---|---|
| Test placebo | *Haloperidol* |

**Table 3.24. Tablet Weight Test Results**

| Time (min) | Average Low Weight (mg) | Average High Weight (mg) | Acceptable (Yes/No) |
|---|---|---|---|
| Start-up | *120.2* | *139.5* | *Yes* |
| 10 | *119.5* | *138.3* | *Yes* |
| 20 | *120.1* | *137.8* | *Yes* |
| 30 | *119.8* | *137.6* | *Yes* |
| Was the average tablet weight held within a ±10% range for each tablet weight setting? (Yes/No) | | | *Yes* |

**Table 3.25. Tablet Hardness Test Results**

| Time (min) | Average Low Hardness (Sc) | Average High Hardness (Sc) | Acceptable (Yes/No) |
|---|---|---|---|
| Start-up | *3.2* | *12.5* | *Yes* |
| 10 | *2.7* | *14.2* | *Yes* |
| 20 | *3.1* | *13.3* | *Yes* |
| 30 | *2.9* | *13.6* | *Yes* |
| Was the average tablet hardness held within a ±10% range for each tablet hardness setting? (Yes/No) | | | *Yes* |

**Table 3.26. Test Instrument Used**

| Test Instrument | Identification Number | Calibration Due Date |
|---|---|---|
| *Weight, hardness, and thickness tester* | *5411* | *10/14/97* |

***Weight calculation:***

| Sample No. | Low Weight-(mg) | | | | High Weight-(mg) | | | |
|---|---|---|---|---|---|---|---|---|
| | Start-up | 10 min | 20 min | 30 min | Start-up | 10 min | 20 min | 30 min |
| 1 | 0.1209 | 0.1196 | 0.1205 | 0.1194 | 0.1403 | 0.1364 | 0.1377 | 0.1361 |
| 2 | 0.1205 | 0.105 | 0.1198 | 0.1190 | 0.1395 | 0.1395 | 0.1379 | 0.1396 |
| 3 | 0.1201 | 0.1187 | 0.1203 | 0.1200 | 0.1397 | 0.1389 | 0.1374 | 0.1379 |
| 4 | 0.1208 | 0.1191 | 0.1198 | 0.1214 | 0.1384 | 0.1387 | 0.1374 | 0.1364 |
| 5 | 0.1189 | 0.1194 | 0.1200 | 0.1194 | 0.1394 | 0.1378 | 0.1388 | 0.1378 |
| Ave. | 0.1202 | 0.1195 | 0.1201 | 0.1198 | 0.1395 | 0.1383 | 0.1378 | 0.1376 |

***Hardness calculation:***

| Sample No. | Low Hardness (Sc) | | | | High Hardness (Sc) | | | |
|---|---|---|---|---|---|---|---|---|
| | Start-up | 10 min | 20 min | 30 min | Start-up | 10 min | 20 min | 30 min |
| 1 | 2.8 | 2.8 | 2.9 | 2.5 | 11.6 | 15.3 | 13.4 | 13.7 |
| 2 | 3.3 | 2.9 | 3.0 | 3.3 | 13.3 | 14.1 | 13.3 | 13.7 |
| 3 | 3.3 | 2.6 | 3.2 | 3.1 | 12.7 | 14.1 | 13.1 | 14.7 |
| 4 | 3.2 | 2.7 | 3.3 | 2.6 | 12.2 | 14.1 | 13.0 | 13.2 |
| 5 | 3.3 | 2.6 | 3.3 | 3.2 | 12.6 | 13.4 | 13.5 | 12.5 |
| Ave. | 3.2 | 2.7 | 3.1 | 2.9 | 12.5 | 14.2 | 13.3 | 13.6 |

### *Tablet Press Operation: Tablet Press Speed Test*

The objective of the speed test is to document the performance of the tablet press using placebo. The objective is also to verify that the tablet press can maintain consistent tableting speeds throughout the tableting speed range. The variable speeds of the tablet press will be calculated and recorded.

The materials and instruments required for this test are a placebo, a scale, and a stopwatch. Compress tablets using placebo granulation and record the placebo used in Table 3.27. Measure the rotor head speed in rpm with a calibrated stopwatch at start-up, 10, 20, and 30 min. Multiply the low and high speed times the number of stations to obtain the tpm and record the results in Table 3.28. Verify that the average speed is held within a 10 percent range for each tablet speed setting. Record the instrument used to measure the speed in Table 3.29.

Low speed tablet press speed calculation:

Start-up: rpm × 45 = 19.77 × 45 = 890 tpm

10 min: rpm × 45 = 19.75 × 45 = 889 tpm

20 min: rpm × 45 = 19.77 × 45 = 890 tpm

**Table 3.27.** ***Test Material***

| Item | Substance |
|---|---|
| Test placebo | *Haloperidol* |

**Table 3.28.** ***Tablet Press Speed Test Results***

| Time (min) | Calculated Low Speed (tpm) | Calculated High Speed (tpm) | Acceptable (Yes/No) |
|---|---|---|---|
| Start-up | *890* | *1,696* | *Yes* |
| 10 | *889* | *1,705* | *Yes* |
| 20 | *890* | *1,707* | *Yes* |
| 30 | *897* | *1,707* | *Yes* |
| Average tpm | *891.5* | *1,707* | *Yes* |
| Was the average speed within ±10% of each speed setting? (Yes/No) | | | *Yes* |

**Table 3.29.** ***Test Instrument Used***

| Test Instrument | Identification Number | Calibration Due Date |
|---|---|---|
| *Stopwatch* | *54039* | *02/16/97* |
| *Weight, hardness, and thickness tester* | *5411* | *10/14/97* |

30 min: rpm × 45 = 19.93 × 45 = 897 tpm

The average high speed of 891.5 is within ± 10% for each tablet speed setting.

High speed tablet press speed calculation:

Start-up: rpm × 45 = 37.68 × 45 = 1,696 tpm

10 min: rpm × 45 = 37.88 × 45 = 1,705 tpm

20 min: rpm × 45 = 37.93 × 45 = 1,707 tpm

30 min: rpm × 45 = 37.93 × 45 = 1,707 tpm

The average high speed of 1,703.7 is within ±10% for each tablet speed setting.

## DRUM LIFT

The validation of a drum lift that is air driven (Figure 3.6) will be described here. The drum lift has a fixed speed that is controlled by up and down push buttons. The drum lift transports a drum full of product from a stationary ground position to a hopper above the equipment. The transport operates at various levels without manual intervention. The transport is accomplished with up, down, forward, reverse, and swivel directions.

The validation protocol for a drum lift is described in the following sections. This protocol example can be used as a cGMP compliance guide, and it should be tailored to meet individual company requirements. Italicized words in the tables represent examples of information that could be recorded in a validation protocol for a drum lift.

### Installation Qualification

An IQ evaluation establishes confidence that the equipment is properly installed (Figure 3.7). The installation must meet the manufacturer's specified guidelines,

***Figure 3.6.*** ***Lifting Column (courtesy of L. B. Bohle, Inc.)***

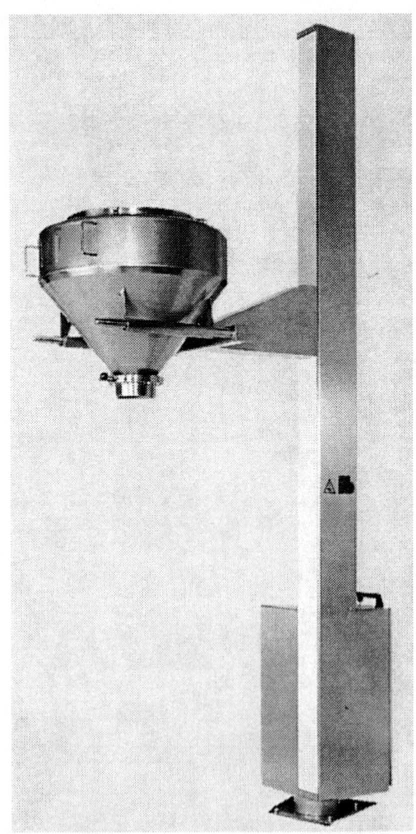

**Figure 3.7. *IQ Elements of a Drum Lift***

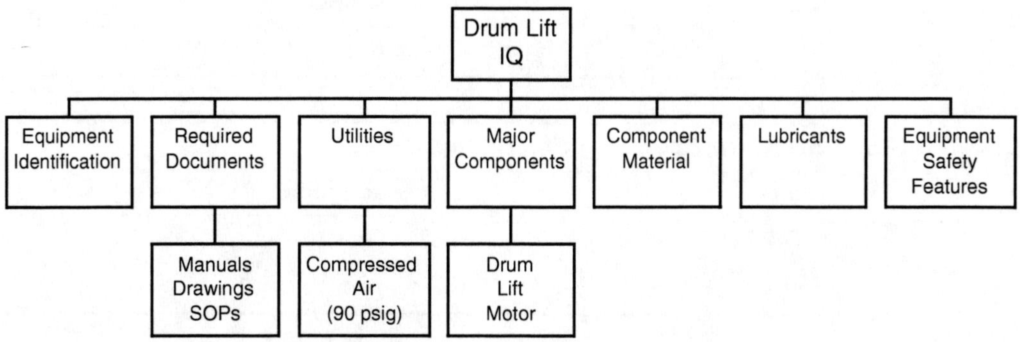


along with design changes at installation. Also, the supporting electrical utilities must meet all electrical codes. The information required for an IQ evaluation is equipment identification, required documentation, equipment utility requirements, major component specifications, component material, lubricants, and equipment safety features.

### *Equipment Identification*

Record the equipment identification numbers in Table 3.30, along with the following information: equipment manufacturer's purchase order number, model number, serial number, company assigned equipment number, and the location of the equipment.

### *Required Documentation*

Record the equipment manufacturer's operation and maintenance manual in Table 3.31. Record the SOPs that cover the setup, operation, and cleaning of the drum lift in Table 3.32.

**Table 3.30. *Equipment Identification***

| Required Information | As-Found Conditions |
|---|---|
| Manufacturer | *Any Drum Lift Co.* |
| Purchase order number | *005519* |
| Model number | *SMD-20* |
| Serial number | *9895* |
| Equipment number | *6921* |
| Location | *Room 423* |

**Table 3.31. Manufacturers' Manuals and Drawings**

| Number | Description | Date |
|---|---|---|
| *None* | *Installing and Maintaining the Drum Lift* | *1987* |

**Table 3.32. Standard Operating Procedures**

| Number | Description | Release Date |
|---|---|---|
| *TAB027* | *Drum Lift Setup and Operation* | *02/08/95* |
| *TAB028* | *Drum Lift Cleaning Procedure* | *03/08/96* |

### *Equipment Utility Requirements*

Compare the manufacturer's specified compressed air requirements to the as-found conditions at the time of qualification testing and record the results in Table 3.33. Record the instrument used to measure the compressed air in Table 3.34.

Compressed air specification = 90 psig

The measured air pressure of 90 psig meets specification (regulated).

### *Major Component Specifications*

The component specifications section of the protocol is designed to verify that the drum lift components purchased were delivered and installed. Record the major components in Table 3.35.

### *Component Material*

Record the material of each component that contacts the product in Table 3.36.

**Table 3.33. Utilities**

| Utility | Specified (psig) | As-Found Condition (psig) | Acceptable (Yes/No) |
|---|---|---|---|
| Compressed air | 90 | *90 (regulated)* | *Yes* |

**Table 3.34. Instrument Used**

| Test Instrument | Identification Number | Calibration Due Date |
|---|---|---|
| *Air pressure gauge* | *540-GP* | *09/15/97* |

**Table 3.35. Major Components**

| Components | As-Found Conditions |
|---|---|
| Drum lift motor | Manufacturer: *Any Drum Lift Co.*<br>Model number: *SMD-20*<br>Serial number: *45621*<br>Air pressure: *90 psig* |
| Drum cage | Part number: *02-0070-DC* |
| Lifting column | Part number: *SMD-21* |
| Control box | Part number: *SMD-22* |

**Table 3.36. Component Material**

| Component | Material |
|---|---|
| Lump breaker | *Stainless steel* |

### *Lubricants*

Record the lubricants used to operate the drum lift in Table 3.37 and indicate if they make contact with the product.

### *Equipment Safety Features*

The objective of testing equipment safety procedures is to verify that the safety features on the drum lift function according to the manufacturer's specifications. This test will be performed with the drum lift full. Verify that all of the safety guards are present and record the results in Table 3.38.

## Operational Qualification

An OQ evaluation should establish that the equipment can operate within specified tolerances and limits (Figure 3.8). The mechanical ranges of the drum lift are

**Table 3.37. Lubricants**

| Where Used | Type | Manufacturer | Product Contact (Yes/No) |
|---|---|---|---|
| *Air motor* | *10W/NR Oil* | *Any Oil Co.* | *No* |
| *Chain and track* | *Lubriplate 630L Grease* | *Any Oil Co.* | *No* |
| Is there a preventive maintenance procedure on file? (Yes/No) | | | *Yes* |

**Table 3.38. Safety Features Test Results**

| Test Function | Expected Results | Acceptable (Yes/No) |
|---|---|---|
| Safety guard | The safety guard was present and functional. | *Yes* |
| Inversion limit switch | The up and down travel of the drum cage was limited. | *Yes* |

**Figure 3.8. OQ Elements of a Drum Lift**

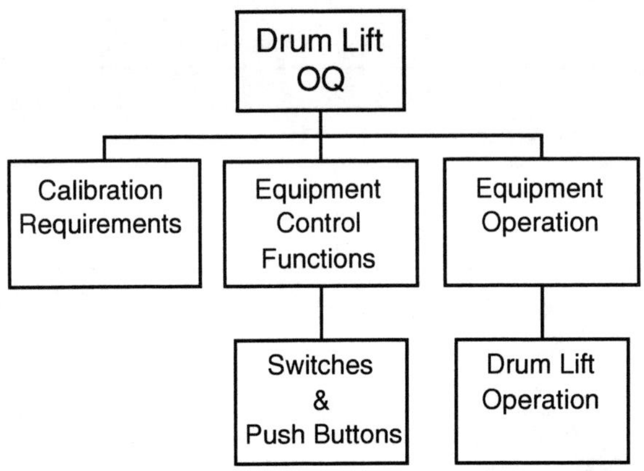


challenged, along with the basic drum lift operations. The drum lift will be validated for its operating ability. The information required for the OQ evaluation is calibration of the instrument used to control the drum lift, equipment control functions (switches and push buttons) and equipment operation.

### *Calibration Requirements*

Verify that all critical instruments on the equipment have been logged into the calibration system, have calibration procedures in place, and are in calibration at the time of qualification testing. Record all information for the calibrated instruments used to control the drum lift in Table 3.39.

### *Equipment Control Functions*

The objective of testing equipment control functions is to verify that the switches and push buttons on the drum lift operate per the manufacturer's specifications.

**Table 3.39. Calibrated and Noncalibrated Instruments**

*There were no calibrated or noncalibrated instruments on the blender.*

Each push button requires testing. Operate each control listed in Table 3.40 and verify its proper operation.

### *Equipment Operation: Drum Lift Operation Test*

The drum lift operation test is designed to verify that the drum lift is operating according to the manufacturer's specification. This test will be performed with the drum lift empty. Perform the functions listed in Table 3.41 and verify their operation.

## Performance Qualification

Once it has been established that the equipment is properly installed and functioning within specified operating parameters, it must be shown that the drum lift can perform reliably under routine, minimum, and maximum operating conditions (Figure 3.9).

### *Drum Lift Operation*

The drum lift operation test is designed to document the performance of the drum lift using placebo. The placebo is used for maximum loading conditions. Fill a drum

***Table 3.40. Equipment Control Function Test Results***

| Test Function | Expected Results | Acceptable (Yes/No) |
|---|---|---|
| On/off push-button operation | When the on/off push button is pressed, the lump breaker starts and stops operating. | *Yes* |
| Open/close push-button operation | When the open/close push button is pressed, the product dumping chute is opened and closed. | *Yes* |
| Up push-button operation | When the up push button is pressed, the drum orientation is controlled. | *Yes* |
| Down push-button operation | When the down push button is pressed, the drum orientation is controlled. | *Yes* |
| Forward push-button operation | When the forward push button is pressed, the drum orientation is controlled. | *Yes* |
| Reverse push-button operation | When the reverse push button is pressed, the drum orientation is controlled. | *Yes* |
| Lock/unlock push-button operation | When the lock/unlock push button is pressed, the lift is locked and unlocked from rotation. | *Yes* |
| Clamp/unclamp push-button operation | When the clamp/unclamp push button is pressed, the cage clamps and unclamps the drum. | *Yes* |
| Reset push-button operation | When the reset push button is pressed while in use, the motion buttons are reset. | *Yes* |

**Table 3.41. Drum Lift Operation Test Results**

| Test Function | Results | Acceptable (Yes/No) |
|---|---|---|
| Press the up push button until the drum cage is approximately 2 ft above the ground. | *The drum cage was approximately 2 ft above the ground.* | *Yes* |
| Press the reverse push button until the drum cage inverts to its stopping point. | *The drum cage was inverted to its stopping point.* | *Yes* |
| Press the up push button until the drum cage reaches its highest limit of travel. | *The drum cage reached its highest limit of travel.* | *Yes* |
| Press the unlock push button and swivel the drum cage until it is over the discharge position. | *The drum cage was swiveled until it was over the discharge position.* | *Yes* |
| Press the down push button until the drum cage stops at the discharge position. | *The drum cage stopped at its discharge position.* | *Yes* |
| Press the open push button and the valve opens. | *The valve opened.* | *Yes* |
| Press the close push button to close the valve before returning to the load position. | *The valve closed.* | *Yes* |
| Press the down push button and the drum cage will go down and stop at the discharge height at approximately 2 ft above the ground. | *The drum cage went down and stopped at the discharge height.* | *Yes* |
| Press the forward push button until the drum cage is inverted back to the upright position. | *The drum cage was inverted back to the upright position.* | *Yes* |
| Press the down push button until the drum cage has reached its lowest limit of travel. | *The drum cage reached its lowest limit of travel.* | *Yes* |
| Press the clamp/unclamp push button and remove the drum. | *The drum was removed.* | *Yes* |

with placebo and load it on the drum lift. Record the placebo used in Table 3.42 Perform the functions listed in Table 3.43 and verify that they operated according to the manufacturer's specifications.

***Figure 3.9.*** ***PQ Elements of a Drum Lift***

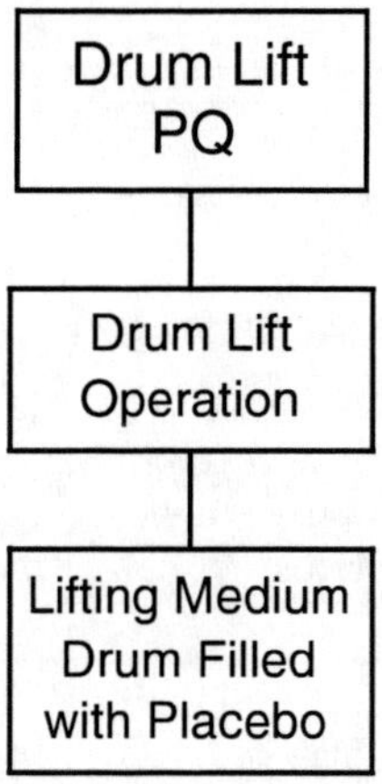


***Table 3.42.*** ***Test Material***

| **Item** | **Substance** |
|---|---|
| Test placebo | *Haloperidol* |

## TABLET DEDUSTER

The validation of a tablet deduster (Figure 3.10) will be described here. A motor drives the deduster, and it operates at variable speeds. The deduster is a stand-alone machine that operates independently when placed into operation with a tablet press. Tablets enter the deduster via an inlet chute. The feed screw assembly spins the tablets at a controlled speed by a potentiometer, while the attached vacuum line removes dust from the tablets. The vacuum level is adjusted by a butterfly valve on the dust outlet nozzle.

The validation protocol for a tablet deduster is given in the following sections. This protocol example can be used as a cGMP compliance guide, and it should be tailored to meet individual company requirements. Italicized words in the tables represent examples of information that could be recorded in a validation protocol for a deduster.

### Installation Qualification

An IQ evaluation establishes confidence that the equipment is properly installed (Figure 3.11). The installation must meet the manufacturer's specified guidelines, along with design changes at installation. Also, the supporting electrical utilities must meet all electrical codes. The information required for an IQ evaluation is equipment identification, required documentation, equipment utility requirements, major component specifications, component material, lubricants, and equipment safety features.

**Table 3.43. Drum Lift Operation Test Results**

| Test Function | Results | Acceptable (Yes/No) |
|---|---|---|
| Place a full drum in the drum cage. Press the clap push button to clamp the drum in place. | *The drum was clamped in place.* | *Yes* |
| Press the up push button until the drum cage is approximately 2 ft above the ground. | *The drum cage was approximately 2 ft above the ground.* | *Yes* |
| Press the reverse push button until the drum cage inverts to its stopping point. | *The drum cage was inverted to its stopping point.* | *Yes* |
| Press the up push button until the drum cage reaches its highest limit of travel. | *The drum cage reached its highest limit of travel.* | *Yes* |
| Press the unlock push button and swivel the drum cage until it is over the discharge position. | *The drum cage was swiveled until it was over the discharge position.* | *Yes* |
| Press the down push button until the drum cage stops at the discharge position. | *The drum cage stopped at its discharge position.* | *Yes* |
| Press the open push button and the valve opens. | *The valve opened.* | *Yes* |
| Press the close push button to close the valve before returning to the load position. | *The valve closed.* | *Yes* |
| Press the down push button and the drum cage will go down and stop at the discharge height at approximately 2 ft above the ground. | *The drum cage went down and stopped at the discharge height.* | *Yes* |
| Press the forward push button until the drum cage has inverted back to the upright position. | *The drum cage was inverted back to the upright position.* | *Yes* |
| Press the down push button until the drum cage has reached its lowest limit of travel. | *The drum cage reached its lowest limit of travel.* | *Yes* |
| Press the clamp/unclamp push button and remove the drum. | *The drum was removed.* | *Yes* |

## *Equipment Identification*

Record the equipment identification numbers in Table 3.44, along with the following information: equipment manufacturer's purchase order number, model number, serial number, company assigned equipment number, and the location of the equipment.

***Figure 3.10. Tablet Deduster (courtesy of BWI Manesty)***

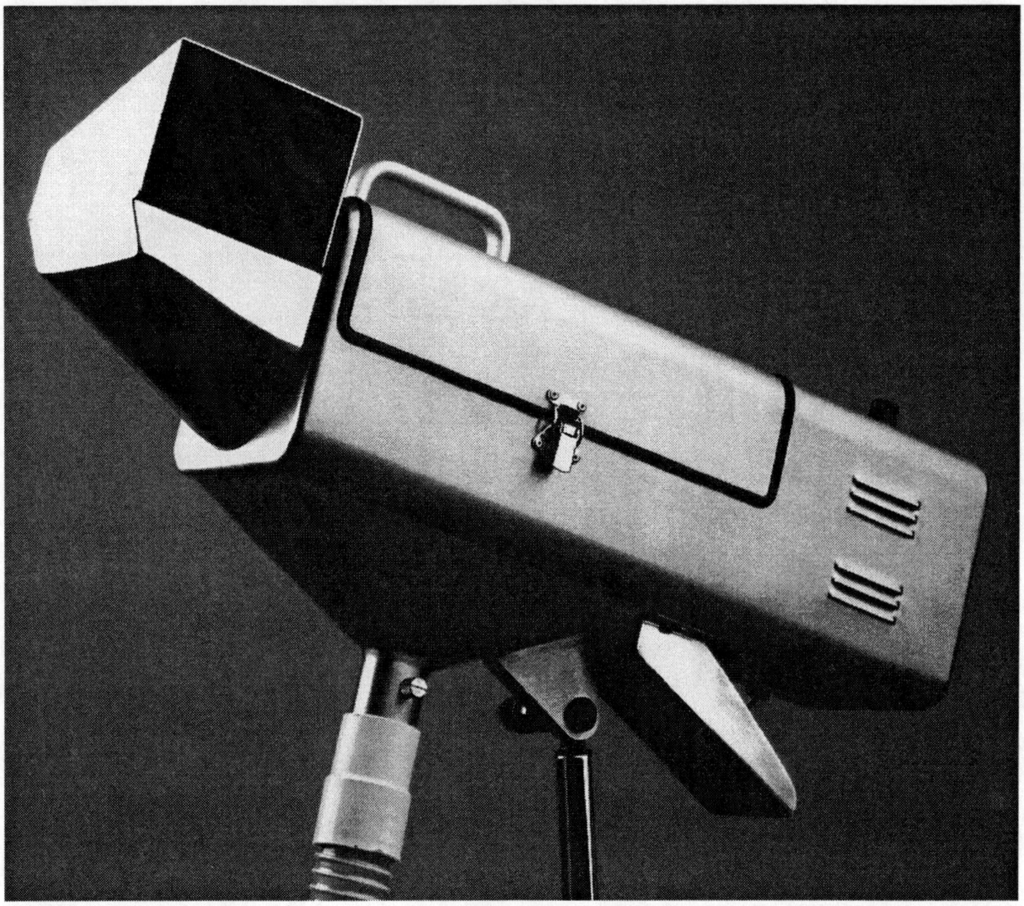

***Figure 3.11. IQ Elements of a Tablet Deduster***

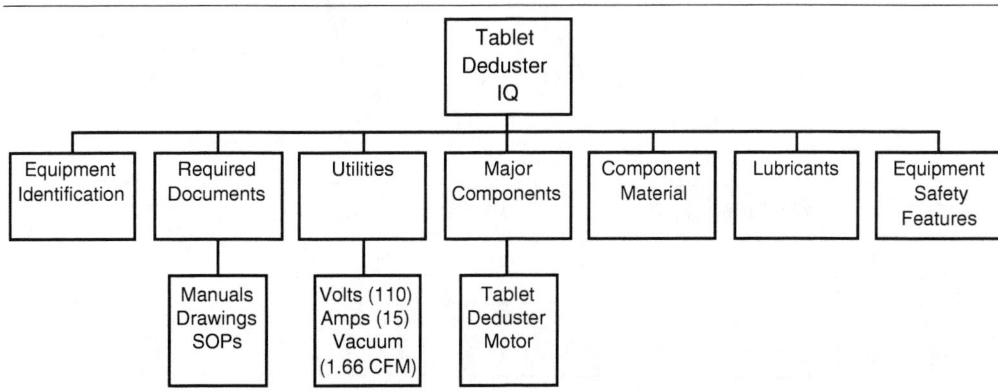

### *Required Documentation*

Record the equipment manufacturer's operation and maintenance manual and drawings in Table 3.45. Record the SOPs that cover the setup, operation, and cleaning of the deduster in Table 3.46.

### *Equipment Utility Requirements*

Compare the manufacturer's specified volts (V), amps (A), and vacuum requirements to their as-found conditions at the time of qualification testing and record the results in Table 3.47. Also record the location of the power supply source. Record the instruments used to measure the volts, amps, and vacuum in Table 3.48.

***Volt calculation:***

Deduster motor volt specification = 110 V ± 10%

± 10% of 110 = ± 11

110 + 11 = 121

110 – 11 = 99

The measured volts of 119 falls within ± 10%.

**Table 3.44. *Equipment Identification***

| Required Information | As-Found Conditions |
|---|---|
| Manufacturer | *Any Deduster Co.* |
| Purchase order number | *0046119* |
| Model number | *25* |
| Serial number | *50W 845671* |
| Equipment number | *113* |
| Location | *Room 349 (This equipment is portable.)* |

**Table 3.45. *Manufacturers' Manuals and Drawings***

| Number | Description | Date |
|---|---|---|
| *None* | *Tablet Deduster Installation, Operation, and Maintenance Manual* | *None* |
| *190100DJA* | *Tablet Deduster Assembly Drawings* | *02/06/93* |

**Table 3.46. *Standard Operating Procedures***

| Number | Description | Release Date |
|---|---|---|
| *TAB021* | *Tablet Deduster Setup and Operation* | *10/23/92* |
| *TAB022* | *Tablet Deduster Cleaning Procedure* | *3/12/96* |

**Table 3.47. *Utilities***

| Utility | Specified | Measured Results | Acceptable (Yes/No) |
|---|---|---|---|
| Volts | Motor = 110 ± 10% | *119* | *Yes* |
| Amps | Motor = 1.2 | 15 circuit rating | *Yes* |
| Vacuum | 1.66 CFM ± 10% | *1.75 CFM* | *Yes* |

Power supply source, breaker box *BB01, wire numbers: 7, 9*

**Table 3.48. *Instruments Used***

| Test Instrument | Identification Number | Calibration Due Date |
|---|---|---|
| *Multimeter* | *ME-025* | *04/19/97* |
| *Vacuum gauge* | *21JP* | *07/10/97* |

***Amp calculation:***

Circuit rating = 15 A

Equipment current draw = 1.2 A

The circuit amp rating of 15 is greater than the maximum current draw of the equipment.

***Vacuum calculation:***

Vacuum specification 1.66 CFM ± 10%

± 10% of 1.66 = ± 0.16

1.66 + 0.16 = 1.82

1.66 – 0.16 = 1.50

The measured CFM of 1.75 falls within ± 10%.

### *Major Component Specifications*

The component specifications section of the protocol is designed to verify that the deduster components purchased were delivered and installed. Record the major components in Table 3.49.

### *Component Material*

Record the material of each component that contacts the product in Table 3.50.

**Table 3.49. *Major Components***

| Components | As-Found Conditions |
|---|---|
| Tablet press motor | Manufacturer: *Any Motor Co.*<br>Model number: *4622005*<br>Serial number: *613W398*<br>Volts: *110*<br>Amperes: *1.2*<br>Phases: *1*<br>Cycles: *60 Hz*<br>hp: *1.6*<br>rpm: *Variable* |

**Table 3.50. *Component Material***

| Component | Material |
|---|---|
| Feed screw | *Stainless steel* |
| Inlet chute | *Stainless steel* |
| Butterfly valve | *Stainless steel* |
| Outlet chute | *Stainless steel* |

### *Lubricants*

Record the lubricants used to operate the deduster in Table 3.51 and indicate if they make contact with the product.

### *Equipment Safety Features*

There are no safety features on this equipment.

## Operational Qualification

An OQ evaluation should establish that the equipment can operate within specified tolerances and limits (Figure 3.12). The mechanical ranges of the deduster are

**Table 3.51. *Lubricants***

| Where Used | Type | Manufacturer | Product Contact (Yes/No) |
|---|---|---|---|
| *The gear box and all bearings were packed at the vendor and required no further attention.* | | | |
| Is there a preventive maintenance procedure on file? (Yes/No) | | | *Yes* |

***Figure 3.12.* *OQ Elements of a Tablet Deduster***

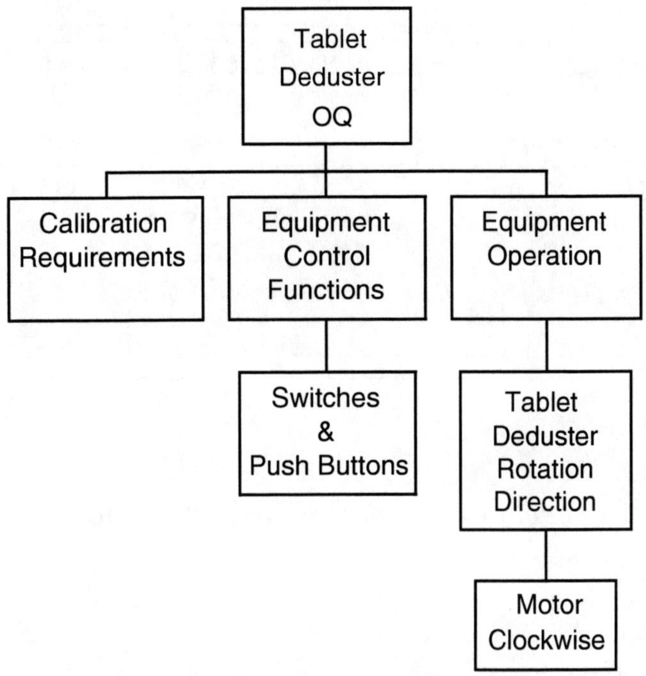


challenged, along with the basic deduster operations. The deduster will be validated for its operating ability, not how well it dedusts tablets. Information required for the OQ evaluation is calibration of the instruments used to control the deduster, equipment control functions (switches and push buttons) and equipment operation (deduster rotation direction).

### *Calibration Requirements*

Verify that all critical instruments on the equipment have been logged into the calibration system, have calibration procedures in place, and are in calibration at the time of qualification testing. Record all information for the calibrated instruments used to control the deduster in Table 3.52.

### *Equipment Control Functions*

The objective of testing equipment control functions is to verify that the switches and push buttons on the deduster operate per the manufacturer's specifications. The tests will be performed with the deduster empty. Operate each control listed in Table 3.53 and verify its proper operation.

**Table 3.52. Calibrated and Noncalibrated Instruments**

*There were no noncalibrated instruments on the blender.*

**Table 3.53. Equipment Control Function Test Results**

| Test Function | Expected Results | Acceptable (Yes/No) |
|---|---|---|
| On/off switch operation | When the on/off switch is pressed, the deduster starts and stops operating, and the LED goes on and off. | *Yes* |
| Potentiometer-operation | When the potentiometer is turned in the direction indicated on the control panel, the motor speed increases and decreases. | *Yes* |

### *Equipment Operation: Deduster Rotation Direction Test*

The objective of the rotation direction test is to verify that the deduster rotates in the proper direction. The tests will be performed with the deduster empty. Press the deduster on switch and observe the direction of rotation of the deduster as viewed from behind the deduster and record the results in Table 3.54.

## Performance Qualification

Once it has been established that the equipment is properly installed and functioning within specified operating parameters, it must be shown that the deduster can operate reliably under routine, minimum, and maximum, operating conditions (Figure 3.13).

### *Tablet Deduster Operation*

The deduster operation test is designed to document the performance of the deduster using placebo tablets. A placebo and stopwatch are required for this test. Feed tablets into the deduster and dedust for 2 min. Perform this operation at 10 percent and 100 percent of the speed and record the results in Table 3.56. Record the placebo used in Table 3.55 and record the instrument used to measure the time in Table 3.57.

**Table 3.54. Deduster Rotation Direction Test Results**

| Item | Expected Results | Results | Acceptable (Yes/No) |
|---|---|---|---|
| Deduster rotation direction | Rotation should be clockwise as viewed from behind the deduster. | *Clockwise rotation was observed.* | *Yes* |

***Figure 3.13. PQ Elements of a Tablet Deduster***

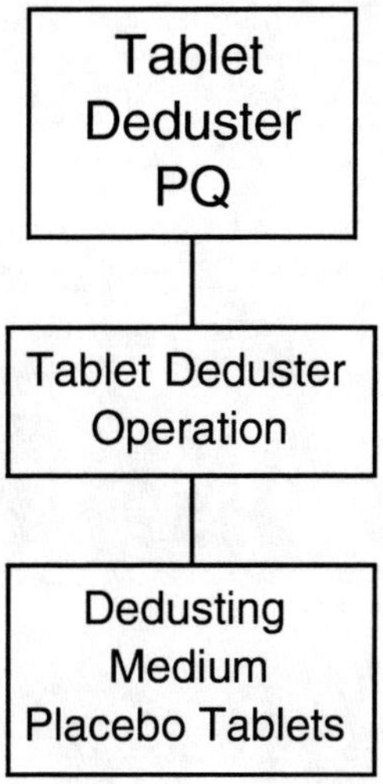


***Table 3.55. Test Materials and Conditions***

| Item | Specifications |
|---|---|
| Test placebo | *Haloperidol* |
| Test time | *2 min each* |

***Table 3.56. Deduster Performance Test results***

*The results of this test were based on observation only.*

| Speed Settings | Expected Results | Results | Time (Min) | Acceptable (Yes/No) |
|---|---|---|---|---|
| 10% | Tablets should be dust free on exiting the deduster. | *Tablets were dust free when they exited the deduster.* | *2* | *Yes* |
| 100% | Tablets should be dust free on exiting the deduster. | *Tablets were dust free when they exited the deduster.* | *2* | *Yes* |

***Table 3.57. Test Instrument Used***

| Test Instrument | Identification Number | Calibration Due Date |
|---|---|---|
| *Stopwatch* | *54039* | *02/16/97* |

## Tablet Hardness Tester

The validation of a tablet hardness tester (Figure 3.14) used to measure the breaking strength of individual tablets is described here. The tablet hardness tester auto-

***Figure 3.14. Tablet Hardness Tester (courtesy of Vector Corporation)***

matically determines the zero value of the load cell before each use. A thrust shaft moves the pressing jaw at a constant speed toward the tablet. When the pressing jaw touches the tablet, the force is measured on a load cell. The pressing jaw continues to push on the tablet until the tablet breaks. Using the principle of a resistance strain gauge, the force at the breaking point is calculated and displayed. Testing can be done as a single operation or as a series.

The validation protocol of a tablet hardness tester is given in the following sections. This protocol example can be used as a cGMP compliance guide, and it should be tailored to meet individual company requirements. Italicized words in the tables represent examples of information that could be recorded in a validation protocol for a tablet hardness tester.

## Installation Qualification

An IQ evaluation establishes confidence that the equipment is properly installed (Figure 3.15). The installation must meet the manufacturer's specified guidelines, along with design changes at installation. Also, the supporting electrical utilities

***Figure 3.15. IQ Elements of a Tablet Hardness Tester***

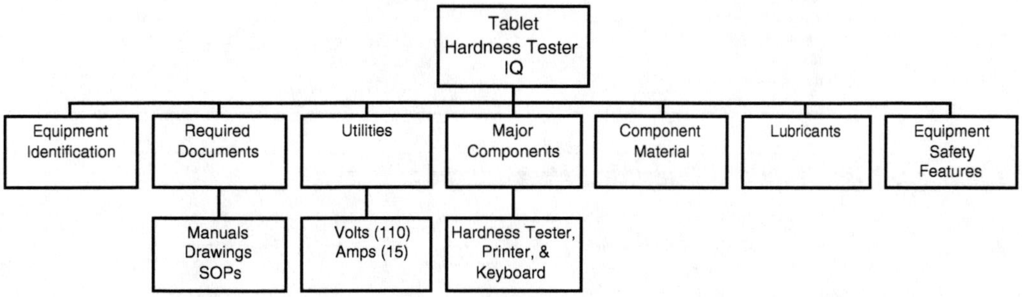


must meet all electrical codes. The information required for an IQ evaluation is equipment identification, required documentation, equipment utility requirements, major component specifications, component material, lubricants, and equipment safety features.

### *Equipment Identification*

Record the equipment identification numbers in Table 3.58, along with the following information: equipment manufacturer's purchase order number, model number, serial number, company assigned equipment number, and the location of the equipment.

### *Required Documentation*

Record the equipment manufacturer's operation and maintenance manual and drawings in Table 3.59. Record the SOPs that cover the setup, operation, and cleaning of the tablet hardness tester in Table 3.60.

***Table 3.58. Equipment Identification***

| **Required Information** | **As-Found Conditions** |
|---|---|
| Manufacturer | *Any Hardness Tester Co.* |
| Purchase order number | *9496300* |
| Model number | *D6 type* |
| Serial number | *363* |
| Equipment number | *1501* |
| Location | *Room 163 (This equipment is portable.)* |

**Table 3.59. Manufacturers' Manuals and Drawings**

| Number | Description | Date |
|---|---|---|
| *None* | *Tablet Hardness Tester Operation & Installation Manual* | *Ver. 3* |
| *SPC-6110* | *Tablet Piping Schematic 0.5–3 Tons, drawing* | *10/91* |

**Table 3.60. Standard Operating Procedures**

| Number | Description | Release Date |
|---|---|---|
| *TAB056* | *Tablet Hardness Tester Setup and Operation* | *01/04/95* |
| *TAB057* | *Tablet Hardness Tester Cleaning Procedure* | *08/11/95* |

### *Equipment Utility Requirements*

Compare the manufacturer's specified volts (V) and amps (A) requirements to their as-found conditions at the time of qualification testing and record the results in Table 3.61. Also record the location of the power supply source. Record the instrument used to measure the volts and amps in Table 3.62.

***Volt calculation:***

Volt specification = 110 V ± 10%

± 10% of 110 = ± 11

110 + 11 = 121

110 – 11 = 99

The measured volts of 116 falls within ± 10%.

**Table 3.61. Utilities**

| Utility | Specified | Measured Results | Acceptable (Yes/No) |
|---|---|---|---|
| Volts | Motor = 110 ± 10% | *116* | *Yes* |
| Amps | 0.01 | 15 circuit rating | *Yes* |
| Power supply source, breaker box *BB47, wire numbers: 19* | | | |

**Table 3.62. Instrument Used**

| Test Instrument | Identification Number | Calibration Due Date |
|---|---|---|
| *Multimeter* | *ME-025* | *04/19/97* |

***Amp calculation:***

Circuit rating = 15 A

Equipment current draw = 0.01 A

The circuit amp rating of 15 is greater than the maximum current draw of the equipment.

### *Major Component Specifications*

The component specifications section of the protocol is designed to verify that the tablet hardness tester components purchased were delivered and installed. Record the major components in Table 3.63.

### *Component Material*

Record the material of each component that contacts the product in Table 3.64.

### *Lubricants*

Record the lubricants used to operate the hardness tester in Table 3.65 and indicate if they make contact with the product.

### *Equipment Safety Features*

There are no safety features on this equipment.

## Operational Qualification

An OQ evaluation should establish that the equipment can operate within specified tolerances and limits (Figure 3.16). The mechanical ranges of the tablet

***Table 3.63. Major Components***

| Components | As-Found Conditions |
|---|---|
| Hardness tester | Manufacturer: *Any Hardness Tester Co.*<br>Model number: *D6 type*<br>Serial number: *363* |
| Printer | Manufacturer: *Any Printer Co.*<br>Model number: *9011-M*<br>Serial number: *49000279 A*<br>Type: *Dot matrix* |
| Keyboard | Manufacturer: *Any Keyboard Co.*<br>Model number: *60/UAH 0003-18G*<br>Serial number: *31C 799610*<br>Type: *Full size with an integrated math pad* |

**Table 3.64. Component Material**

| Component | Material |
|---|---|
| Fixed and pressing jaws | *Stainless steel* |

**Table 3.65. Lubricants**

| Where Used | Type | Manufacturer | Product Contact (Yes/No) |
|---|---|---|---|
| *None* | | | |
| Is there a preventive maintenance procedure on file? (Yes/No) | | | *Yes* |

**Figure 3.16. OQ Elements of a Tablet Hardness Tester**

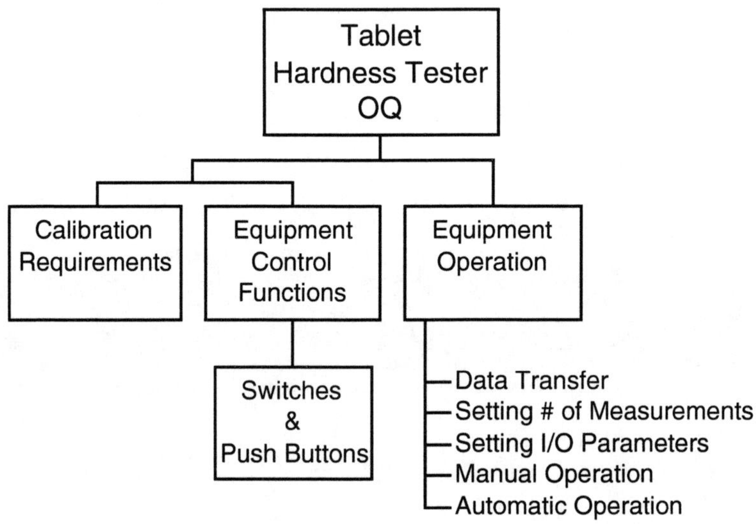


hardness tester are challenged, along with the basic tablet hardness tester operations. The tablet hardness tester will be validated for its operating ability. Information required for the OQ evaluation is calibration of the instruments used to control the tablet hardness tester, equipment control functions (switches and push buttons) and equipment operation (data transfer, setting the number of valid measurements, setting input/output (I/O) parameters, manual operation, automatic operation).

### *Calibration Requirements*

Verify that all critical instruments on the equipment have been logged into the calibration system, have calibration procedures in place, and are in calibration at the time of qualification testing. Record all information for the calibrated instruments used to control the tablet hardness tester in Table 3.66.

**Table 3.66. Calibrated and Noncalibrated Instruments**

*There were no noncalibrated instruments on the blender.*

| Calibrated Instrument | As-Found Conditions |
|---|---|
| Identification number | *2279* |
| Type | *Load cell* |
| Manufacturer | *Any Load Cell Co.* |
| Model number | *D7* |
| Serial number | *463* |
| Range | *0.3–55 Sc* |
| Scale division | *0.1 Sc* |
| Location | *Inside the unit* |
| Use | *Measures the hardness of tablets* |
| Calibration due date | *12/28/95* |
| Critical or not critical | *Not critical to the operation of the tablet hardness tester* |

### *Equipment Control Functions*

The objective of testing equipment control functions is to verify that the switches and push buttons on the tablet hardness tester operate per the manufacturer's specifications. This test will be performed with the tablet hardness tester empty. Operate each control listed in Table 3.67 and verify its proper operation.

### *Equipment Operation: Data Transfer Test*

The objective of the data transfer test is to verify that the data displayed is correctly transferred to the printer. Placebo tablets are required for this test. Set the unit force to Strong Cobbs (Sc) and operate the tablet hardness tester in the automatic mode for 15 tablets. Read the breakage force on the display for each tablet and record the force value in Table 3.68. Print out the test results and record the results in Table 3.68. Verify that the displayed breakage force is the same as the printout breakage force. Record the placebo tablets used in Table 3.73.

### *Equipment Operation: Setting the Number of Valid Measurements Test*

The objective of setting the number of measurements is to verify that the number of valid measurements can be set. This test will be performed with the tablet hardness tester empty. Perform each operation listed in Table 3.69 and verify its proper operation.

**Table 3.67. Equipment Control Function Test Results**

| Test Function | Expected Results | Acceptable (Yes/No) |
|---|---|---|
| Power start/stop push-button operation | When the start/stop push button is pressed with the power on, the tester starts operating. | *Yes* |
| | When the start/stop push button is pressed with the tester in automatic mode and operating, the tester stops at the completion of the test cycle operating. | *Yes* |
| Auto/calibration push-button operation | When the auto/calibration push button is pressed with the power on, the "Auto" LED goes on and the tester performs in the automatic mode. | *Yes* |
| | When the push button is pressed again, the "Auto" LED goes off and the tester operates in the manual mode. | *Yes* |
| | With the power off, press and hold the auto/calibration push button. When the power is turned on, "7.00" is displayed. Release the auto/calibration push button and "0.00" is displayed. | *Yes* |
| Statistics/time-date push-button operation | With the tester operating in automatic mode, press the stop push button. Press the statistics push button, and the tablet hardness data is printed out. | *Yes* |
| | With the power off, press and hold the time-date push button; then turn on the power and release the push button. "1 XX" is displayed, and the time and date are set. | *Yes* |
| Reset push-button operation | With the power on, press and hold the reset push button. Press and hold the time-date push button. Release the reset push button and "1 XX" is displayed. | *Yes* |
| Main power switch operation | When the main power switch is turned on or off, the power goes on or off. | *Yes* |
| Line voltage selector operation | When the line voltage selector is set to 110 V, the equipment will operate at this voltage. | *Yes* |
| Unit of force selector operation | When one of the unit of force options is selected, the corresponding LED goes on. | *Yes* |

### *Equipment Operation: Setting the Input/Output Parameters Test*

The objective of the I/O parameters test is to verify that the input/output parameters can be set. This test will be performed with the tablet hardness tester empty. Perform each operation listed in Table 3.70 and verify its proper operation.

### *Equipment Operation: Manual Test*

The objective of the manual test is to verify the manual operation of the tablet hardness tester. Perform each operation listed in Table 3.71 and verify its proper operation. Record the placebo tablets used in Table 3.73.

**Table 3.68. *Data Transfer Test Results***

| Tablet Number | Displayed Force (Sc) | Printout (Sc) | Acceptable (Yes/No) |
|---|---|---|---|
| 1 | *8.9* | *8.9* | *Yes* |
| 2 | *8.2* | *8.2* | *Yes* |
| 3 | *9.7* | *9.7* | *Yes* |
| 4 | *10.7* | *10.7* | *Yes* |
| 5 | *10.8* | *10.8* | *Yes* |
| 6 | *9.6* | *9.6* | *Yes* |
| 7 | *9.1* | *9.1* | *Yes* |
| 8 | *7.7* | *7.7* | *Yes* |
| 9 | *7.9* | *7.9* | *Yes* |
| 10 | *8.4* | *8.4* | *Yes* |
| 11 | *9.1* | *9.1* | *Yes* |
| 12 | *8.4* | *8.4* | *Yes* |
| 13 | *10.0* | *10.0* | *Yes* |
| 14 | *10.7* | *10.7* | *Yes* |
| 15 | *8.1* | *8.1* | *Yes* |

**Table 3.69. *Setting the Number of Valid Measurements Test Results***

| Test Function | Results | Acceptable (Yes/No) |
|---|---|---|
| Turn the tester on. | *The display goes on and the LED for the force unit that was chosen goes out.* | *Yes* |
| Press and hold the reset push button. Press and hold the start/stop push button; then release. | *"8 XX" is displayed.* | *Yes* |
| Set the measurement (XX) by pressing the start/stop push button. Enter 5. | *The display changes when the start/stop push button is pressed. The value 5 is displayed.* | *Yes* |
| Press the statistics push button. | *The input mode is ended.* | *Yes* |
| After the automatic test run, determine the total number of valid measurements printed on the printout. | *The total number of valid measurements was 5.* | *Yes* |

**Table 3.70. Setting the Input/Output Parameters Test Results**

| Test Function | Results | Acceptable (Yes/No) |
|---|---|---|
| Press the ESC key to enter the program mode. Press the key combination of CONTROL_C on the keyboard to delete the last parameters. Use the keys on the math keyboard to switch from one of the six input parameters to another. Enter the following values for input parameters: | *The program mode was entered.* | *Yes* |
| Product = TEST 1 | | |
| Product Number = 1234 | | |
| Batch = T4567 | | |
| Comment = HARDNESS TEST | | |
| Low Limit = 2 | | |
| High Limit = 14 | *The six parameters were entered.* | *Yes* |
| After each entry, press the ENTER key. End the program mode by leaving the next input blank and pressing the ENTER key. | *The program mode ended.* | *Yes* |
| After the automatic run is performed, print out the results. Review the printout and compare the printout with the one inputted. | *The six parameters entered are the same as the ones printed.* | *Yes* |

**Table 3.71. Manual Operation Test Results**

| Test Function | Expected Results | Acceptable (Yes/No) |
|---|---|---|
| If the auto LED is on, press the auto push button to change to the manual mode. | *If the auto LED is on, the LED goes out when the auto push button is pressed.* | *Yes* |
| Insert a tablet and press and hold the start push button until the moving jaw comes in contact with the tablet. Release the button. | *The jaw moves toward the tablet.* | *Yes* |
| The jaw continues until the tablet is broken. | *The breakage force is displayed, and the jaw returns to the position where the start push button was released.* | *Yes* |
| Remove the broken tablet. Insert another tablet and press the start push button. | *The tester begins another hardness test cycle.* | *Yes* |

### *Equipment Operation: Automatic Test*

The objective of the automatic test is to verify automatic operation of the tablet hardness tester. Perform each operation listed in Table 3.72 and verify its proper operation. Record the placebo tablets used in Table 3.73.

## Performance Qualification

Once it has been established that the equipment is properly installed and functioning within specified operating parameters, it must be shown that the tablet hardness tester can perform reliably under routine, minimum, and maximum operating conditions (Figure 3.17).

**Table 3.72. *Automatic Operation Test Results***

| Test Function | Results | Acceptable (Yes/No) |
|---|---|---|
| If the auto LED is not lit, press the auto push button to change to the automatic mode. | *If the auto LED is not on, the LED goes on when the auto push button is pressed.* | *Yes* |
| Insert a tablet and press and hold the start push button until the moving jaw comes in contact with the tablet. Release the start push button. | *The jaw moves toward the tablet.* | *Yes* |
| The jaw continues until the tablet is broken. | *The breaking force is displayed, and the jaw returns to the position where the start push button was released.* | *Yes* |
| Remove the broken tablet and insert a new one. After approximately 3 sec, the tester cycle restarts. | *The jaw moves toward the tablet, and the breakage force is determined again.* | *Yes* |
| After the assigned number of valid measurements is obtained, the statistics for this run are automatically printed. | *A printout was obtained.* | *Yes* |

**Table 3.73. *Test Materials and Conditions***

| Item | Specifications |
|---|---|
| Test placebo | *Haloperidol* |
| Operating mode | *Automatic* |
| Independent calculator | *Any PC database software* |

**Figure 3.17. PQ Elements of a Tablet Hardness Tester**

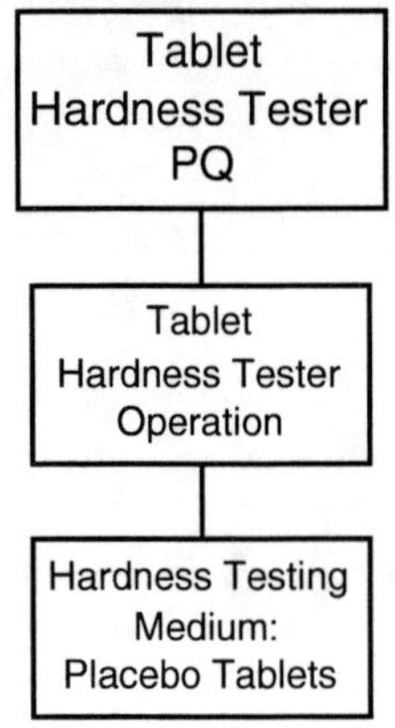


### *Tablet Hardness Tester Operation*

The objective of testing the tablet hardness test is to compare the displayed data with an independent calculation of the displayed data. Placebo tablets and a calculator or personal computer are required for this test.

Set the unit force to Strong Cobbs and operate the tablet hardness tester in the automatic mode for 100 tablets. Read the breakage force on the display for each tablet and record the force values in Table 3.74. Use an independent calculator to verify that the tablet hardness tester's breaking force calculations are the same. Record the results in Table 3.74. Record placebo and independent calculator used in Table 3.73.

**Table 3.74. Tablet Hardness Tester Test Results**

| Item | Hardness Tester Calculation | Independent Calculation | Acceptable (Yes/No) |
|---|---|---|---|
| Mean | *9.47 Sc* | *9.47 Sc* | *Yes* |
| Maximum value measured | *11.20 Sc* | *11.20 Sc* | *Yes* |
| Minimum value measured | *6.5 Sc* | *6.5 Sc* | *Yes* |
| Difference between maximum & minimum | *4.70 Sc* | *4.70 Sc* | *Yes* |
| Standard deviation absolute | *1.09* | *1.09* | *Yes* |
| Standard deviation relative | *11.6%* | *11.6%* | *Yes* |

## Tablet Sorter

The validation of a tablet sorter (Figure 3.18) that uses diverging rollers to sort tablets of varying thickness into three bins is described here. There are three sets of rollers on the unit which are connected by single adjustment handle. With one adjustment, the same upper and lower gap settings can made on each roller set. The counterrotating rollers orient the tablets and move them along the diverging gap until the space is large enough for the tablet to fall through. Tablets that are not thick enough fall into the first bin, where they are rejected. Tablets of the target thickness range fall into the second bin and are retained. Tablets that are too thick do not fall through the rollers until the third bin, where they are rejected. Feed bin vibration and roller speeds are adjusted per product for optimum performance.

The validation protocol for a tablet sorter is described in the following sections. This protocol example can be used as a cGMP compliance guide and it should be tailored to meet individual company requirements. Italicized words in the tables represent examples of information that could be recorded in a validation protocol for a tablet sorter.

***Figure 3.18. Tablet Sorter (courtesy of Modern Controls, Inc.)***

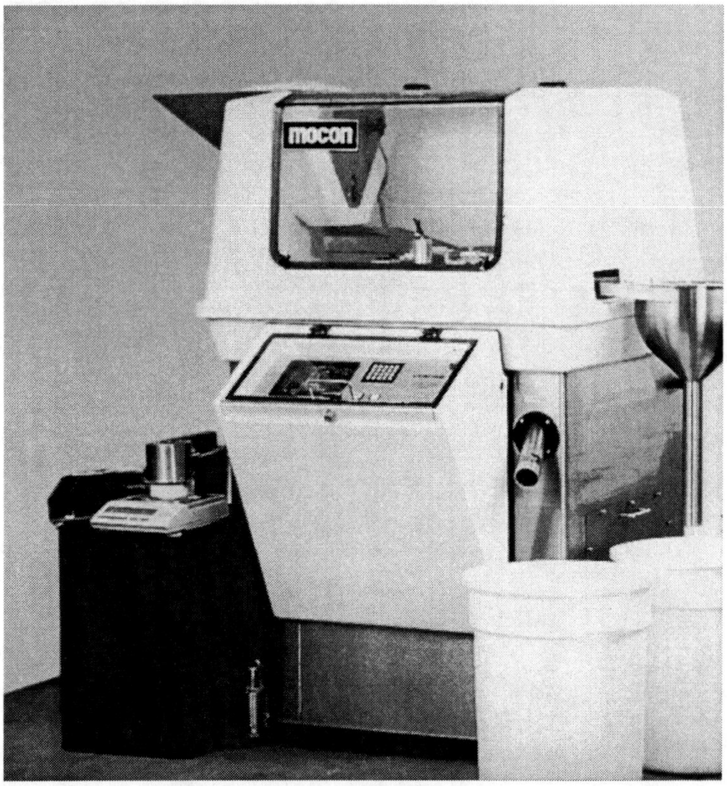

## Installation Qualification

An IQ evaluation establishes confidence that the equipment is properly installed (Figure 3.19). The installation must meet the manufacturer's specified guidelines, along with design changes at installation. Also, the supporting electrical utilities must meet all electrical codes. The information required for an IQ evaluation is equipment identification, required documentation, equipment utility requirements, major component specifications, component material, lubricants, and equipment safety features.

### *Equipment Identification*

Record the equipment identification numbers in Table 3.75, along with the following information: equipment manufacturer's purchase order number, model number, serial number, company assigned equipment number, and the location of the equipment.

### *Required Documentation*

Record the equipment manufacturer's operation and maintenance manual and drawings in Table 3.76. Record the SOPs that cover the setup, operation, and cleaning of the tablet sorter in Table 3.77.

**Figure 3.19. IQ Elements of a Tablet Sorter**

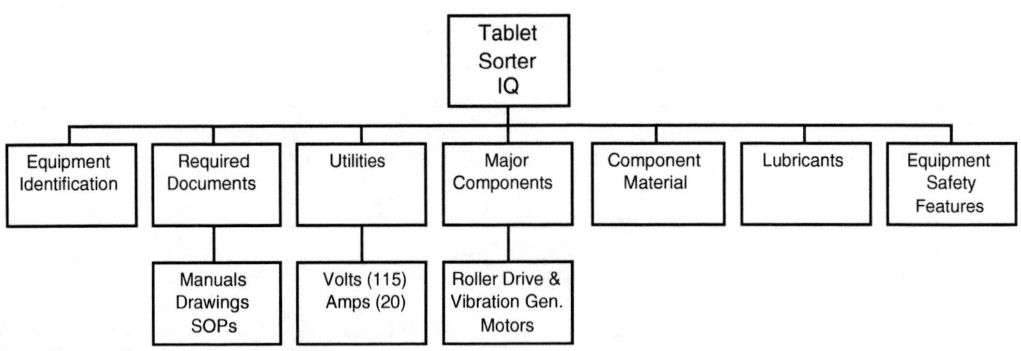


**Table 3.75. Equipment Identification**

| Required Information | As-Found Conditions |
|---|---|
| Manufacturer | *Any Tablet Sorter Co.* |
| Purchase order number | *0033610* |
| Model number | *4-SCT* |
| Serial number | *060-68* |
| Equipment number | *8230* |
| Location | *Room 400 (This equipment is portable.)* |

**Table 3.76. *Manufacturers' Manuals and Drawings***

| Number | Description | Date |
|---|---|---|
| *None* | *Tablet Sorter Installation, Operation, and Maintenance Manual* | *96* |
| *12357* | *Four-Channel Roller Sorter Assembly Drawing* | *04/09/88* |

**Table 3.77. *Standard Operating Procedures***

| Number | Description | Release Date |
|---|---|---|
| *TAB110* | *Tablet Sorter Setup and Operation* | *10/04/91* |
| *TAB111* | *Tablet Sorter Cleaning Procedure* | *10/10/94* |

### *Equipment Utility Requirements*

Compare the manufacturer's specified volts (V) and amps (A) requirements to their as-found conditions at the time of qualification testing and record the results in Table 3.78. Also record the location of the power supply source. Record the instrument used to measure the volts and amps in Table 3.79.

***Volt calculation:***

Volt specification = 115 V ± 10%

± 10% of 115 = ± 11.5

115 + 11.5 = 126.5

115 – 11.5 = 103.5

The measured volts of 114 falls within ± 10%.

**Table 3.78. *Utilities***

| Utility | Specified | Measured Results | Acceptable (Yes/No) |
|---|---|---|---|
| Volts | 115 ± 10% | *114* | *Yes* |
| Amps | Roller drive motor = 1.8 A<br>Vibration generation motor = 1.8 A | 20 circuit rating | *Yes* |

Power supply source, breaker box *BB11, wire number: 9*

**Table 3.79. *Instrument Used***

| Test Instrument | Identification Number | Calibration Due Date |
|---|---|---|
| *Multimeter* | *ME-025* | *04/19/97* |

***Amp calculation:***

Amp specification = 20 A

Roller drive motor = 1.8 A

Vibration generator motor = 1.8 A

Equipment current draw = roller drive motor A + vibration generator motor A = 1.8 + 1.8 = 3.6 A

The circuit amp rating of 20 is greater than the maximum current draw of the equipment.

### *Major Component Specifications*

The component specifications section of the protocol is designed to verify that the tablet sorter components purchased were delivered and installed. Record the major components in Table 3.80.

### *Component Material*

Record the material of each component that contacts the product in Table 3.81.

**Table 3.80. Major Components**

| Components | As-Found Conditions |
|---|---|
| Roller drive motor | Manufacturer: *Any Motor Co.*<br>Model number: *3BE07414*<br>Serial number: *85 9 82736 RP*<br>Volts: *115*<br>Amperes: *1.8*<br>Phases: *1*<br>Cycles: *60 Hz*<br>hp: *1/4*<br>rpm: *165* |
| Vibration generator motor | Manufacturer: *Any Motor Co.*<br>Model number: *80551137*<br>Serial number: *F594*<br>Volts: *115*<br>Amperes: *1.8*<br>Phases: *1*<br>Cycles: *60 Hz*<br>hp: *1/4*<br>rpm: *165* |

**Table 3.81. Component Material**

| Component | Material |
|---|---|
| Sorting rollers | *Stainless steel* |
| Vibratory linear channel | *Stainless steel* |

### *Lubricants*

Record the lubricants used to operate the tablet sorter in Table 3.82 and indicate if they make contact with the product.

### *Equipment Safety Features*

There are no safety features on this equipment.

## Operational Qualification

An OQ evaluation should establish that the equipment can operate within specified tolerances and limits (Figure 3.20). The mechanical ranges of the tablet sorter are challenged along with the basic tablet sorter operations that the operator will use. The tablet sorter will be validated for its operating ability. Information required for the OQ evaluation is calibration of the instruments used to control the tablet sorter, equipment control functions (switches and push buttons) and equipment operation (roller drive and vibration generator motor rotation direction).

### *Calibration Requirements*

Verify that all critical instruments on the equipment have been logged into the calibration system, have calibration procedures in place, and are in calibration at the time of qualification testing. Record all information for the calibrated instruments used to control the tablet sorter in Table 3.83.

### *Equipment Control Functions*

The objective of testing equipment control functions is to verify that the switches and push buttons on the tablet sorter operate according to the manufacturer's

**Table 3.82. Lubricants**

| Where Used | Type | Manufacturer | Product Contact (Yes/No) |
|---|---|---|---|
| *Drive motors* | *APG 90 Oil* | *Any Oil Co.* | *No* |
| Is there a preventive maintenance procedure on file? (Yes/No) | | | *Yes* |

**Figure 3.20. OQ Elements of a Tablet Sorter**

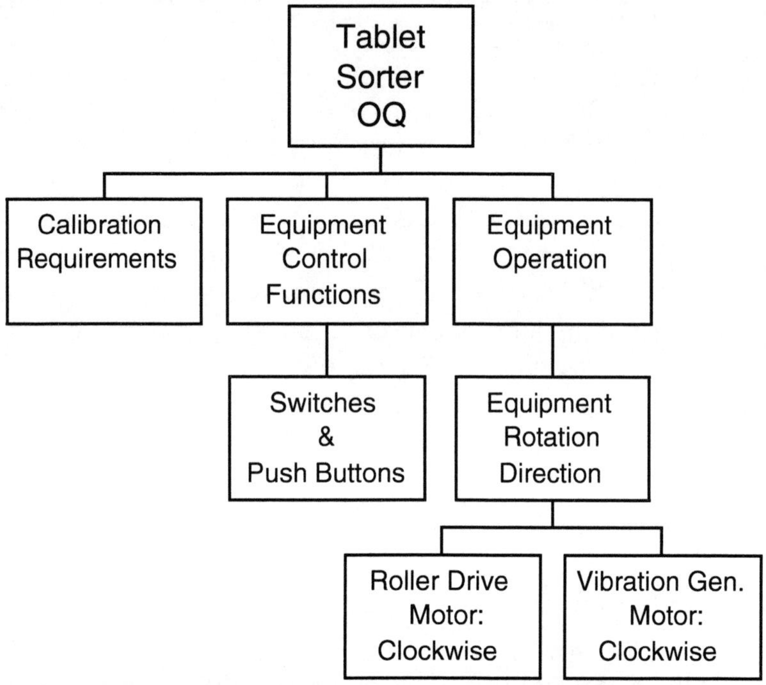


**Table 3.83. Calibrated and Noncalibrated Instruments**

*There were no calibrated or noncalibrated instruments on the blender.*

specifications. The test will be performed with the tablet sorter empty. Operate the controls listed in Table 3.84 and verify its proper operation.

### *Equipment Operation: Roller Drive and Vibration Generator Motor Rotation Direction Test*

The objective of the rotation direction test is to verify that the roller drive and vibration generator motor rotate in the proper direction. The tests will be performed with the tablet sorter empty. With the equipment operating, observe the direction of rotation of the roller drive and the vibration generator motors as viewed from behind the motors and record the results in Table 3.85.

## Performance Qualification

Once it has been established that the equipment is properly installed and functioning within specified operating parameters, it must be shown that the tablet sorter can operate reliably under routine, minimum, and maximum operating conditions (Figure 3.21).

**Table 3.84. *Equipment Control Function Test Results***

| Test Function | Expected Results | Acceptable (Yes/No) |
|---|---|---|
| Roller potentiometer operation | When the roller potentiometer is rotated from 0 to 100, the roller speed is increased. | *Yes* |
| | When the roller potentiometer is rotated from 100 to 0, the roller speed is decreased. | *Yes* |
| Vibratory feeder potentiometer operation | When the vibratory feeder potentiometer is rotated from 0 to 100, the vibratory feeder speed is increased. | *Yes* |
| | When the vibratory feeder potentiometer is rotated from 100 to 0, the vibratory feeder speed is decreased. | *Yes* |

**Table 3.85. *Rotation Direction Test Results***

| Item | Expected Results | Results | Acceptable (Yes/No) |
|---|---|---|---|
| Roller drive motor rotation direction | Rotation should be clockwise as viewed from behind the motor. | *Clockwise rotation was observed.* | *Yes* |
| Vibration generator motor rotation direction | Rotation should be clockwise as viewed from behind the motor. | *Clockwise rotation was observed.* | *Yes* |

**Figure 3.21. *PQ Elements of a Tablet Sorter***

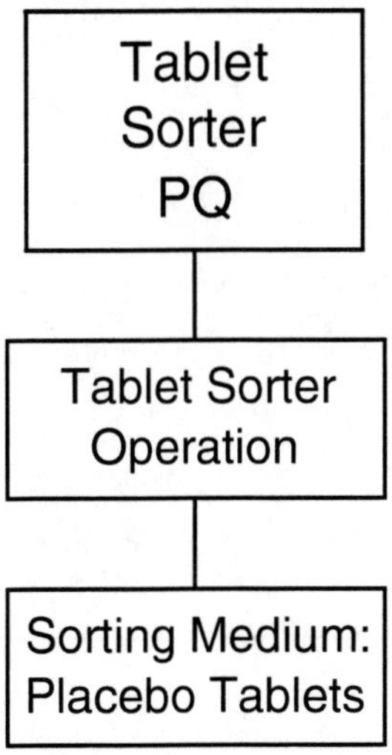

### *Tablet Sorter Operation*

The operation of the tablet sorter is tested to document its performance with placebo tablets. Using the following procedure to test the tablet sorter:

- Setup the equipment for normal operation and fill the infeed hopper half full with a mixture of placebo (50 undersize, marked red; 50 oversize, marked blue; and the balance of acceptable size, unmarked). Record the speed dial setting in Table 3.86.
- With the equipment ready to process tablets, record the size sorter parameters in Table 3.87.
- Process the placebo tablets until the hopper is empty, inspect the contents of the three receiving containers, and record the information in the following table. Measure a sample of the unmarked samples in the undersized and oversized receivers and include the results of measurement in Table 3.88.

**Table 3.86. Average Representative Size Parameters**

| Test Function | Measured/Recorded Setpoint |
|---|---|
| Infeed vibrator potentiometer setpoint (recorded) | *6* |
| Hopper gates gap (measured) | *0.264 in.* |
| Channel 1 roller gap (measured)—top end | *0.075 in.* |
| Channel 1 roller gap (measured)—bottom end | *0.121 in.* |
| Channel 2 roller gap (measured)—top end | *0.075 in.* |
| Channel 2 roller gap (measured)—bottom end | *0.121 in.* |
| Channel 3 roller gap (measured)—top end | *0.075 in.* |
| Channel 3 roller gap (measured)—top end | *0.123 in.* |
| Channel 4 roller gap (measured)—bottom end | *0.073 in.* |
| Channel 4 roller gap (measured)—bottom end | *0.124 in.* |
| Roller potentiometer setpoint (recorded) | *45 rpm* |
| Lower limit category divider plate position (measured) | *9 in. from upper roller ends* |
| Upper limit category divider plate position (measured) | *12-1/8 in. from the lower limit category, divider plate* |

**Table 3.87. Test Materials and Conditions**

| Item | Specifications |
|---|---|
| Undersize placebo tablets | *Desipramine 10*<br>*Size range = 0.066 in. to 0.070 in.*<br>*Average tablet size = 0.068 in.* |
| Target placebo tablets | *Desipramine 10*<br>*Target size = 0.096 in.* |
| Oversize placebo tablets | *Desipramine 10*<br>*Size range = 0.109 in. to 0.127 in.*<br>*Average tablet size = 0.115 in.* |

**Table 3.88. Representative of Average Production Size**

| Tablet Placebo Size | Number of Placebo Tablets with Marking | Total Number of Placebo Tablets in Receiving | Acceptable (Yes/No) |
|---|---|---|---|
| Undersize (red) | *50* | *50* | *Yes* |
| Acceptable size (no mark) | *99* | *99* | *Yes* |
| Oversize (blue) | *51* | *51* | *Yes* |

# Chapter 4

# Validation of Coating Equipment

This chapter will cover the qualification testing of equipment used to perform operations in the coating department (Figure 4.1).

## Tablet Coating Pan

The validation of a tablet coating pan (Figure 4.2) that is motor driven and operates at variable speeds is described here. Tablets are placed in a 24 in. diameter perforated stainless steel drum. The pan rotates on its horizontal axes about a spray system that coats the tablets as they are tumbling. The pan is equipped with two built-in baffles that, along with the rotation of the pan, promote uniform coating

***Figure 4.1. Coating Equipment***

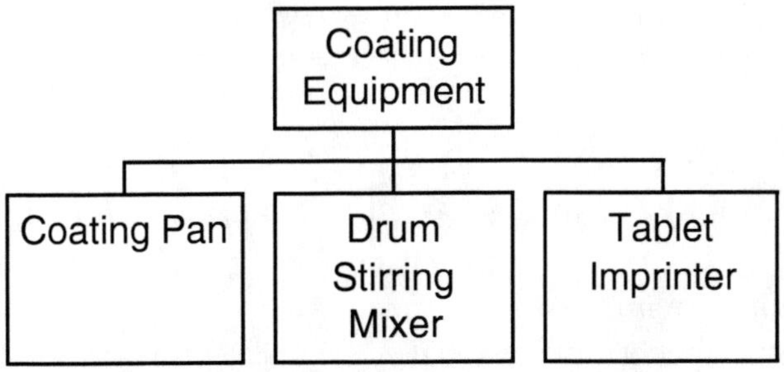

***Figure 4.2. Tablet Coating Pan (courtesy of Thomas Engineering, Inc.)***

and prevent the tablets from sticking together during the coating process. The tablets are coated by the spray distribution system. Preconditioned air circulated through the pan dries the coated tablets. The air is exhausted at a higher rate than it is supplied in order to reduce vapor emissions by creating a negative pressure in the coating chamber.

The validation protocol for a tablet coating pan is described in the following sections. Following are the major elements of a validation protocol for a tablet coating pan. This protocol example can be used as a cGMP compliance guide, and it should be tailored to meet individual company requirements. Italicized words in the tables represent examples of information that could be recorded in a validation protocol for a tablet coating pan.

## Installation Qualification

An IQ evaluation establishes confidence that the equipment is properly installed (Figure 4.3). The installation must meet the manufacturer's specified guidelines, along with design changes at installation. Also, the supporting electrical utilities must meet all electrical codes. The information required for an IQ evaluation is equipment identification, required documentation, equipment utility requirements,

***Figure 4.3. IQ Elements of a Tablet Coating Pan***

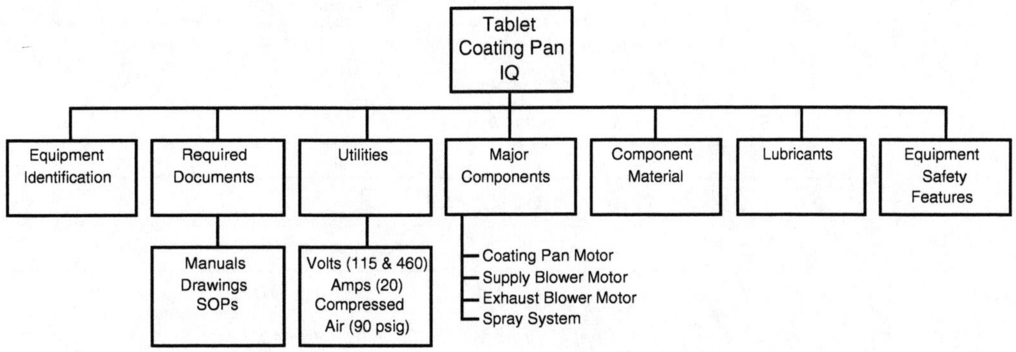


major component specifications, component material, lubricants, and equipment safety features.

### *Equipment Identification*

Record the equipment identification numbers in Table 4.1, along with the following information: equipment manufacturer's purchase order number, model number, serial number, company assigned equipment number, and the location of the equipment.

### *Required Documentation*

Record the equipment manufacturer's operation and maintenance manual and drawings in Table 4.2. Record the SOPs that cover the setup, operation, and cleaning of the tablet coating pan in Table 4.3.

***Table 4.1. Equipment Identification***

| Required Information | As-Found Conditions |
|---|---|
| Manufacturer | *Any Tablet Coating Pan Co.* |
| Purchase order number | *008332* |
| Model number | *24" COTER* |
| Serial number | *381 42* |
| Equipment number | *8570* |
| Location | *Room 727* |

**Table 4.2. *Manufacturers' Manuals and Drawings***

| Number | Description | Date |
|---|---|---|
| *None* | *Tablet Coating Pan Instructions, Installation, Operation, Maintenance Manual* | *None* |
| *A47-042* | *Electrical Diagram Drawing* | *03/19/74* |

**Table 4.3. *Standard Operating Procedures***

| Number | Description | Release Date |
|---|---|---|
| *COA059* | *Setup and Operation of a Tablet Coating Pan* | *09/08/94* |
| *COA060* | *Coating Dept.—Equipment Cleaning Procedures* | *03/22/96* |

### *Equipment Utility Requirements*

Compare the manufacturer's specified volts (V), amps (A), and compressed air pressure requirements to their as-found conditions at the time of qualification testing and record the results in Table 4.4. Also record the location of the power supply source. Record the instruments used to measure the volts, amps, and compressed air in Table 4.5.

**Table 4.4. *Utilities***

| Utility | Specified | Measured Results | Acceptable (Yes/No) |
|---|---|---|---|
| Volts | Spray system = 115 ± 10%<br>Coating pan = 460 ± 10% | *120*<br>*A-B 461; A-C 468; B-C 466* | *Yes* |
| Amps | Coating pan motor = 4.8<br>Supply blower = 4.8<br>Exhaust blower = 4.8 | 20 circuit rating | *Yes* |
| Compressed air | 90 ± 10% psig | *90 psig (regulated)* | *Yes* |
| Power supply source, breaker box *BB65, wire numbers: 21 and 13, 15, 17 respectively* | | | |

**Table 4.5. *Instruments Used***

| Test Instrument | Identification Number | Calibration Due Date |
|---|---|---|
| *Multimeter* | *ME-025* | *04/19/97* |
| *Air pressure gauge* | *P-102* | *02/10/97* |

***Volt calculation:***

Spray system motor volt specification = 115 V ± 10%

±10% of 115 = ±11.5

115 + 11.5 = 126.5

115 – 11.5 = 103.5

The measured volts of 120 falls within the ±10%.

Coating pan motor volt specification = 460 V ± 10%

±10% of 460 = ±46

460 + 46 = 506

460 – 46 = 414

The measured volts of 461/468/466 fall within the ±10%.

***Amp calculation:***

Circuit rating = 20 A

Coating pan motor = 4.8 A

Supply blower motor = 4.8 A

Exhaust blower motor = 4.8 A

Equipment current draw = coating pan A + supply blower A + exhaust blower A = 4.8 + 4.8 + 4.8 = 14.4 A

The circuit amp rating of 20 is greater than the maximum current draw of the equipment.

***Compressed air calculation:***

Compressed air specification = 90 psig ± 10%

The measured air pressure of 90 psig falls within the ±10% (regulated).

### *Major Component Specifications*

The component specifications section of the protocol verifies that the tablet coating pan components purchased were delivered and installed. Record the major components in Table 4.6.

### *Component Material*

Record the material of each component that contacts the product in Table 4.7.

**Table 4.6. Major Components**

| Component | As-found Conditions |
|---|---|
| Coating pan motor | Manufacturer: *Any Motor Co.*<br>Model number: *MN2261*<br>Serial number: *66450*<br>Volts: *460*<br>Amperes: *4.8*<br>Phases: *3*<br>Cycles: *60 Hz*<br>hp: *3/4*<br>rpm: *1,200* |
| Supply blower motor | Manufacturer: *Any Motor Co.*<br>Model number: *MN2262*<br>Serial number: *66451*<br>Volts: *460*<br>Amperes: *4.8*<br>Phases: *3*<br>Cycles: *60 Hz*<br>hp: *1.5*<br>rpm: *3,500* |
| Exhaust blower motor | Manufacturer: *Any Motor Co.*<br>Model number: *MN2263*<br>Serial number: *66452*<br>Volts: *460*<br>Amperes: *4.8*<br>Phases: *3*<br>Cycles: *60 Hz*<br>hp: *3*<br>rpm: *3,500* |
| Spray system | Manufacturer: *Any Pump Co.*<br>Model number: *306*<br>Serial number: *67150049B*<br>Type: *Peristaltic* |

**Table 4.7. Component Material**

| Component | Material |
|---|---|
| Coating pan | *304 stainless steel* |
| Spray system | *Stainless steel* |

### *Lubricants*

Record the lubricants used to operate the tablet coating pan in Table 4.8 and indicate if they make contact with the product.

### *Equipment Safety Features*

There are no safety features on this equipment.

**Table 4.8. Lubricants**

| Where Used | Type | Manufacturer | Product Contact (Yes/No) |
|---|---|---|---|
| *Drive chain* | *Lubriplate APG 90* | *Any Oil Co.* | *No* |
| Is there a preventive maintenance procedure on file? (Yes/No) | | | *Yes* |

## Operational Qualification

An OQ evaluation should establish that the equipment can operate within specified tolerances and limits (Figure 4.4). The mechanical ranges of the tablet coating pan are challenged, along with the basic tablet coating pan operations that the operator will use. The tablet coating pan will be validated for its operating ability, not how well it coats tablets. The information required for the OQ evaluation is calibration of the instrument used to control the tablet coating pan, equipment control functions (switches and push buttons) and equipment operation (pan motor rotation direction, pan motor speed, pan supply temperature control, pan differential air pressure control, pan airflow control, spray system general operation).

### *Calibration Requirements*

Verify that all critical instruments on the equipment have been logged into the calibration system, have calibration procedures in place, and are currently in calibration at the time of qualification testing. Record all information for the calibrated instruments used to control the tablet coating pan in Table 4.9.

**Figure 4.4. OQ Elements of a Tablet Coating Pan**

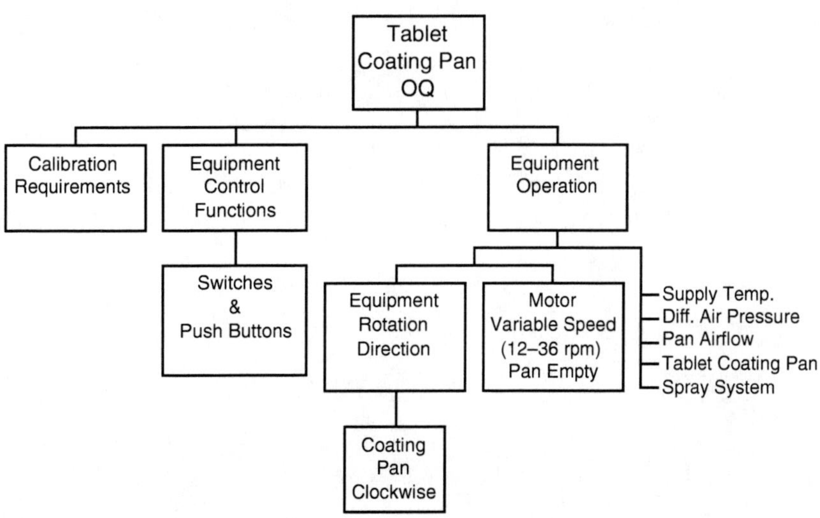

### *Equipment Control Functions*

The objective of testing equipment control functions is to verify that the push buttons on the tablet coating pan operate per manufacturer specifications. The tablet coating pan will be operated with the pan empty. Operate each control listed in Table 4.10 and verify its proper operation.

**Table 4.9. Calibrated and Noncalibrated Instruments**

*There were no noncalibrated instruments on this equipment.*

| Calibrated Instrument | As-Found Conditions |
|---|---|
| Identification number | *GPD-100-0* |
| Type | *Pressure gauge, analog* |
| Manufacturer | *Any Gauge Co.* |
| Model number | *C 3002* |
| Serial number | *722E011139R* |
| Range | *0–3 in. of* $H_2O$ |
| Scale division | *0.1 in. of* $H_2O$ |
| Location | *Left wall in room 627* |
| Use | *Prefilter 012-IP* |
| Calibration due date | *01/10/97* |
| Critical or not critical | *Critical to the operation of the coating pan* |
| **Calibrated Instrument** | **As-Found Conditions** |
| Identification number | *GPD-200-0* |
| Type | *Pressure gauge, analog* |
| Manufacturer | *Any Gauge Co.* |
| Model number | *3002* |
| Serial number | *6726011139R* |
| Range | *0–3 in. of* $H_2O$ |
| Scale division | *0.1 in. of* $H_2O$ |
| Location | *Left wall in room 627* |
| Use | *HEPA filter 022-IP* |
| Calibration due date | *01/10/97* |
| Critical or not critical | *Critical to the operation of the coating pan* |

*Continued on next page.*

*Continued from previous page.*

| **Calibrated Instrument** | **As-Found Conditions** |
|---|---|
| Identification number | *GPD-100-0* |
| Type | *Pressure gauge, analog* |
| Manufacturer | *Any Gauge Co.* |
| Model number | C *3002* |
| Serial number | *765011139R* |
| Range | *0–3 in. of* $H_2O$ |
| Scale division | *0.1 in. of* $H_2O$ |
| Location | *Left wall in room 627* |
| Use | *Pan pressure 006-IP* |
| Calibration due date | *01/10/97* |
| Critical or not critical | *Critical to the operation of the coating pan* |
| **Calibrated Instrument** | **As-Found Conditions** |
| Identification number | *DRP-100-00* |
| Type | *Temperature gauge, digital* |
| Manufacturer | *Any Gauge Co.* |
| Model number | *521CM* |
| Serial number | *3000* |
| Range | *0.0–120.0°C* |
| Scale division | *0.1°C* |
| Location | *Left wall in room 627* |
| Use | *Supply temperature controller* |
| Calibration due date | *01/10/97* |
| Critical or not critical | *Critical to the operation of the coating pan* |
| **Calibrated Instrument** | **As-Found Conditions** |
| Identification number | *DRP-200-00* |
| Type | *Temperature gauge, digital* |
| Manufacturer | *Any Gauge Co.* |
| Model number | *100* |
| Serial number | *65330339* |
| Range | *0.0–120.0°C* |
| Scale division | *1°C* |
| Location | *Left wall in room 627* |
| Use | *Exhaust temperature controller 005-IT* |
| Calibration due date | *01/10/97* |
| Critical or not critical | *Critical to the operation of the coating pan* |

*Continued on next page.*

*Continued from previous page.*

| Calibrated Instrument | As-Found Conditions |
|---|---|
| Identification number | *DRP-100-00-MFA* |
| Type | *Air volume gauge, digital* |
| Manufacturer | *Any Gauge Co.* |
| Model number | *100* |
| Serial number | *Not specified* |
| Range | *0–500 CFM* |
| Scale division | *1 CFM* |
| Location | *Left wall in room 627* |
| Use | *Air volume: 1–200 CFM* |
| Calibration due date | *01/10/97* |
| Critical or not critical | *Critical to the operation of the coating pan* |

**Table 4.10. Control Function Test Results**

| Test Function | Expected Results | Acceptable (Yes/No) |
|---|---|---|
| Power start/stop push-button | When the pan start/stop push button is pressed, the pan starts and rotates continuously. | *Yes* |
| | When the pan start/stop push button is pressed again, the pan stops rotating. | *Yes* |
| Pan jog push button | When the pan jog push button is pressed, the pan rotates clockwise; then it stops rotating when the pan jog push button is released. | *Yes* |
| Heater on/off push button | When the heater on push button is pressed, the heater starts. | *Yes* |
| | When the heater off push button is pressed, the fan heater stops. | *Yes* |
| Fan on/off push button | When the fan start push button is pressed, the fan starts and rotates continuously. | *Yes* |
| | When the fan stop push button is pressed, the fan stops. | *Yes* |
| Exhaust plenum clamp switch | When the exhaust plenum clamp switch is rotated counterclockwise, the plenum is clamped into place. | *Yes* |
| Pan speed control knob | When the pan speed control knob is rotated counter-clockwise, the pan speed increases. | *Yes* |
| | Slow speed: 11 rpm<br>Increased speed: 18 rpm | |
| | When the pan speed control knob is rotated clockwise, the pan speed decreases. | *Yes* |
| | Fast speed: 18 rpm<br>Decreased speed: 13 rpm | |

### *Equipment Operation: Tablet Coating Pan Rotation Direction Test*

The objective of the rotation direction test is to verify that the tablet coating pan rotates in the proper direction. The tablet coating pan will be operated with the pan empty. Press the start push button and observe the direction of rotation of the tablet coating pan as viewed from the front and record the results in Table 4.11.

### *Equipment Operation: Tablet Coating Pan Speed Test*

The objective of the pan speed test is to document the speed of the tablet coating pan with the pan empty. A tachometer will be required for this test. Use the following procedure to test the speed of the tablet coating pan:

- Measure the speed of the coating pan with a calibrated tachometer and record the results in Table 4.12. Verify that the measured speed is within ±10 percent of the variable speed of 12 to 36 rpm. Record the instrument used to measure the speed in Table 4.13.

  Coating pan speed specification = 12 to 36 rpm

  The measured minimum rpm of 12 falls within the specification.

  The measured maximum rpm of 35 falls within the specification.

### *Equipment Operation: Pan Supply Temperature Control Test*

The objective of the temperature control test is to verify that the pan supply temperature controller operates according to the manufacturer's specifications. Set the

**Table 4.11. *Tablet Coating Rotation Direction Test Results***

| Item | Expected Results | Results | Acceptable (Yes/No) |
|---|---|---|---|
| Tablet coating pan rotation direction | Rotation should be clockwise as viewed from the front of the tablet coating pan. | *Clockwise rotation was observed.* | *Yes* |

**Table 4.12. *Tablet Coating Pan Speed Test Results***

| Item | Specification (rpm ± 10%) | Results | Acceptable (Yes/No) |
|---|---|---|---|
| Minimum speed | 12 | *12* | *Yes* |
| Maximum speed | 36 | *35* | *Yes* |

**Table 4.13. *Instrument Used***

| Test Instrument | Identification Number | Calibration Due Date |
|---|---|---|
| *Tachometer* | *64020* | *06/21/97* |

pan supply temperature control to a low, medium, and high setting and verify that the setpoint temperatures are within ±10 percent of the indicated temperatures. Record the results in Table 4.14.

### *Equipment Operation: Pan Differential Air Pressure Control Test*

The objective of the air pressure control test is to verify that the pan differential air pressure controller operates according to the manufacturer's specifications. Set the pan differential air pressure control to a low, medium, and high setting and verify that the pan differential air pressure increases. Record the results in Table 4.15.

### *Equipment Operation: Pan Airflow Control Test*

The objective of the airflow control test is to verify that the pan airflow controller operates according to the manufacturer's specifications. Set the pan airflow control to a low, medium, and high setting and verify that the pan airflow increases. Record the results in Table 4.16.

**Table 4.14. Pan Supply Temperature Control Test Results**

| Temperature Setting | Inlet Temperature (Setpoint ± 10%) | Indicated Supply Temperature | Acceptable (Yes/No) |
|---|---|---|---|
| Low | 0°C | 0°C | *Yes* |
| Medium | 60°C | 60°C | *Yes* |
| High | 120°C | 120°C | *Yes* |

**Table 4.15. Pan Differential Air Pressure Control Test Results**

| Differential Air Pressure Setting | Differential Air Pressure Dial Setting | Indicated Differential Air Pressure | Acceptable (Yes/No) |
|---|---|---|---|
| Low | 1 | *0* | *Yes* |
| Medium | 5 | *-1.15 in. of* $H_2 0$ | *Yes* |
| High | 10 | *-0.55 in. of* $H_2 0$ | *Yes* |

**Table 4.16. Airflow Control Test Results**

| Differential Air Pressure Setting | Airflow Dial Setting | Indicated Airflow (CFM) | Acceptable (Yes/No) |
|---|---|---|---|
| Low | 1 | *2* | *Yes* |
| Medium | 5 | *262* | *Yes* |
| High | 10 | *562* | *Yes* |

### *Equipment Operation: Tablet Coating Pan Operation*

The operation of the tablet coating pan is tested to document its performance using placebo tablets but without the spray system. Placebo will be used for maximum loading conditions. The following procedure should be used:

- Fill the pan with placebo tablets to the appropriate level and record the test material used in Table 4.17.
- Set the pan speed to 11 rpm and run 5 tests for 5 min each after the steady state is achieved. Enter the setpoints and conduct the tests. Record the results in Table 4.18.

### *Equipment Operation: Spray System Operation Test*

The operation of the spray system is tested to verify that it operates according to the manufacturer's specifications. Water will be used as a spray medium. A graduated cylinder and a stopwatch will be required.

- Start the spray system and select a low flow rate. Capture the water from the spray system with a graduated cylinder for 1 min. Repeat this test for medium and high flow rates. Record the results in Table 4.19. Record the instruments used to measure the volume of the water and the operating time in Table 4.20.

**Table 4.17. *Required Materials***

| Item | Specifications |
|---|---|
| Test material | *Placebo tablets* |
| Batch size | *1 kg* |

**Table 4.18. *Tablet Coating Pan Operation Test Results***

| Test Number | Inlet Temperature Setpoint (°C) | Indicated Inlet Temperature (±2°C) | Inlet Air Volume Target (CFM) | Indicated Air Volume (±40 CFM) | Acceptable Acceptable (Yes/No) |
|---|---|---|---|---|---|
| 1 | 40 | *38.9–40.1* | 200 | *198–200* | *Yes* |
| 2 | 40 | *40.1–40.2* | 350 | *349–351* | *Yes* |
| 3 | 40 | *40.0–40.1* | 450 | *450–452* | *Yes* |
| 4 | 80 | *79.8–80.0* | 500 | *500–501* | *Yes* |
| 5 | 60 | *59.9–60.0* | 350 | *349–351* | *Yes* |

**Table 4.19. *Spray System Test Results***

| Flow Rate | Selected Delivery Flow Rate (mL/min 2 spray head) | Measure Delivery Flow Rate (mL/min 2 spray head) | Acceptable (Yes/No) |
|---|---|---|---|
| Low | 50 | 50 | *Yes* |
| Medium | 100 | 110 | *Yes* |
| High | 200 | 240 | *Yes* |

**Table 4.20. Instruments Used**

| Test Instrument | Identification Number | Calibration Due Date |
|---|---|---|
| *Graduated cylinder* | *00-770-ETM* | *01/16/97* |
| *Stopwatch* | *54039* | *02/16/97* |

## Performance Qualification

Once it has been established that the equipment is properly installed and functioning within specified operating parameters, it must be shown that the tablet coating pan can operate reliably under routine, minimum, and maximum operating conditions (Figure 4.5).

### *Tablet Coating Pan Operation*

The pan operation is tested to document the performance and speed of the tablet coating pan using placebo tablets and orange opadry coating medium. A tachometer will be required. Use the following procedure:

- Fill the pan with placebo tablets and record the placebo and spraying materials used in Table 4.21.
- Enter the setpoints and start spraying the tablets after the steady state is achieved. Record the results in Table 4.22.
- Measure the speed of the coating pan with a calibrated tachometer and record the results in Table 4.23. Verify that the measured speed is within ±10 percent

**Figure 4.5. PQ Elements of a Tablet Coating Pan**

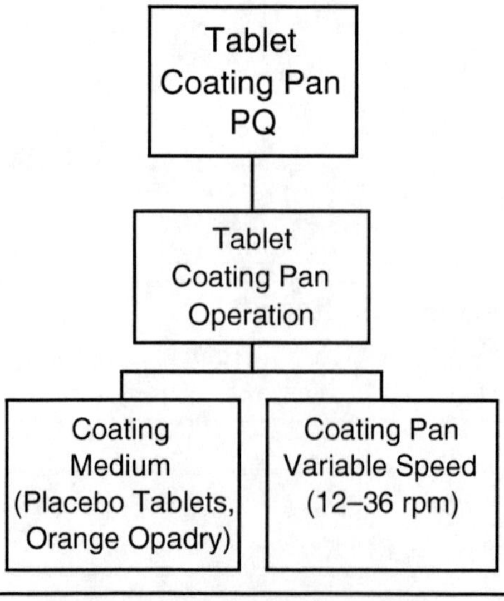

of the variable speed from 12 to 36 rpm. Record the instrument used to measure the speed in Table 4.24.

Coating pan speed specification = 12 to 36 rpm

The measured minimum rpm of 12 falls within the specification.

The measured maximum rpm of 35 falls within the specification.

**Table 4.21. Required Materials**

| Required Materials | Specifications |
|---|---|
| Test material | *Placebo tablets* |
| Coating medium | *Orange opadry* |
| Batch size | *1 kg* |

**Table 4.22. Tablet Coating Pan Performance Test Results**

| Test Function | Results | Acceptable (Yes/No) |
|---|---|---|
| Supply temperature | *75°C* | *Yes* |
| Exhaust temperature | *41°C* | *Yes* |
| Supply airflow rate | *363 CFM* | *Yes* |
| Pan differential air pressure | *0.05 in. of* $H_2O$ | *Yes* |
| Pan speed | *9 rpm* | *Yes* |
| Spray delivery rate | *30 mL/min* | *Yes* |
| Were the tablets coated adequately? (Yes/No) | | *Yes* |

**Table 4.23. Tablet Coating Pan Speed Test Results**

| Item | Specification (rpm) | Measured Speed (rpm) | Acceptable (Yes/No) |
|---|---|---|---|
| Minimum speed | *12* | *12* | *Yes* |
| Maximum speed | *36* | *35* | *Yes* |

**Table 4.24. Instrument Used**

| Test Instrument | Identification Number | Calibration Due Date |
|---|---|---|
| *Tachometer* | *64020* | *06/21/97* |

## DRUM STIRRING MIXER

The validation of a drum stirring mixer (Figure 4.6) that is motor driven and remotely controlled will be described here. The mixer is used for stirring tablet coating solutions during the spraying process. The operator uses the mixer's variable speed capability and digital speed display to ensure proper stirring of the spray solution. The mixer cannot be operated without the impeller being submerged in water or other representative fluid or powder.

The validation protocol of a drum stirring mixer is described in the following sections. This protocol example can be used as a cGMP compliance guide, and it should be tailored to meet individual company requirements. Italicized words in the tables represent examples of information that could be recorded in a validation protocol for a mixer.

### Installation Qualification

An IQ evaluation establishes confidence that the equipment is properly installed (Figure 4.7). The installation must meet the manufacturer's specified guidelines, along with design changes at installation. Also, the supporting electrical utilities must meet all electrical codes. The information required for an IQ evaluation is equipment identification, required documentation, equipment utility requirements, major component specifications, component material, lubricants, and equipment safety features.

#### *Equipment Identification*

Record the equipment identification numbers in Table 4.25, along with the following information: equipment manufacturers' purchase order number, model number, serial number, company assigned equipment number, and the location of the equipment.

#### *Required Documentation*

Record the equipment manufacturer's operation and maintenance manual and drawings in Table 4.26. Record the SOPs that cover the setup, operation, and cleaning of the mixer in Table 4.27.

#### *Equipment Utility Requirements*

Compare the manufacturer's specified volts (V) and amps (A) requirements to their as-found conditions at the time of qualification testing and record the results in Table 4.28. Also record the location of the power supply source. Record the instrument used to measure the volts and amps in Table 4.29.

**Figure 4.6 High-Shear Mixer (courtesy of Lightnin)**

**Figure 4.7. IQ Elements of a Drum Stirring Mixer**

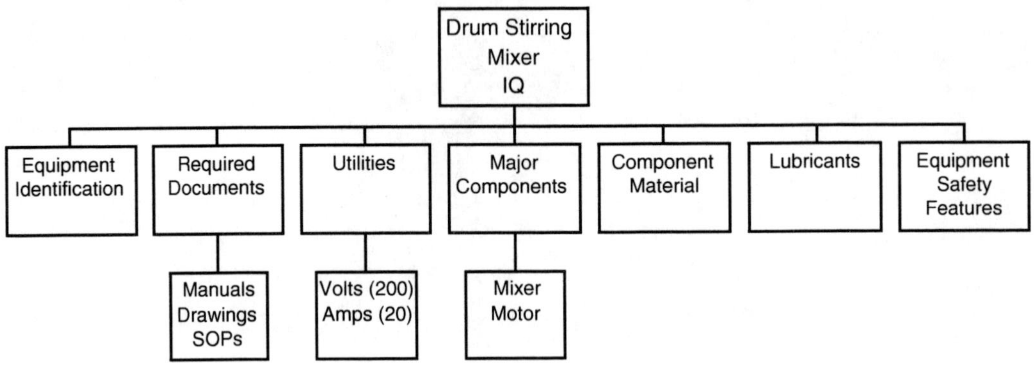


**Table 4.25. Equipment Identification**

| Required Information | As-Found Conditions |
|---|---|
| Manufacturer | *Any Mixer Co.* |
| Purchase order number | *005524* |
| Model number | *PC1322NX* |
| Serial number | *45584-02* |
| Equipment number | *35678* |
| Location | *Room 253 (This equipment is portable.)* |

**Table 4.26. Manufacturers' Manuals and Drawings**

| Number | Description | Date |
|---|---|---|
| *None* | *Mixer Instructions, Installation, Operation, Maintenance Manual* | *1996* |
| *PD-P-106* | *Electronic Controls for Variable Speed Drives* | *1993* |

**Table 4.27. Standard Operating Procedures**

| Number | Description | Release Date |
|---|---|---|
| *CAOA009* | *Setup and Operation of a Drum Mixer* | *10/23/92* |
| *COA010* | *Coating Dept.—Equipment Cleaning Procedures* | *03/12/96* |

**Table 4.28. Utilities**

| Utility | Specified | Measured | Acceptable (Yes/No) |
|---|---|---|---|
| Volts | 200 ± 10% | *A-B 199; A-C 197; B-C 201* | *Yes* |
| Amps | Motor = 1.2<br>Controller = 6.4 | *20 circuit rating* | *Yes* |

Power supply source, breaker box *BB23, wire numbers: 1, 3, 5*

**Table 4.29. Instrument Used**

| Test Instrument | Identification Number | Calibration Due Date |
|---|---|---|
| *Multimeter* | *ME-025* | *04/19/97* |

***Volt calculation:***

Mixer motor volt specification = 200 V ± 10%

±10% of 200 = ±20

200 + 20 = 220

200 – 20 = 180

The measured volts of 199/197/201 fall within ±10%.

***Amp calculation:***

Circuit rating = 20 A

Motor = 1.2 A

Controller = 6.4 A

Equipment current draw = motor A + controller A = 1.2 + 6.4 = 7.6 A

The circuit amp rating of 20 is greater than the maximum current draw of the equipment.

### *Major Component Specifications*

The component specifications of the protocol verifies that the mixer components purchased were delivered and installed. Record the major components in Table 4.30.

### *Component Material*

Record the material of each component that contacts the product in Table 4.31.

### *Lubricants*

Record the lubricants used to operate the mixer in Table 4.32 and indicate if they make contact with the product.

### *Equipment Safety Features*

There are no safety features on this equipment. Nevertheless, there are some general safety precautions that should be followed when operating a mixer.

**Table 4.30. Major Components**

| Components | As-Found Conditions |
|---|---|
| Mixer motor | Manufacturer: *Any Motor Co.*<br>Model number: *3BE07414*<br>Serial number: *LX*<br>Volts: *200*<br>Amperes: *1.2*<br>Phases: *3*<br>Cycles: *60 Hz*<br>hp: *1*<br>rpm: *149–1,785* |
| Shaft | Part number: *14263*<br>Size: *3/8 in. diameter by 36 in. long*<br>Material: *316 stainless steel* |
| Impeller | Part number:<br>Size: *3/8 in. bore by 5 in. diameter*<br>Material: *316 stainless steel* |
| A-C controller | Manufacturer: *Any Controller Co.*<br>Model number: *SP5004*<br>Serial number: *01GN033875*<br>Volts: *200–230*<br>Amperes: *6.4*<br>Phases: *3*<br>Cycles: *60 Hz*<br>hp: *1/4–1*<br>rpm: *149–1,785* |

**Table 4.31. Component Material**

| Component | Material |
|---|---|
| Shaft | *316 stainless steel* |
| Impeller | *316 stainless steel* |

**Table 4.32. Lubricants**

| Where Used | Type | Manufacturer | Product Contact (Yes/No) |
|---|---|---|---|
| *The mixer bearings were packed at the vendor and require no further attention.* | | | |
| Is there a preventive maintenance procedure on file? (Yes/No) | | | *Yes* |

- Never touch a mixer, which has an electric motor, or any part of an electrical service line cord with wet hands or wet feet or if standing on a wet surface.
- Never attempt to move or adjust a mixer while it is running.
- Never touch any rotating part of a mixer with bare hands, gloved hands, or any hand-held object. Rotating parts include, but are not limited to, the mixer shaft, impeller(s), mechanical seals, and motor fans.
- Do not touch a mixer motor until it cools. The motor temperature may be high enough to cause burns to the hands.

## Operational Qualification

An OQ evaluation should establish that the equipment can operate within specified tolerances and limits (Figure 4.8). The mechanical ranges of the mixer are challenged, along with the basic mixer operations that the operator will use. The mixer will be validated for its operating ability, not how well it mixes liquids. The information required for the OQ evaluation is calibration of the instrument used to control the mixer, equipment control functions (switches and push buttons) and equipment operation (mixer motor rotation direction and speed test).

***Figure 4.8*** ***OQ Elements of a Drum Stirring Mixer***

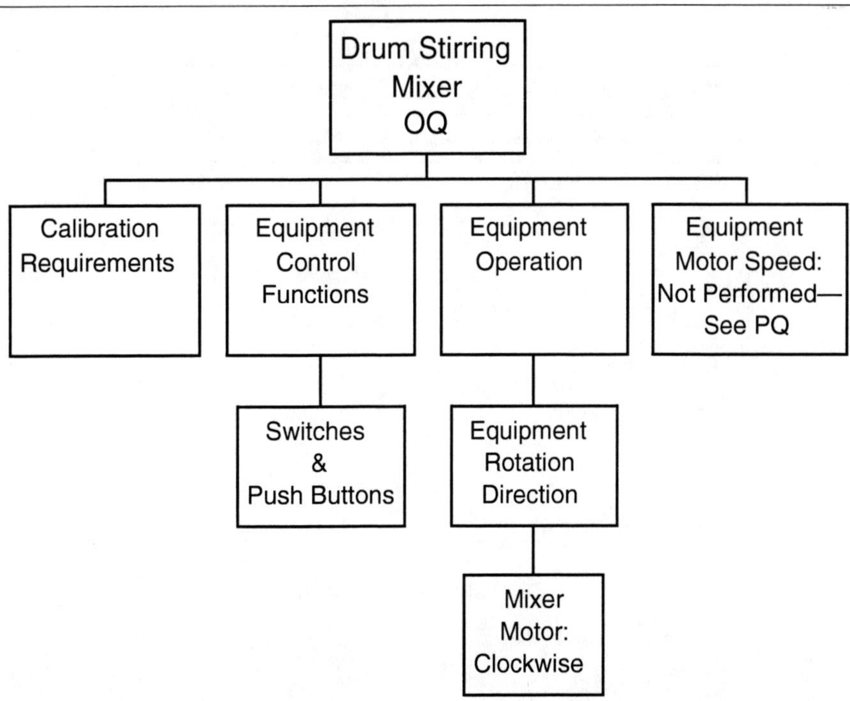

### *Calibration Requirements*

Verify that all critical instruments on the equipment have been logged into the calibration system, have calibration procedures in place, and are currently in calibration at the time of qualification testing. Record all information for the calibrated instruments used to control the mixer in Table 4.33.

### *Equipment Control Functions*

The objective of testing equipment control functions is to verify that the push buttons on the mixer operate per manufacturer specifications. The mixer will be operated with the impeller submerged in water. Record the results in Table 4.35. Using a mixing container, the procedure is as follows:

- Fill the mixing container with water to its maximum working volume of 130 L and record the amount used in Table 4.34. When operating the mixer, the head must be submerged in the water to prevent damage to the mixer.

  Test fluid volume = 75% of 173 L = 129.75 L (rounded to 130 L)

- Press the start key and verify that the mixer starts operating. Record the results in Table 4.35.

- Press the up/down arrow key and verify that the rpm increases and decreases according to the appropriate key. Record the initial speed, increased speed, and decreased speed in Table 4.35.

- Press the stop/reset key and verify that the mixer stops operating. Record the results in Table 4.35.

***Table 4.33. Calibrated and Noncalibrated Instruments***

*There were no noncalibrated instruments on the mixer.*

| Calibrated Instrument | As-Found Conditions |
|---|---|
| Identification number | *OPS/VFD-001* |
| Type | *Speed controller, digital* |
| Manufacturer | *Any Gauge Co.* |
| Model number | *SP500* |
| Serial number | *01GN033875* |
| Range | *149–1,785 rpm* |
| Scale division | *1 rpm* |
| Location | *Mounted on the wall next to the mixer* |
| Use | *Controls the speed of the mixer* |
| Calibration due date | *07/08/97* |
| Critical or not critical | *Critical to the operation of the mixer* |

**Table 4.34. *Required Materials***

| Required Materials | Specifications |
|---|---|
| Mixing container | *173 L* |
| Test fluid | *Water* |
| Test fluid volume | *130 L* |

**Table 4.35. *Control Function Test Results***

| Test Function | Expected Results | Acceptable (Yes/No) |
|---|---|---|
| Start key operation | When the start key is pressed, the mixer starts. | *Yes* |
| Up/down arrow key operation | When the up/down arrow keys are pressed, the mixer rpm increases and decreases.<br>Initial speed: *252 rpm*<br>Increased speed: *275 rpm*<br>Decreased speed: *175 rpm* | *Yes* |
| Stop/reset key operation | When the stop/reset key is pressed, the mixer stops. | *Yes* |
| | The corresponding LED goes on. | *Yes* |

### *Equipment Operation: Mixer Motor Rotation Direction Test*

The objective of the rotation direction test is to verify that the mixer motor rotates in the proper direction. The mixer will be operated with the impeller submerged in water. Using the same materials as for the previous test, press the start key and observe the direction of rotation of the mixer motor as viewed from the top of the mixer and record the results in Table 4.36.

### *Equipment Operation: Mixer Speed Test*

The speed test without water will not be performed because the mixer cannot be operated without the impeller submerged in water or other representative fluid or powder.

**Table 4.36. *Mixer Motor Rotation Direction Test Results***

| Item | Expected Results | Results | Acceptable (Yes/No) |
|---|---|---|---|
| Mixer motor rotation direction | Rotation should be clockwise as viewed from the top of the mixer. | *Clockwise rotation was observed.* | *Yes* |

## Performance Qualification

Once it has been established that the equipment is properly installed and functioning within specified operating parameters, it must be shown that the mixer can operate reliably under routine, minimum, and maximum operating conditions (Figure 4.9).

### *Drum Stirring Mixer Operation*

The operation of the drum stirring mixer is tested to document its performance with water and a colored dye. Water will be used for maximum loading conditions. The objective also includes verification that the mixer can move the fluid about the container (i.e., mixing). The variable speed of the mixer will be measured and recorded.

The materials and instruments required for the test are a mixing container, test fluid, dye, and a tachometer. The operation test is performed as follows:

- Fill a mixing container with water to its maximum working volume of 130 L and record the material used in Table 4.37. When operating the mixer, the head must be submerged in the water.

  Test fluid volume = 75% of 173 L = 129.75 L (rounded to 130 L)

- Turn the mixer on, observe the motion of the fluid, and record the results in Table 4.38. The results of this test are qualitative only and are based on observation.

- Add colored dye to the water, observe the mixing action of the mixer, and record the results in Table 4.38. Verify that the added dye is distributed uniformly throughout the container. Record the dye that was used in Table 4.37.

***Figure 4.9. PQ Elements of a Drum Stirring Mixer***

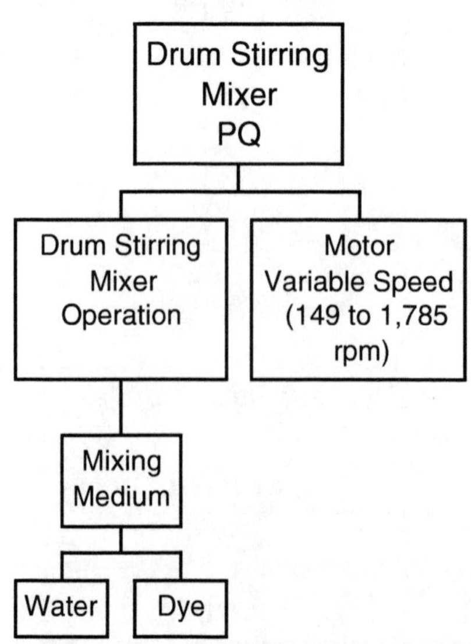

**Table 4.37. Required Materials**

| Required Materials | Specifications |
|---|---|
| Mixing container | *173 L* |
| Test fluid | *Water* |
| Test dye | *FDC blue #1* |
| Test fluid volume | *130 L* |

**Table 4.38. Mixer Performance Test Results**

| Test Function | Expected Results | Results* | Acceptable (Yes/No) |
|---|---|---|---|
| Fluid Mixing capabilities | The mixer should move the fluid about the container. | *Fluid motion and vortex action was observed. There was increased motion with increased speed.* | *Yes* |
| | The added dye should be distributed uniformly throughout the container. | *The dye was distributed uniformly throughout the container.* | *Yes* |

*The results of this test are qualitative only and are based on observation.

- Set the mixer to the minimum speed of 149 rpm and measure the rpm of the mixer with a calibrated tachometer. Verify that the rpm reading displayed on the controller is within ±10 percent of the tachometer measurement. Record the mixer speed measurement in Table 4.39. Record the instrument used to measure the speed in Table 4.40.
- Set the mixer to the maximum speed of 1,785 rpm and measure the rpm of the mixer with a calibrated tachometer. Verify that the rpm reading displayed on the controller is within ±10 percent of the tachometer measurement. Record the mixer speed measurement in Table 4.39. Record the instrument used to measure the speed in Table 4.40.

**Table 4.39. Mixer Speed Test Results**

| Specification (rpm) | Controller Display (rpm) | Measured Speed (rpm) | Acceptable (Yes/No) |
|---|---|---|---|
| 149 ± 10% | *149* | *146* | *Yes* |
| 1,785 ± 10% | *1,785* | *1,787* | *Yes* |

**Table 4.40. Instrument Used**

| Test Instrument | Identification Number | Calibration Due Date |
|---|---|---|
| *Tachometer* | *64020* | *04/21/97* |

***Mixer slow speed calculation:***

Mixer speed specification = 149 rpm ± 10%

±10% of 149 = ±14.9

149 + 14.9 = 163.9

149 – 14.9 = 134.1

The measured rpm of 146 falls within ±10%.

***Mixer fast speed calculation:***

Mixer speed specification = 1,785 rpm ± 10%.

±10% of 1,785 = ±178.3

1785 + 178.3 = 1961.3

1785 – 178.3 = 1604.7

The measured rpm of 1,787 falls within ±10%

## Tablet Imprinter

The validation of a tablet imprinter (Figure 4.10) that is motor driven and operates at variable speeds will be described here. The imprinter stamps tablets with an inked impression. A vibrator places tablets in a hopper; then they are distributed on a conveyor that moves the tablets between two rubber grippers. One or both grippers are in contact with an imprinting wheel that rolls through an ink well. Ink is deposited on engraved labels that are on the imprinting wheel. The ink is transferred from the imprinting wheel to the gripper and then to the tablets. After imprinting, the tablets pass through an air-driven drying area and then on to a collection bucket.

The validation protocol for a tablet imprinter is given in the following sections. This protocol example can be used as a cGMP compliance guide, and it should be tailored to meet individual company requirements. Italicized words in the tables represent examples of information that could be recorded in a validation protocol for a tablet imprinter.

### Installation Qualification

An IQ evaluation establishes confidence that the equipment is properly installed (Figure 4.11). The installation must meet the manufacturer's specified guidelines, along with design changes at installation. Also, the supporting electrical utilities must meet all electrical codes. The information required for an IQ evaluation is equipment identification, required documentation, equipment utility requirements, major component specifications, component material, lubricants, and equipment safety features.

***Figure 4.10. Tablet Imprinter (courtesy of The Hartnett Co.)***

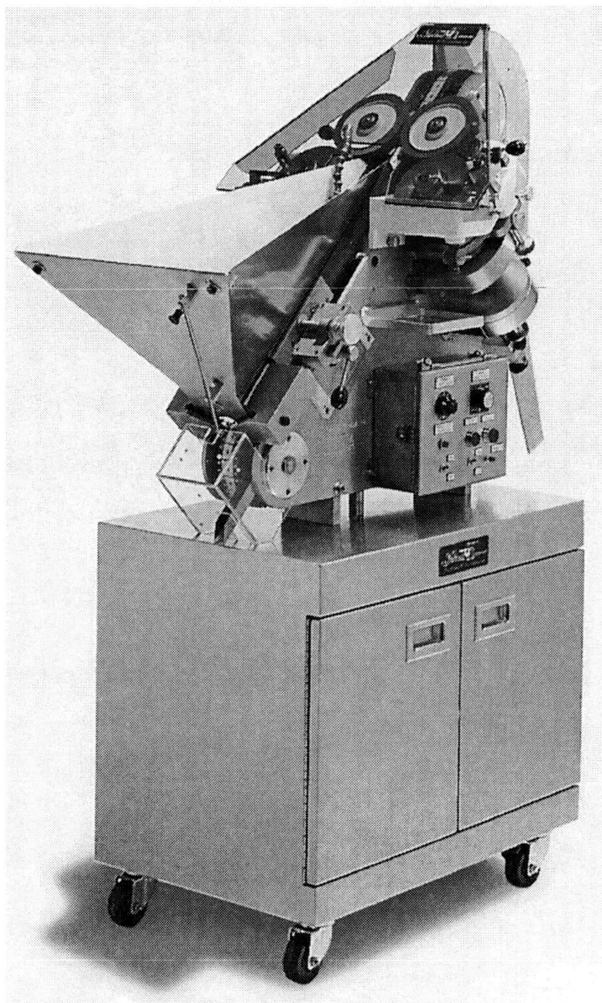

***Figure 4.11. IQ Elements of a Tablet Imprinter***

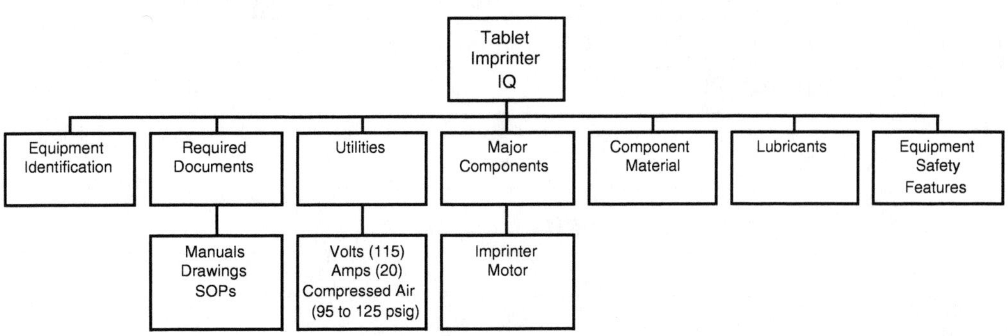

### *Equipment Identification*

Record the equipment identification numbers in Table 4.41, along with the following information: equipment manufacturer's purchase order number, model number, serial number, company assigned equipment number, and the location of the equipment.

### *Required Documentation*

Record the equipment manufacturer's operation and maintenance manual and drawings in Table 4.42. Record the SOPs that cover the setup, operation, and cleaning of the tablet imprinter in Table 4.43.

### *Equipment Utility Requirements*

Compare the manufacturer's specified volts (V), amps (A), and compressed air pressure requirements to their as-found conditions at the time of qualification testing and record the results in Table 4.44. Also record the location of the power supply source. Record the instrument used to measure the volts, amps, and compressed air in Table 4.45.

**Table 4.41. *Equipment Identification***

| Required Information | As-Found Conditions |
|---|---|
| Manufacturer | *Any Imprinter Co.* |
| Purchase order number | *0077521* |
| Model number | *Delta* |
| Serial number | *291D* |
| Equipment number | *0341* |
| Location | *Room 401* |

**Table 4.42. *Manufacturers' Manuals and Drawings***

| Number | Description | Date |
|---|---|---|
| *None* | *Imprinter Instructions, Installation, Operation, and Maintenance Manual* | *None* |

**Table 4.43. *Standard Operating Procedures***

| Number | Description | Release Date |
|---|---|---|
| *IMP009* | *Setup and Operation of a Tablet Imprinter* | *04/13/94* |
| *IMP010* | *Imprinting Dept.—Equipment Cleaning Procedures* | *09/13/94* |

**Table 4.44. Utilities**

| Utility | Specified | Measured Results | Acceptable (Yes/No) |
|---|---|---|---|
| Volts | 115 ± 10% | *119* | *Yes* |
| Amps | 3 | 20 circuit rating | *Yes* |
| Compressed air | 95 to 125 psig | *110 psig* | *Yes* |

Power supply source, breaker box *BB11, wire number: 6*

**Table 4.45. Instrument Used**

| Test Instrument | Identification Number | Calibration Due Date |
|---|---|---|
| *Multimeter* | *ME-025* | *04/19/97* |

***Volt calculation:***

Imprinter motor volt specification = 115 V ± 10%

±10% of 115 = ±11.5

115 + 11.5 = 126.5

115 – 11.5 = 103.5

The measured voltage of 119 falls within ±10%.

***Amp calculation:***

Circuit rating = 20 A

Equipment current draw = 3 A

The circuit amp rating of 20 is greater than the maximum current draw of the equipment.

***Compressed air calculation:***

Compressed air = 95 to 125 psig

The measured air pressure of 110 psig falls within the specification.

### *Major Component Specifications*

The component specifications section of the protocol verifies that the tablet imprinter components purchased were delivered and installed. Record the major components in Table 4.46.

**Table 4.46. Major Components**

| Components | As-Found Conditions |
|---|---|
| Imprinter motor | Manufacturer: *Any Motor Co.*<br>Model number: *C761M2*<br>Serial number: *651009*<br>Volts: *115*<br>Amperes: *3*<br>Phases: *1*<br>Cycles: *60 Hz*<br>hp: *1/4*<br>rpm: *0–150* |

### *Component Material*

Record the material of each component that contacts the product in Table 4.47.

### *Lubricants*

Record the lubricants used to operate the tablet imprinter in Table 4.48 and indicate if they make contact with the product.

### *Equipment Safety Features*

There are no safety features on this equipment.

**Table 4.47. Component Material**

| Component | Material |
|---|---|
| Hopper | *Stainless steel* |
| Rubber grippers | *Rubber 45 durometer* |
| Imprinter wheel | *Rubber 45 durometer* |
| Carrier link | *Stainless steel* |
| Dryer/ejector assembly | *Stainless steel* |
| Printing ink | *Food grade* |

**Table 4.48. Lubricants**

| Where Used | Type | Manufacturer | Product Contact (Yes/No) |
|---|---|---|---|
| *Main chain* | *SAE 20-30* | *Any Oil Co.* | *No* |
| Is there a preventive maintenance procedure on file? (Yes/No) | | | *Yes* |

## Operational Qualification

An OQ evaluation should establish that the equipment can operate within specified tolerances and limits (Figure 4.12). The mechanical ranges of the tablet imprinter are challenged, along with the basic tablet imprinter operations that the operator will use. The tablet imprinter will be validated for its operating ability, not how well it imprints tablets. The information required for the OQ evaluation is calibration of the instrument used to control the tablet imprinter, equipment control functions (switches and push buttons) and equipment operation (tablet imprinter motor rotation direction, tablet imprinter motor speed control, dryer jet operation, conveyor and vibrator operation).

### *Calibration Requirements*

Verify that all critical instruments on the equipment have been logged into the calibration system, have calibration procedures in place, and are currently in calibration at the time of qualification testing. Record all information for the calibrated instruments used to control the tablet imprinter in Table 4.49.

### *Equipment Control Functions*

The objective of testing equipment control functions is to verify that the switches and push buttons on the tablet imprinter operate according to the manufacturer's

**Figure 4.12. OQ Elements of a Tablet Imprinter**

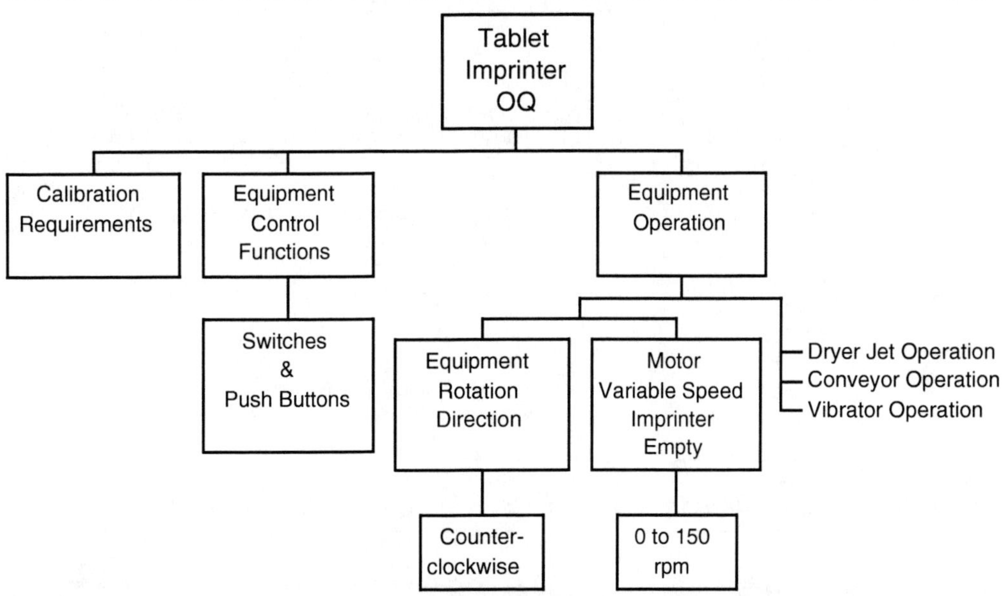


**Table 4.49. Calibrated and Noncalibrated Instruments**

*There were no calibrated or noncalibrated instruments on this equipment.*

specifications. The tablet imprinter will be operated without tablets. Operate each control listed in Table 4.50 and verify its proper operation.

### *Equipment Operation: Tablet Imprinter Rotation Direction Test*

The objective of the rotation direction test is to verify that the tablet imprinter motor rotates in the proper direction. The tablet imprinter will be operated with the imprinter empty. Press the start push button, observe the direction of rotation of the tablet imprinter motor as viewed from the front, and record the results in Table 4.51.

### *Equipment Operation: Tablet Imprinter Motor Speed Control Test*

The objective of the speed control test is to document the speed of the tablet imprinter motor with the imprinter empty. A tachometer is required for the following test procedure:

- Measure the speed of the imprinter motor with a calibrated tachometer and record the results in Table 4.52. Verify that the measured speed is within ±10 percent of the variable speed from 0 to 150 rpm. Record the instrument used to measure the speed in Table 4.53.

**Table 4.50. Control Function Test Results**

| Test Function | Expected Results | Acceptable (Yes/No) |
|---|---|---|
| Conveyor speed control knob operation | When the conveyor speed control knob is rotated, the speed of the conveyor increases and decreases. | *Yes* |
| On/off toggle switch operation | When the on/off toggle switch is flipped, the imprinter goes on and off. | *Yes* |
| Vibrator toggle pushbutton operation | When the vibration on/off push button is pressed, the vibrator goes on and off. | *Yes* |
| Imprinter speed potentiometer knob operation | When the imprinter speed potentiometer knob is rotated, the speed increases and decreases. | *Yes* |

**Table 4.51. Tablet Imprinter Motor Rotation Direction Test Results**

| Item | Expected Results | Results | Acceptable (Yes/No) |
|---|---|---|---|
| Tablet imprinter motor rotation direction | Rotation should be counterclockwise as viewed from the front of the tablet imprinter. | *Counterclockwise rotation was observed.* | *Yes* |

**Table 4.52. *Tablet Imprinter Speed Control Test Results***

| Speed | Specification (rpm) | Measured Speed (rpm) | Acceptable (Yes/No) |
|---|---|---|---|
| Minimum | 0 | *0* | *Yes* |
| Maximum | 150 | *150* | *Yes* |

**Table 4.53. *Instrument Used***

| Test Instrument | Identification Number | Calibration Due Date |
|---|---|---|
| *Tachometer* | *64020* | *06/21/97* |

Tablet imprinter motor speed specification = 0 to 150 rpm

The measured minimum rpm of 0 falls within the specification.

The measured maximum rpm of 150 falls within the specification.

### *Equipment Operation: Dryer Jet Operation Test*

The dryer jets are tested to verify that they operate according to the manufacturer's specifications. Remove the rear guard and verify that the air is blowing out the dryer jets when the unit is operating. Record the results in Table 4.54.

### *Equipment Operation: Conveyor and Vibrator Operation Test*

The conveyor and vibrator are tested to verify that they operate according to the manufacturer's specifications. Observe the conveyor for 3 min at the beginning and end of a run. Record the number of tablets picked up and the number of times the conveyor slot was empty in Table 4.55. Record the instrument used to measure the time in Table 4.56.

**Table 4.54. *Dryer Jet Test Results***

| Item | Expected Results | Results | Acceptable (Yes/No) |
|---|---|---|---|
| Dryer jet test | Air should be blowing out the dryer jets. | *Air was blowing out the dryer jets.* | *Yes* |

**Table 4.55. *Conveyor and Vibrator Test Results***

| Run | Number of Tablets Picked Up | Number of Times Conveyor Was Empty | Acceptable (Yes/No) |
|---|---|---|---|
| Beginning | 342 | *0* | *Yes* |
| End | 342 | *0* | *Yes* |

**Table 4.56. Instrument Used**

| Test Instrument | Identification Number | Calibration Due Date |
|---|---|---|
| *Stopwatch* | *54039* | *02/16/97* |

## Performance Qualification

Once it has been established that the equipment is properly installed and functioning within specified operating parameters, it must be shown that the tablet imprinter can operate reliably under routine, minimum, and maximum operating conditions (Figure 4.13).

### *Tablet Imprinter Operation*

Operation of the tablet imprinter is tested to document its performance and speed using placebo tablets. A tachometer will be required. The test is performed as follows:

- Visually inspect 100 tablets at the beginning and the end of a production run to verify that the tablets were imprinted. The acceptable quality level (AQL) is 100 percent. Record the results in Table 4.58. Record the materials used in Table 4.57.
- Measure the speed of the imprinter with a calibrated tachometer and record the results in Table 4.59. Verify that the measured speed is within ±10 percent

**Figure 4.13. PQ Elements of a Tablet Imprinter**

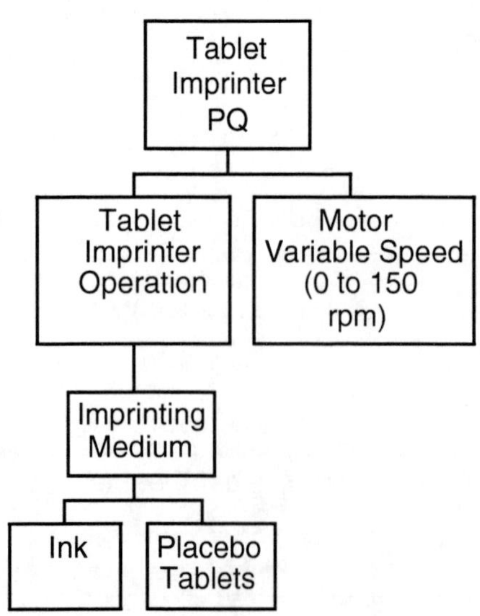

**Table 4.57.** ***Required Material***

| Item | Specifications |
|---|---|
| Test material | *Placebo tablets* |

**Table 4.58.** ***Tablet Imprinter Performance Test Results***

| Production Run | Number of Tablets | Number of Tablets (AQL=100%) | Acceptable (Yes/No) |
|---|---|---|---|
| Beginning | 100 | *100* | *Yes* |
| End | 100 | *100* | *Yes* |

**Table 4.59.** ***Tablet Imprinter Speed Test Results***

| Speed | Specification (rpm) | Measured Speed (rpm) | Acceptable (Yes/No) |
|---|---|---|---|
| Minimum | 0 | *0* | *Yes* |
| Maximum | 150 | *148* | *Yes* |

of the variable speed from 0 to 150 rpm. Record the instrument used to measure the speed in Table 4.60.

Tablet imprinter speed specification = 0 to 150 rpm

The measured minimum rpm of 0 falls within the specification.

The measured maximum rpm of 148 falls within the specification.

**Table 4.60.** ***Instrument Used***

| Test Instrument | Identification Number | Calibration Due Date |
|---|---|---|
| *Tachometer* | *64020* | *06/21/97* |

# Chapter 5

# Validation of Encapsulation Equipment

This chapter will cover the qualification testing of the equipment used to perform operations in the encapsulation department (Figure 5.1).

## Capsule Filler

The validation of a capsule filler (Figure 5.2) will be described here. The unit is motor driven and operates at variable speeds. The capsule filler can dose powder, tablets, and beads into hard shell gelatin capsules. The machine is connected to a control (PLC) cabinet that houses all of the electronics and a programmable logic

***Figure 5.1. Encapsulation Equipment***

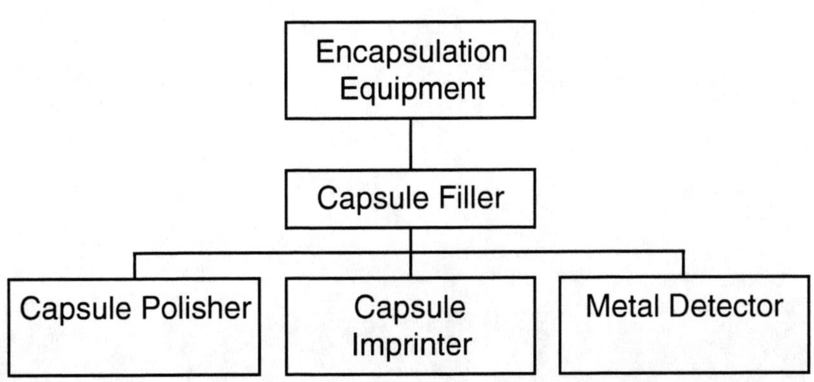

**Figure 5.2. MG Futura Capsule Filler (courtesy of MG America, Inc.)**

controller (PLC). On the powder dosing unit, the powder chamber, compression, and layer values are modified by drives integrated in the machine logic and controlled by the PLC. The bead dosing unit is entirely integrated in the machine logic and is also controlled by the PLC.

There are seven product hoppers: one for capsules, one for powders, three for tablets, and two for beads. Capsules are fed into the capsule hopper connected to the orientation station. At this station, the capsule bodies are separate from the capsule lids. The lids attach themselves to the lid transfer station and travel directly to the closing/ejection station. The bodies travel on the capsule carrier belt to the powder dosing station where they are filled by dosing fingers with product from the powder hopper; with one, two, or three tablets from the tablet dosing disk; or with beads from dosing chambers one and/or two. Once the bodies have been filled, they travel to the closing/ejection station where they match up with the lids that have been circulated on the lid transfer station. The capsules are then closed and ejected from the machine.

Provided with the machine is an exhaust and vacuum cabinet that is located in the room with the equipment. The vacuum system provides the necessary vacuum required for the operation, opening, transferring, and closing of the capsules as they pass through the various stations of the machine. The vacuum cabinet also contains vacuum canisters for product tailing.

The validation protocol for a capsule filler is described in the following sections. This protocol example can be used as cGMP compliance guide, and it should be tailored to meet individual company requirements. Italicized words in the tables represent examples of information that could be recorded in a validation protocol for a capsule filler.

## Installation Qualification

An IQ evaluation establishes confidence that the equipment is properly installed (Figure 5.3). The installation must meet the manufacturer's specified guidelines, along with design changes at installation. Also, the supporting electrical utilities must meet all electrical codes. The information required for an IQ evaluation is: equipment identification, required documentation, equipment utility requirements, major component specifications, component material, lubricants, and equipment safety features.

### *Equipment Identification*

Record the equipment identification numbers in Table 5.1, along with the following information: equipment manufacturer's purchase order number, model number, serial number, company assigned equipment number, and the location of the equipment.

### *Required Documentation*

Record the equipment manufacturer's operation and maintenance manuals and drawings in Table 5.2. Record the SOPs that cover the setup, operation, and cleaning of the capsule filler in Table 5.3.

**Figure 5.3. IQ Elements of a Capsule Filler**

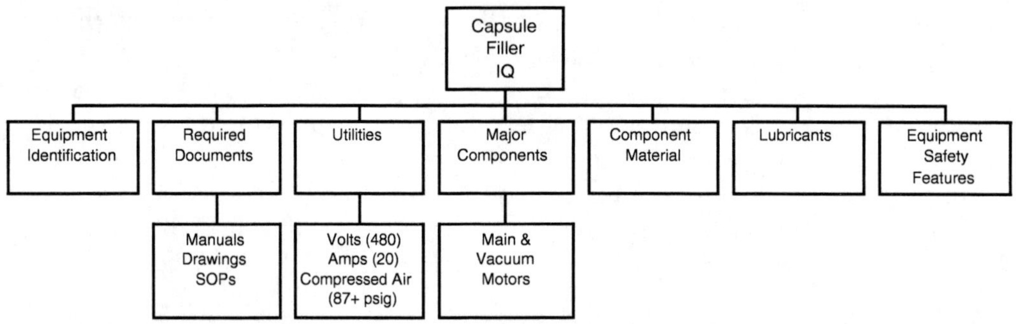


**Table 5.1. Equipment Identification**

| Required Information | As-Found Conditions |
|---|---|
| Manufacturer | *Any Capsule Filler Co.* |
| Purchase order number | *5637100* |
| Model number | *2GM* |
| Serial number | *8615* |
| Equipment number | *0173* |
| Location | *Room 147* |

**Table 5.2. Manufacturers' Manuals and Drawings**

| Number | Description | Date |
|---|---|---|
| *None* | *Capsule Filler, Instructions for Use and Maintenance Manual* | *1997* |
| *720000* | *Powder Dosing Unit, Instruction Manual* | *09/12/96* |
| *730000* | *Tablet Dosing Unit, Instruction Manual* | *06/30/96* |
| *930000* | *Bread Dosing Unit, Instruction Manual* | *09/12/96* |
| *None* | *Automatic Capsule Filling Machine, Spare Parts Manual* | *None* |
| *None* | *Drawings are too numerous to list. They are bound into a manual.* | *None* |

**Table 5.3. Standard Operating Procedures**

| Number | Description | Date |
|---|---|---|
| *ENC001* | *Capsule Filler Setup and Operation* | *05/20/97* |
| *ENC002* | *Capsule Filler Cleaning Procedure* | *05/26/97* |

## *Equipment Utility Requirements*

Compare the manufacturer's specified volts (V), amps (A), and compressed air pressure requirements to their as-found conditions at the time of qualification testing and record the results in Table 5.4. Also record the location of the power supply source. Record the instruments used to measure the volts, amps, and compressed air in Table 5.5.

***Volt calculation:***

Capsule filler motor volt specification = 480 V ± 10%

± 10% of 480 = ± 48

480 + 48 = 528

480 – 48 = 432

The measured voltage of 477/475/472 falls within ± 10%.

***Amp calculation:***

Circuit rating = 20 A

Main motor = 6.5 A

Vacuum motor = 1.7 A

Equipment current draw = main motor A + vacuum motor A = 6.5 A + 1.7 A = 8.2 A

The circuit amp rating of 20 is greater than the maximum current draw of the equipment.

***Compressed air calculation:***

Compressed air ≥ 87 psig

The measured air pressure of 110 psig is greater than 87 psig.

**Table 5.4. Utilities**

| Utility | Specified | Measured | Acceptable (Yes/No) |
|---|---|---|---|
| Volts | 480 ± 10% | *A-B 477; A-C 475; B-C 472* | *Yes* |
| Amps | Main motor = 6.5<br>Vacuum = 1.7 | 20 circuit rating | *Yes* |
| Compressed air | ≥ 87 psig | *110 psig* | *Yes* |

Power supply source, breaker box *BB47, wire numbers: 17, 19, 21*

**Table 5.5. Instruments Used**

| Test Instrument | Identification Number | Calibration Due Date |
|---|---|---|
| *Multimeter* | *ME-025* | *04/19/97* |
| *Air pressure gauge* | *P-102* | *02/10/97* |

### *Major Component Specifications*

The component specifications section of the protocol verifies that the capsule filler components purchased were delivered and installed. Record the major components in Table 5.6.

### *Component Material*

Record the material of each component that contacts the product in Table 5.7.

### *Lubricants*

Record the lubricants used to operate the capsule filler in Table 5.8 and state whether they make contact with the product.

**Table 5.6. Major Components**

| Components | As-Found Conditions |
|---|---|
| Main motor | Manufacturer: *Any Motor Co.*<br>Model number: *3B 4AL09 KZ*<br>Serial number: *Not specified*<br>Volts: *480*<br>Amperes: *6.5*<br>Phases: *3*<br>Cycles: *60 Hz*<br>hp: *2*<br>rpm: *1,715* |
| Control cabinet software | Programmable controller:<br>Manufacturer: *Any Motor Co.*<br>Model number: *2V-11cCPU-1MVC*<br>Serial number: *8615*<br><br>Memory:<br>Manufacturer: *Any EPROM Co.*<br>Model Number: *EEPROM 27993*<br>Version: *1.03* |
| Vacuum cabinet motor | Manufacturer: *Any Motor Co.*<br>Model number: *2095141 F*<br>Serial number: *Not specified*<br>Volts: *480*<br>Amperes: *1.7*<br>Phases: *3*<br>Cycles: *60 Hz*<br>hp: *1*<br>rpm: *1,420–1,700* |

**Table 5.7. *Component Material***

| Component | Material |
|---|---|
| Too numerous to list | *304 or 316 stainless or biocompatible materials* |

**Table 5.8. *Lubricants***

| Where Used | Type | Manufacturer | Product Contact (Yes/No) |
|---|---|---|---|
| *Gears* | *Kluber 02201HU Oil* | *Any Oil Co.* | *No* |
| *Motor & reducers* | *SAE 80W/140 Oil* | *Any Oil Co.* | *No* |
| Is there is a preventive maintenance procedure on file? (Yes/No) | | | *Yes* |

### *Equipment Safety Features*

The objective of testing equipment safety procedures is to verify that the safety features on the capsule filler function according to manufacturer specifications. With the machine switched on and the equipment parts stationary, verify that the safety features operate as specified. Record the results in Table 5.9.

## Operational Qualification

An OQ evaluation should establish that the equipment can operate within specified tolerances and limits (Figure 5.4). The mechanical ranges of the capsule filler are challenged, along with the basic capsule filler operations that the operator will use. The capsule filler will be validated for its operating ability, not how well it fills capsules. Black box software testing is performed, and it evaluates the PLC against the documented functional requirements or intended uses of the equipment.

**Table 5.9. *Safety Features Test Results***

| Test Function | Expected Results | Acceptable (Yes/No) |
|---|---|---|
| Emergency stop push-button operation | When the emergency stop push button is pressed, the machine stops. | *Yes* |
| Protective hood operation | When each hood is opened, the machine stops. | *Yes* |
| Acoustic signal operation | When an error occurs, the buzzer sounds. | *Yes* |
| Protective guards operation | When the protective guards are removed, the machine stops. | *Yes* |

**Figure 5.4. OQ Elements of a Capsule Filler**

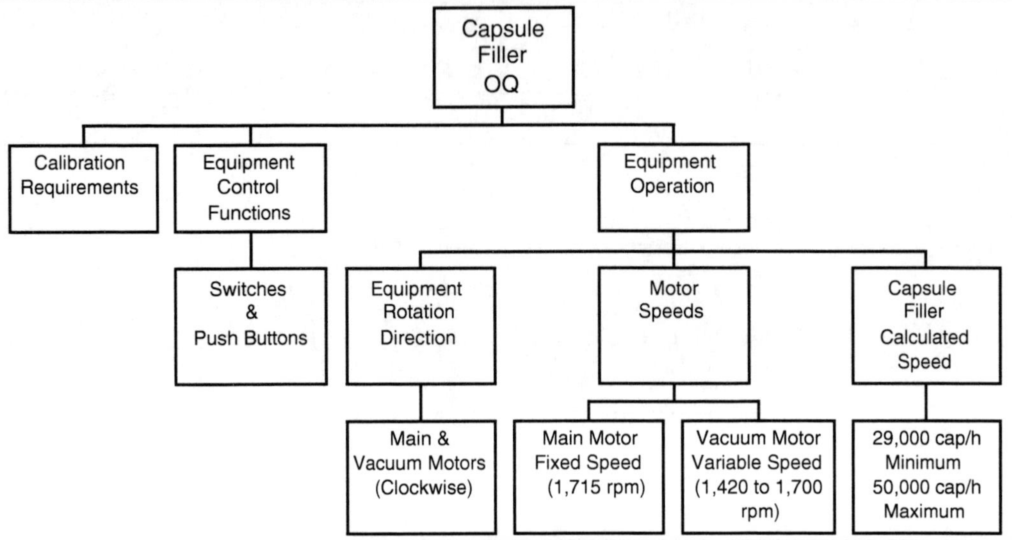


Information required for the OQ evaluation is: calibration of the instruments used to control the capsule filler, equipment control functions (switches and push buttons) and equipment operation (main and vacuum motor rotation direction and speed and capsule filler speed).

### *Calibration Requirements*

Verify that all critical instruments on the equipment, have been logged into the calibration system, have calibration procedures in place, and are in calibration at the time of qualification testing. Record all information for the calibrated instruments used to control the capsule filler in Table 5.10.

### *Equipment Control Functions*

The objective of testing equipment control functions is to verify that the switches and push buttons on the capsule filler operate per the manufacturer's specifications. The tests will be performed with the capsule filler empty. Operate each control listed in Table 5.11 and verify its proper operation.

### *Equipment Operation: Main and Vacuum Motor Rotation Direction Test*

The objective of the rotation direction test is to verify that the main and vacuum motors rotate in the proper direction. The tests will be performed with the capsule filler empty. With the capsule filler operating, verify that the main and vacuum motors are rotating in the proper direction as viewed from behind the motors. Record the results in Table 5.12.

**Table 5.10. Calibrated and Noncalibrated Instruments**

*There were no noncalibrated instruments on the blender.*

| Calibrated Instrument | As-Found Conditions |
|---|---|
| Identification number | *PRD0712* |
| Type | *Display gauge, digital* |
| Manufacturer | *Any Gauge Co.* |
| Model number | *Not specified* |
| Serial number | *Not specified* |
| Range | *30–50 caps/h × 1000* |
| Scale division | *1,000 caps* |
| Location | *Front of the control cabinet* |
| Use | *Displays the machine operations* |
| Calibration due date | *10/17/97* |
| Critical or not critical | *Critical to the operation of the capsule filler* |
| **Calibrated Instrument** | **As-Found Conditions** |
| Identification number | *PRD0713* |
| Type | *Vacuum gauge, analog* |
| Manufacturer | *Any Gauge Co.* |
| Model number | *Not specified* |
| Serial number | *Not specified* |
| Range | *0–1,200 mbar* |
| Scale division | *50 mbar* |
| Location | *Lower right side of the dosing station* |
| Use | *Measures the vacuum* |
| Calibration due date | *11/12/97* |
| Critical or not critical | *Not critical to the operation of the capsule filler* |
| **Calibrated Instrument** | **As-Found Conditions** |
| Identification number | *PRDVG0731* |
| Type | *Air pressure gauge, analog* |
| Manufacturer | *Any Gauge Co.* |
| Model number | *1002* |
| Serial number | *Not specified* |
| Range | *0.00–1" w.c.* |
| Scale division | *0.020" w.c.* |
| Location | *Left side of the control cabinet* |
| Use | *Measures air pressure* |
| Calibration due date | *11/12/97* |
| Critical or not critical | *Not critical to the operation of the capsule filler* |

**Table 5.11. Control Function Test Results**

| Test Function | Expected Results | Acceptable (Yes/No) |
|---|---|---|
| Because this equipment is controlled by a PLC, the control functions are too numerous to list. | | |

**Table 5.12. Main and Vacuum Motor Rotation Direction Test Results**

| Item | Expected Results | Results | Acceptable (Yes/No) |
|---|---|---|---|
| Main motor rotation direction | Rotation should be clockwise as viewed from behind the motor. | *Clockwise rotation was observed.* | *Yes* |
| Vacuum motor rotation direction | Rotation should be clockwise as viewed from behind the motor. | *Clockwise rotation was observed.* | *Yes* |

### *Equipment Operation: Main and Vacuum Motors Speed Test*

The objective of the motor speed test is to document the speed of the main and vacuum motors with the capsule filler empty. A tachometer is used with the following test procedure:

- Measure the speed of the main motor with a calibrated tachometer. Verify that the measured speed is within ±10 percent of the fixed speed of 1,715 rpm. Record the results in Table 5.13. Record the instrument used to measure the speed in Table 5.16.

  Main motor speed specification = 1,715 rpm ± 10%

  ±10% of 1,715 = ±171.5

  1,715 + 171.5 = 1,886.5

  1,715 – 171.5 = 1,543.5

  The measured rpm of 1,767 falls within ±10%.

- Measure the speed of the vacuum motor with a calibrated tachometer. Verify that the measured speed is within ±10 percent of the variable speeds of

**Table 5.13. Main and Vacuum Motor Speed Test Results**

| Motor | Speed Specifications (rpm ± 10%) | Measured Speed (rpm) | Acceptable (Yes/No) |
|---|---|---|---|
| Main | 1,715 | *1,767* | *Yes* |
| Vacuum | 1,420 to 1,700 | *1,451 to 1,726* | *Yes* |

1,420 to 1,700 rpm. Record the results in Table 5.13. Record the instrument used to measure the speed in Table 5.15.

Vacuum motor minimum speed specification = 1,420 rpm ± 10%

±10% of 1,420 = ±142

1,420 + 142 = 1,562

1,420 – 142 = 1,278

The measured rpm of 1,451 falls within ±10%.

Vacuum motor maximum speed specification = 1,700 rpm ± 10%

±10% of 1,700 = ±170

1,700 + 170 = 1,870

1,700 – 170 = 1,530

The measured rpm of 1,726 falls within ±10%.

### *Equipment Operation: Capsule Filler Speed Test*

The objective of the capsule filler speed test is to document the variable speed of the capsule filler with the equipment empty. The manufacturer's speed specification is 29,000 to 50,000 cap/h. Using a tachometer, perform the following procedure:

- Set the equipment to 29,000 cap/h and verify that the calculated speed is within ±10 percent of the indicated speed. Measure the speed of the feeding unit with a calibrated tachometer and record the measurement in Table 5.14. Record the instrument used to measure the speed in Table 5.15.

  Measured speed = 30 rpm

  Number of stations = 16

  cap/h = rpm(60)(number of stations)

  cap/h = 30(60)(16) = 28,800

**Table 5.14. Capsule Filler Speed Test Results**

| Speed | Displayed Speed (cap/h) | Calculated Speed (cap/h) | Acceptable (Yes/No) |
|---|---|---|---|
| Minimum | *29,000* | *28,800* | *Yes* |
| Maximum | *50,000* | *49,920* | *Yes* |

**Table 5.15. Instrument Used**

| Test Instrument | Identification Number | Calibration Due Date |
|---|---|---|
| *Tachometer* | *64020* | *06/21/97* |

- Set the equipment to 50,000 cap/h and verify that the calculated speed is within ±10 of the indicated speed. Measure the speed of the feeding unit with a calibrated tachometer and record the measurement in Table 5.14. Record the instrument used to measure the speed in Table 5.15.

  Measured speed = 52 rpm

  Number of stations = 16

  cap/h = rpm(60)(number of stations)

  cap/h = 52(60)(16) = 49,920

## Performance Qualification

Once it has been established that the equipment is properly installed and functioning within specified operating parameters, it must be shown that the capsule filler can perform reliably under routine, minimum, and maximum operating conditions (Figure 5.5).

### *Capsule Filler Operation (Tablets/Capsules)*

The operation of the capsule filler is tested to verify that the capsule filler is capable of inserting a specified number of tablets into capsules. The machine will be operated from 30,000 to 35,000 cap/h. Record the results of each test in Table 5.17. Record the materials used in Table 5.16.

***Figure 5.5. PQ Elements of a Capsule Filler***

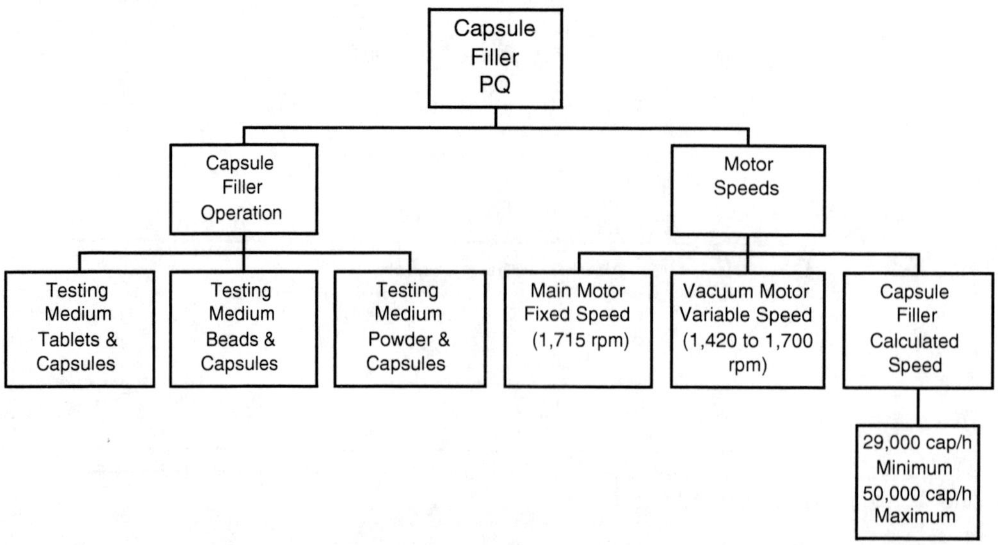

**Table 5.16. *Test Materials***

| Item | Specifications |
|---|---|
| Tablets | *Lactose placebo* |
| Capsules | *Gelatin capsules size 3* |

**Table 5.17. *Capsule Filler Performance Test Results (Tablets/Capsules)***

**Position C**

| Capsule Size | # of Caps | # Filled 1 Tablet (AQL = 100%) | # Filled 2 Tablets (AQL = 100%) | # Filled 3 Tablets (AQL = 100%) | Acceptable (Yes/No) |
|---|---|---|---|---|---|
| 3 | *100* | *100* | — | — | *Yes* |
| 3 | *100* | — | *100* | — | *Yes* |
| 3 | *100* | — | — | *100* | *Yes* |

**Position E**

| Capsule Size | # of Caps | # Filled 1 Tablet (AQL = 100%) | # Filled 2 Tablets (AQL = 100%) | # Filled 3 Tablets (AQL = 100%) | Acceptable (Yes/No) |
|---|---|---|---|---|---|
| 3 | *100* | *100* | — | — | *Yes* |
| 3 | *100* | — | *100* | — | *Yes* |
| 3 | *100* | — | — | *100* | *Yes* |

- Set up the machine to insert one tablet into size 3 capsules, with the tableting dosing unit in position "C" and "E." Sample 100 capsules of each size and verify that there is one tablet in each capsule.
- Set up the machine to insert two tablets into each size 3 capsules, with the tableting dosing unit in position "C" and "E." Sample 100 capsules of each size and verify that there are two tablets in each capsule.
- Set up the machine to insert three tablets into each size 3 capsules, with the tableting dosing unit in position "C" and "E." Sample 100 capsules of each size and verify that there are three tablets in each capsule.

### *Capsule Filler Operation (Beads/Capsules)*

Operation of the capsule filler is tested to verify that it is capable of filling capsules with beads. Record the results in Table 5.19. Record the materials used in Table 5.18. Record the instrument used to weigh the materials in Table 5.22.

Set up the machine to accept size 3 capsules and perform the following tests for dosing chamber 1 in position "C" and then "E."

**Table 5.18. *Test Materials***

| Item | Specifications |
|---|---|
| Beads | *Placebo blend* |
| Capsules | *Gelatin capsules size 3* |
| Beads (2 colors) | *White sugar spheres, orange diltazem* |

**Table 5.19. *Capsule Filler Performance Test Results (Beads/Capsules)***

**Beads: Chamber, 1 Position C, Capsule Size 3**

| Test Functions | Expected Results | Observed Results | (Yes/No) |
|---|---|---|---|
| Machine speed | 48,000 caps/h | *48,000 caps/h* | *Yes* |
| Material flow | No significant powder and capsule jams or failures | *None* | *Yes* |
| Target weight | — | *0.190 g* | *Yes* |

Capsule weight variation, 95% to 105% of the target weight, RSD ± 6% for 10 capsules.

| Weights Obtained | 10 min | 20 min | 30 min | Average |
|---|---|---|---|---|
| Mean weight | *0.1902 g* | *0.1889 g* | *0.1893 g* | *0.1894 g* |
| Standard deviation | *0.0027 g* | *0.0024 g* | *0.0032 g* | *0.0027 g* |
| % RSD | *1.42* | *1.26* | *1.69* | *1.45* |

**Beads: Chamber, 1 Position E, Capsule Size 3**

| Test Functions | Expected Results | Observed Results | (Yes/No) |
|---|---|---|---|
| Machine speed | 48,000 caps/h | *48,000 caps/h* | *Yes* |
| Material flow | No significant powder and capsule jams or failures | *None* | *Yes* |
| Target weight | — | *0.190 g* | *Yes* |

Capsule weight variation, 95% to 105% of the target weight, RSD ± 6% for 10 capsules.

| Weights Obtained | 10 min | 20 min | 30 min | Average |
|---|---|---|---|---|
| Mean weight | *0.1935 g* | *0.1935 g* | *0.1903 g* | *0.1924 g* |
| Standard deviation | *0.0038 g* | *0.0033 g* | *0.0035 g* | *0.0035 g* |
| % RSD | *1.94* | *1.71* | *1.84* | *1.83* |

*Continued on next page.*

*Continued from previous page.*

**Beads: Chamber, 1 Position C, Capsule Size 3**

| Test Functions | Expected Results | Observed Results | (Yes/No) |
|---|---|---|---|
| Machine speed | 48,000 caps/h | *48,000 caps/h* | *Yes* |
| Material flow | No significant powder and capsule jams or failures | *None* | *Yes* |
| Target weight | — | *0.190 g* | *Yes* |

Capsule weight variation, 95% to 105% of the target weight, RSD ± 6% for 10 capsules.

| Weights Obtained | 10 min | 20 min | 30 min | Average |
|---|---|---|---|---|
| Mean weight | *0.1884 g* | *0.1889 g* | *0.1888 g* | *0.1887 g* |
| Standard deviation | *0.0023 g* | *0.0026 g* | *0.0014 g* | *0.0021 g* |
| % RSD | *1.21* | *1.38* | *0.74* | *1.11* |

**Beads: Chamber, 1 Position E, Capsule Size 3**

| Test Functions | Expected Results | Observed Results | (Yes/No) |
|---|---|---|---|
| Machine speed | 48,000 caps/h | *48,000 caps/h* | *Yes* |
| Material flow | No significant powder and capsule jams or failures | *None* | *Yes* |
| Target weight | — | *0.190 g* | *Yes* |

Capsule weight variation, 95% to 105% of the target weight, RSD ± 6% for 10 capsules.

| Weights Obtained | 10 min | 20 min | 30 min | Average |
|---|---|---|---|---|
| Mean weight | *0.1909 g* | *0.1921 g* | *0.1903 g* | *0.1911 g* |
| Standard deviation | *0.0047 g* | *0.0035 g* | *0.0034 g* | *0.0386 g* |
| % RSD | *2.46* | *1.81* | *1.77* | *1.99* |

Were the capsules, from chamber 1 and 2, position "C" filled with two different colors? (Yes/No) *Yes*

Were the capsules, from chamber 1 and 2, position "E," filled with two different colors? (Yes/No) *Yes*

- Operate the machine at the speed of 48,000 caps/h and obtain a target weight by sampling 10 bead-filled capsules periodically until the mean target fill weight is reached ±5 percent. Operate the machine for 30 min. Repeat the test for each chamber 1 in position "C" and then in "E."
- Put one color of beads in dosing chamber 1 and a different color of beads in dosing chamber 2 in position "C" and then in "E." Sample 100 capsules and visually verify that the capsules are filled with the both colors.

### *Capsule Filler Operation (Powder/Capsules)*

Operation of the capsule filler is tested to verify that it is capable of filling capsules with powder. Record the results in Table 5.21. Record the materials used in Table 5.20. Record the instrument used to weigh the materials in Table 5.22.

- Set-up the machine to accept size 3 capsules and perform the following tests for dosing chamber 1 in position "C" and then in "E".
- Operate the machine at the speed of 48,000 caps/h and obtain a target weight by sampling 10 powder-filled capsules periodically until the mean target fill weight is reached ±5 percent. Operate the machine for 30 min. Repeat the test for each chamber 1 in position "C" and then in "E".

### *Main and Vacuum Motors Speed Test*

The objective of the motor speed test is to document the speed of the main and vacuum motors with the capsule filler full of capsules and placebo powder. A tachometer will be required for the following tests. Record the results in Table 5.23. Record the instrument used to measure the speed in Table 5.25. Record the material used in Table 5.21.

**Table 5.20. Test Materials**

| Item | Specifications |
|---|---|
| Powder | *Lactose placebo* |
| Capsules | *Gelatin capsules size 3* |

**Table 5.21. Capsule Filler Performance Test Results (Powder/Capsules)**

**Beads: Chamber, 1 Position C, Capsule Size 3**

| Test Functions | Expected Results | Observed Results | (Yes/No) |
|---|---|---|---|
| Machine speed | 48,000 caps/h | *48,000 caps/h* | *Yes* |
| Material flow | No significant powder and capsule jams or failures | *None* | *Yes* |
| Target weight | — | *0.377 g* | *Yes* |

Capsule weight variation, 95% to 105% of the target weight, RSD ± 6% for 10 capsules.

| Weights Obtained | 10 min | 20 min | 30 min | Average |
|---|---|---|---|---|
| Mean weight | *0.3749 g* | *0.3786 g* | *0.3773 g* | *0.3769 g* |
| Standard deviation | *0.0039 g* | *0.0047 g* | *0.0032 g* | *0.0039 g* |
| % RSD | *1.05* | *1.23* | *0.84* | *1.04* |

*Continued on next page.*

*Continued from previous page.*

**Beads: Chamber, 1 Position C, Capsule Size 3**

| Test Functions | Expected Results | Observed Results | (Yes/No) |
|---|---|---|---|
| Machine speed | 48,000 caps/h | *48,000 caps/h* | *Yes* |
| Material flow | No significant powder and capsule jams or failures | *None* | *Yes* |
| Target weight | — | *0.377 g* | *Yes* |

Capsule weight variation, 95% to 105% of the target weight, RSD ± 6% for 10 capsules.

| Weights Obtained | 10 min | 20 min | 30 min | Average |
|---|---|---|---|---|
| Mean weight | *0.3749 g* | *0.3786 g* | *0.3773 g* | *0.3769 g* |
| Standard deviation | *0.0039 g* | *0.0047 g* | *0.0032 g* | *0.0039 g* |
| % RSD | *1.05* | *1.23* | *0.84* | *1.04* |

**Beads: Chamber, 1 Position E, Capsule Size 3**

| Test Functions | Expected Results | Observed Results | (Yes/No) |
|---|---|---|---|
| Machine speed | 48,000 caps/h | *48,000 caps/h* | *Yes* |
| Material flow | No significant powder and capsule jams or failures | *None* | *Yes* |
| Target weight | — | *0.377 g* | *Yes* |

Capsule weight variation, 95% to 105% of the target weight, RSD ± 6% for 10 capsules.

| Weights Obtained | 10 min | 20 min | 30 min | Average |
|---|---|---|---|---|
| Mean weight | *0.3777 g* | *0.3779 g* | *0.3767 g* | *0.3774 g* |
| Standard deviation | *0.0067 g* | *0.0056 g* | *0.0044 g* | *0.0055 g* |
| % RSD | *1.77* | *1.48* | *1.15* | *1.46* |

**Beads: Chamber, 2 Position C, Capsule Size 3**

| Test Functions | Expected Results | Observed Results | (Yes/No) |
|---|---|---|---|
| Machine speed | 48,000 caps/h | *48,000 caps/h* | *Yes* |
| Material flow | No significant powder and capsule jams or failures | *None* | *Yes* |
| Target weight | — | *0.225 g* | *Yes* |

Capsule weight variation, 95% to 105% of the target weight, RSD ± 6% for 10 capsules.

| Weights Obtained | 10 min | 20 min | 30 min | Average |
|---|---|---|---|---|
| Mean weight | *0.2253 g* | *0.2260 g* | *0.2242 g* | *0.2225 g* |
| Standard deviation | *0.0032 g* | *0.0022 g* | *0.0034 g* | *0.0029 g* |
| % RSD | *1.44* | *0.98* | *1.51* | *1.31* |

*Continued on next page.*

*Continued from previous page.*

| Beads: Chamber, 2 Position E, Capsule Size 3 | | | |
|---|---|---|---|
| **Test Functions** | **Expected Results** | **Observed Results** | **(Yes/No)** |
| Machine speed | 48,000 caps/h | *48,000 caps/h* | *Yes* |
| Material flow | No significant powder and capsule jams or failures | *None* | *Yes* |
| Target weight | — | *0.225 g* | *Yes* |

Capsule weight variation, 95% to 105% of the target weight, RSD ± 6% for 10 capsules.

| Weights Obtained | 10 min | 20 min | 30 min | Average |
|---|---|---|---|---|
| Mean weight | *0.2253 g* | *0.2262 g* | *0.2285 g* | *0.2266 g* |
| Standard deviation | *0.0021 g* | *0.0028 g* | *0.0032 g* | *0.0027 g* |
| % RSD | *0.94* | *1.23* | *1.40* | *1.19* |

**Table 5.22. *Instrument Used***

| Test Instrument | Identification Number | Calibration Due Date |
|---|---|---|
| *Scale* | *PER600* | *07/18/97* |

**Table 5.23. *Main and Vacuum Motor Speed Test Results***

| Motor | Speed Specification (rpm ± 10%) | Measured Speed (rpm) | Acceptable (Yes/No) |
|---|---|---|---|
| Main | 1,715 | *1,725* | *Yes* |
| Vacuum | 1,420 to 1,700 | *1,409 to 1,721* | *Yes* |

- With the machine operating, measure the speed of the main motor with a calibrated tachometer. Verify that the measured speed is within ±10 percent of the fixed speed of 1,715 rpm.

  Main motor speed specification = 1,715 rpm ±10%

  ±10% of 1,715 = ±171.5

  1,715 + 171.5 = 1,886.5

  1,715 – 171.5 = 1,543.5

  The measured rpm of 1,725 falls within ±10%.

- Measure the speed of the vacuum motor with a calibrated tachometer. Verify that the measured speed is within ±10 percent of the variable speeds of 1,420 to 1,700 rpm.

Vacuum motor minimum speed specification = 1,420 rpm ± 10%

±10% of 1,420 = ±142

1,420 + 142 = 1,562

1,420 – 142 = 1,278

The measured rpm of 1,409 falls within ±10%.

Vacuum motor maximum speed specification = 1,700 rpm ± 10%

±10% of 1,700 = ±170

1,700 + 170 = 1,870

1,700 – 170 = 1,530

The measured rpm of 1,721 falls within ±10%.

### *Capsule Filler Speed Test*

The objective of the capsule filler speed test is to document the variable speed of the capsule filler with the equipment full of capsules and placebo powder. The manufacturer's speed specification is 29,000 to 50,000 caps/h. A tachometer will be required for this test. Record the results in Table 5.24. Record the instrument used to measure the speed in Table 5.25. Record the materials used in Table 5.20.

- Set the equipment to 29,000 caps/h and verify that the calculated speed is within ±10 percent of the indicated speed. Measure the speed of the feeding unit with a calibrated tachometer and record the measurement below.

  Measured speed = 30 rpm

  Number of stations = 16

  caps/h = rpm(60)(number of stations)

  caps/h = 30(60)(16) = 28,787

***Table 5.24. Capsule Filler Speed Test Results***

| Speed | Displayed Speed (cap/h) | Calculated Speed (cap/h) | Acceptable (Yes/No) |
|---|---|---|---|
| Minimum | *29,000* | *28,787* | *Yes* |
| Maximum | *50,000* | *49,915* | *Yes* |

***Table 5.25. Instrument Used***

| Test Instrument | Identification Number | Calibration Due Date |
|---|---|---|
| *Tachometer* | *64020* | *06/21/97* |

- Set the equipment to 50,000 cap/h and verify that the calculated speed is within ±10 percent of the indicated speed. Measure the speed of the feeding unit with a calibrated tachometer and record the measurement below.

  Measured speed = 52 rpm

  Number of stations = 16

  caps/h = rpm(60)(number of stations)

  caps/h = 52(60)(16) = 49,915

## Capsule Polisher

The validation of a capsule polisher (Figure 5.6) will be described here. A motor drives the capsule polisher and it operates at variable speeds. The capsule polisher is a stand alone machine that operates independently when placed into operation with capsules. Capsules enter the capsule polisher via the inlet chute. The feed screw assembly spins the capsules at a speed controlled by a potentiometer while the attached vacuum line removes dust from the capsules. The vacuum level is also user adjustable via a butterfly valve on the dust outlet nozzle.

The validation for a capsule polisher is described in the following sections. This protocol example can be used as cGMP compliance guide, and it is tailored to meet individual company requirements. All italicized words in the tables represent examples of information that could be recorded in a validation protocol for a capsule polisher.

***Figure 5.6. Capsule Polisher (courtesy of Stokes)***

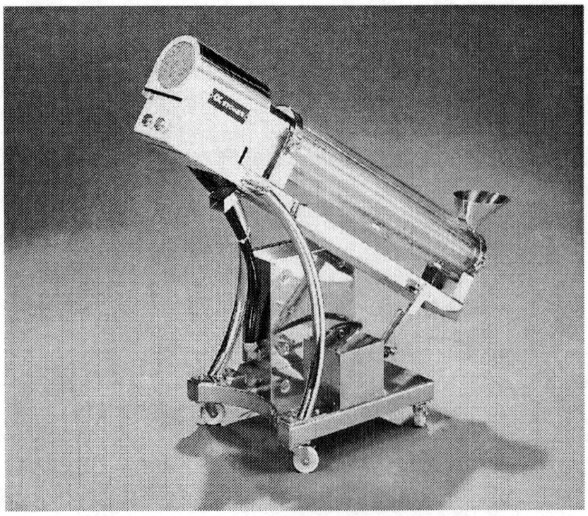

## Installation Qualification

An IQ evaluation will establish confidence that the equipment is properly installed (Figure 5.7). The installation must meet the manufacturer's specified guidelines along with design changes at installation. Also, the supporting electrical utilities must meet all electrical codes. The information required for an IQ evaluation should be: equipment identification, required documentation, equipment utility requirements, major component specifications, component material, lubricants, and equipment safety features.

### *Equipment Identification*

Record the equipment identification numbers in Table 5.27, along with the following information: equipment manufacturers' purchase order number, model number, serial number, company assigned equipment number, and the location of the equipment.

### *Required Documentation*

Record the equipment manufacturer's operation and maintenance manual and drawings in Table 5.28. Record the standard operating procedures that cover the setup, operation and cleaning of the capsule polisher in Table 5.29.

### *Equipment Utility Requirements*

Compare the manufacturer's specified volts (V), amps (A) and vacuum requirements to their as-found conditions at the time of qualification testing and record the results in Table 5.30. Also record the location of the power supply source. Record the instruments used to measure the volts, amps, and vacuum in Table 5.31.

***Figure 5.7. IQ Elements of a Capsule Polisher***

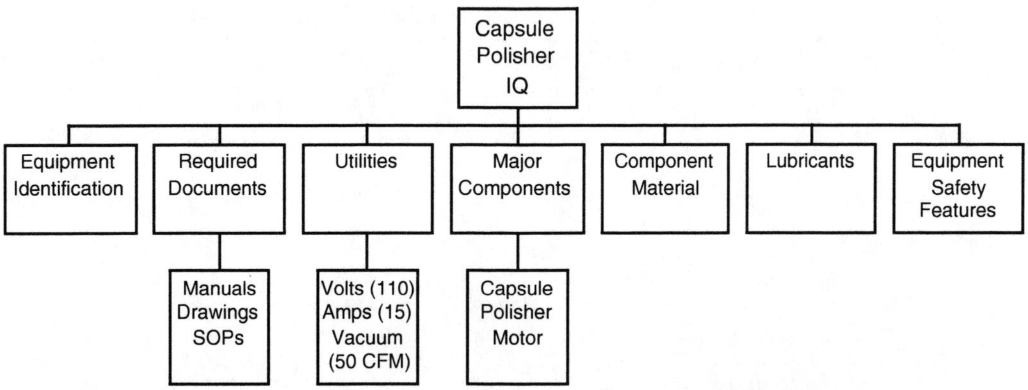

**Table 5.27. Equipment Identification**

| Required Information | As-Found Conditions |
|---|---|
| Manufacturer | *Any Capsule Polisher Co.* |
| Purchase order number | *0046118* |
| Model number | *8-081* |
| Serial number | *030086* |
| Equipment number | *5410* |
| Location | *Room 210 (This equipment is portable.)* |

**Table 5.28. Manufacturers' Manuals and Drawings**

| Number | Description | Date |
|---|---|---|
| *None* | *Capsule Polisher Operation Manual Model 8-081* | *96* |
| *3-8-081-IC* | *Wiring Diagram Drawing* | *None* |

**Table 5.29. Standard Operating Procedures**

| Number | Description | Release Date |
|---|---|---|
| *ENC011* | *Capsule Polisher Setup, Operation and Disassembly* | *03/20/96* |
| *ENC012* | *Capsule Polisher Cleaning Procedure* | *03/20/96* |

**Table 5.30. Utilities**

| Utility | Specified | Measured Results | Acceptable (Yes/No) |
|---|---|---|---|
| Volts | 110 ± 10% | *118* | *Yes* |
| Amps | 1.2 | 15 circuit rating | *Yes* |
| Vacuum | 50 cfm ± 10% | *52 cfm* | Yes |
| Power supply source, breaker box *BB16, wire number: 25* | | | |

**Table 5.31. Instruments Used**

| Test Instrument | Identification Number | Calibration Due Date |
|---|---|---|
| Multimeter | ME-025 | 04/19/97 |
| Vacuum gauge | 21JP | 07/10/97 |

***Volt calculation:***

Capsule polisher motor volt specification = 110 V ± 10%

±10% of 110 = ±11

110 + 11 = 121

110 – 11 = 99

The measured volts of 118 falls within ± 10%.

***Amp calculation:***

Circuit rating = 15 A

Equipment current draw = 1.2 A

The circuit amp rating of 15 is greater than the maximum current draw of the equipment.

***Vacuum calculation:***

Vacuum specification 50 cfm ± 10%.

±10% of 50 = ±5

50 + 5 = 55

50 – 5 = 45

The measured cfm of 52 falls within ±10%.

### *Major Component Specifications*

This section is used to verify that the capsule polisher components purchased were delivered and installed. Record the major components in Table 5.32.

### *Component Material*

Record the material of each component that contacts the product in Table 5.33.

**Table 5.32. Major Components**

| Components | As-Found Conditions |
|---|---|
| Roller drive motor | Manufacturer: *Any Motor Co.*<br>Model number: *4622005*<br>Serial number: *613W398*<br>Volts: *110*<br>Amperes: *1.2*<br>Phases: *1*<br>Cycles: *60 Hz*<br>hp: *1.6*<br>rpm: *Variable* |

**Table 5.33. Component Material**

| Component | Material |
|---|---|
| Inlet chute | *Stainless steel* |
| Brush | *Nylon* |
| Brush housing | *Stainless steel* |
| Outlet chute | *Stainless steel* |

### *Lubricants*

Record the lubricants used to operate the capsule polisher in Table 5.34. And indicate if they make contact with the product.

### *Equipment Safety Features*

There are no safety features on this equipment.

## Operational Qualification

An OQ evaluation should establish that the equipment can operate within specified tolerances and limits (Figure 5.8.). The mechanical ranges of the capsule polisher is challenged along with the basic capsule polisher operations. The capsule polisher will be validated for its operating ability, not how well it polishes capsules. Information required for the OQ evaluation should be: calibration of the instruments used to control the capsule polisher, equipment control functions (switches and push buttons), and equipment operation (capsule polisher motor rotation direction).

### *Calibration Requirements*

Verify that all critical instruments on the equipment are logged into the calibration system, have calibration procedures in place and are in calibration at the time of qualification testing. Record all of the necessary information for the calibrated instruments used to control the capsule polisher in Table 5.35.

**Table 5.34. Lubricants**

| Where Used | Type | Manufacturer | Product Contact (Yes/No) |
|---|---|---|---|
| *The gearbox and all bearings were packed at the vendor and required no further attention.* | | | |
| Is there is a preventive maintenance procedure on file? (Yes/No) | | | *Yes* |

***Figure 5.8*** ***OQ Elements of a Capsule Polisher***

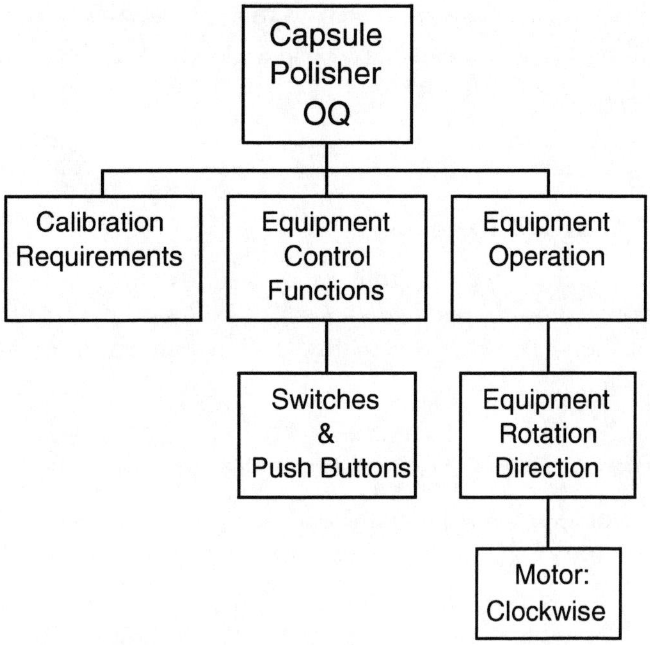


***Table 5.35.*** ***Calibrated and Noncalibrated Instruments***

*There were no calibrated or noncalibrated instruments on the capsule polisher.*

## *Equipment Control Functions*

The objective of equipment control function test is to verify that the switches and push buttons on the capsule polisher operate per manufacturer's specifications. Record the results in Table 5.36. The tests will be performed with the capsule polisher empty. Operate each control listed in Table 5.36 and verify their proper operation.

***Table 5.36.*** ***Equipment Control Function Test Results***

| Test Function | Expected Results | Acceptable (Yes/No) |
|---|---|---|
| ON/OFF switch | When the ON/OFF switch is pressed, the motor goes on and off. | *Yes* |
| Air pressure regulator operation | When the air pressure regulator is rotated, the broken and empty capsules are diverted from the finished product container. | *Yes* |

### *Equipment Operation: Capsule Polisher Motor Rotation Direction Test*

The objective of the motor rotation direction test is to verify that the motor rotates in the proper direction. The tests will be performed with the capsule polisher empty. Press the capsule polisher On switch and observe the direction of rotation of the motor when facing the shaft end of the motor and record the results in Table 5.37.

## Performance Qualification

Once it has been established that the equipment is properly installed and functioning within specified operating parameters, it must be shown that the capsule polisher can operate reliably under routine, minimum and maximum operating conditions (Figure 5.9).

### *Capsule Polisher Operation*

The objective of the polisher operation test is to verify that dust can be removed form the capsules. Capsules filled with placebo powder will be required as follows:

- Put 200 capsules into a clean container and pour placebo powder into the container. Feed the capsules into an operating capsule polisher, for 2 min at 10 percent speed and 2 min at 100 percent speed, and verify that when they exit they are free of dust. Record the results in Table 5.39. Record the capsules and placebo powder used in Table 5.38. Record the instrument used to measure the time in Table 5.40.

***Table 5.37. Capsule Polisher Motor Rotation Direction***

| Item | Expected Results | Results | Acceptable (Yes/No) |
|---|---|---|---|
| Capsule polisher motor rotation direction | Rotation should be clockwise when facing the shaft end of the motor. | *Clockwise rotation was observed.* | *Yes* |

***Figure 5.9. PQ Elements of a Capsule Polisher***

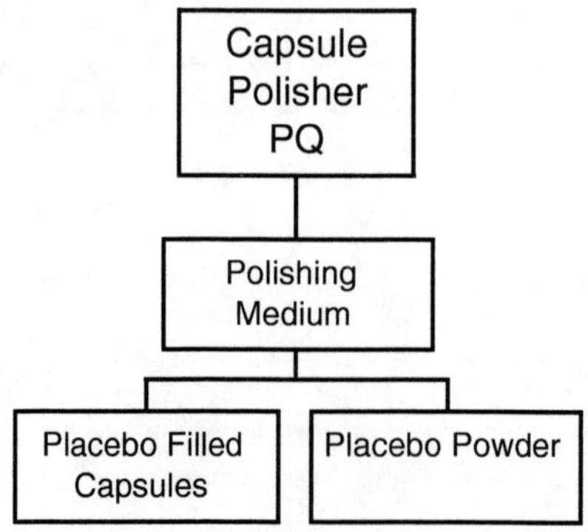

**Table 5.38. Test Materials and Conditions**

| Item | Specifications |
|---|---|
| Test capsules | *Gelatin, size 3* |
| Test placebo powder | *Haloperidol 5* |

**Table 5.39. Capsule Polisher Test Results**

The results of this test were based on observation only.

| Speed Settings | Expected Results | Results | Time (Min) | Acceptable (Yes/No) |
|---|---|---|---|---|
| 10% | Capsules should be dust free upon exiting the capsule polisher. | *Capsules were dust free when they exited the capsule polisher.* | *2* | *Yes* |
| 100% | Capsules should be dust free upon exiting the capsule polisher. | *Capsules were dust free when they exited the capsule polisher.* | *2* | *Yes* |

**Table 5.40. Instrument Used**

| Test Instrument | Identification Number | Calibration Due Date |
|---|---|---|
| *Stopwatch* | *54039* | *02/16/97* |

## Capsule Imprinter

The validation of a capsule imprinter (Figure 5.10) will be described here the motor driven and operates at variable speeds. The capsule imprinter is bowl fed, with a gravure offset process printer designed to print on one or both sides of capsules. Printing is performed with materials that meet the requirements for a printing ink for use on food and pharmaceutical products.

The validation protocol for a capsule imprinter is described in the following sections. This protocol example can be used as cGMP compliance guide, and it should be tailored to meet individual company requirements. Italicized words in the tables represent examples of information that could be recorded in a validation protocol for a capsule imprinter.

### Installation Qualification

An IQ evaluation establishes confidence that the equipment is properly installed (Figure 5.11). The installation must meet the manufacturer's specified guidelines along with design changes at installation. Also, the supporting electrical utilities must meet all electrical codes. The information required for an IQ evaluation is: equipment identification, required documentation, equipment utility requirements, major component specifications, component material, lubricants, and equipment safety features.

**Figure 5.10. Capsule Imprinter (courtesy of Markum)**

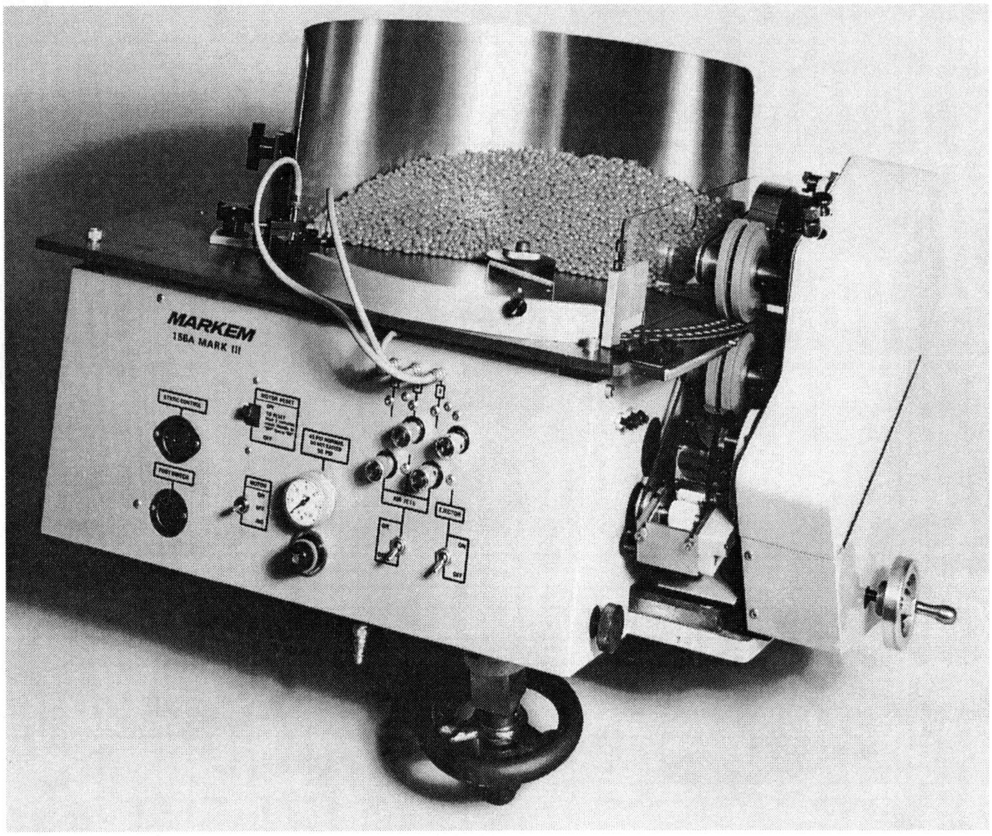


**Figure 5.11. IQ Elements of a Capsule Imprinter**

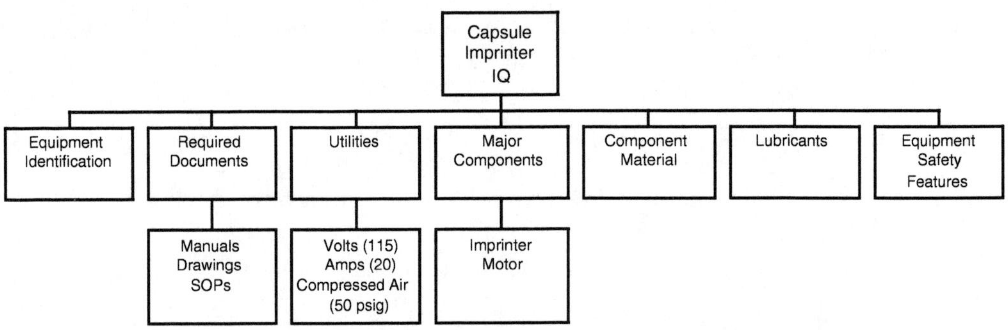


## *Equipment Identification*

Record the equipment identification numbers in Table 5.41, along with the following information: equipment manufacturer's purchase order number, model number, serial number, company assigned equipment number, and the location of the equipment.

**Table 5.41. *Equipment Identification***

| Required Information | As-Found Conditions |
|---|---|
| Manufacturer | *Any Imprinter Co.* |
| Purchase order number | *078127* |
| Model number | *IIIKM A651* |
| Serial number | *894809* |
| Equipment number | *5403* |
| Location | *Room 367 (This equipment is portable.)* |

### *Required Documentation*

Record the equipment manufacturer's operation and maintenance manual and drawings in Table 5.42. Record the SOPs that cover the setup, operation, and cleaning of the capsule imprinter in Table 5.43.

### *Equipment Utility Requirements*

Compare the manufacturer's specified volts (V), amps (A), and compressed air requirements to their as-found conditions at the time of qualification testing and record the results in Table 5.44. Also record the location of the power supply source. Record the instrument used to measure the volts, amps, and compressed air in Table 5.45.

**Table 5.42. *Manufacturers' Manuals and Drawings***

| Number | Description | Date |
|---|---|---|
| *0405580* | *Imprinter, Operator and Service Manual* | *10/84* |
| *None* | *Drawings are too numerous to list.* | *None* |

**Table 5.43. *Standard Operating Procedures***

| Number | Description | Date |
|---|---|---|
| *IMP011* | *Setup and Operation of a Capsule Imprinter* | *10/13/95* |
| *IMP012* | *Imprinter Cleaning Procedure* | *05/26/96* |
| *IMP023* | *Operations Acceptable Quality Level (AQL) Testing Procedure* | *08/02/96* |

**Table 5.44. Utilities**

| Utility | Specified | Measured Results | Acceptable (Yes/No) |
|---|---|---|---|
| Volts | 115 ± 10% | *117* | *Yes* |
| Amps | 2.5 | 20 circuit rating | *Yes* |
| Compressed air | 50 psig ± 10% | 55 psig | *Yes* |

Power supply source, breaker box *BB11, wire numbers: 6*

**Table 5.45. Instrument Used**

| Test Instrument | Identification Number | Calibration Due Date |
|---|---|---|
| *Multimeter* | *ME-025* | *04/19/97* |

***Volt calculation:***

Imprinter motor volt specification = 115 V ± 10%

±10% of 115 = ±11.5

115 + 11.5 = 126.5

115 – 11.5 = 103.5

The measured voltage of 117 falls within ± 10%.

***Amp calculations:***

Circuit rating = 20 A

Equipment current draw = 2.5 A

The circuit amp rating of 20 is greater than the maximum current draw of the equipment

***Compressed air calculation:***

Compressed air specification = 50 psig ±10%

±10% of 50 = ±5

50 + 5 = 55

50 – 5 = 45

The measured air pressure of 55 psig falls within ±10%

### *Major Component Specifications*

The component specification section of this protocol verifies that the capsule imprinter components purchased were delivered and installed. Record the major com-

**Table 5.46. Major Components**

| Components | As-Found Conditions |
|---|---|
| Imprinter motor | Manufacturer: *Any Motor Co.*<br>Model number: *FGH23799*<br>Serial number: *AM45190*<br>Volts: *115*<br>Amperes: *2.5*<br>Phases: *1*<br>Cycles: *60 Hz*<br>hp: *1/4*<br>rpm: *1,750* |

ponents in Table 5.46.

### *Component Material*

Record the material of each component that contacts the product in Table 5.47.

### *Lubricants*

Record the lubricants used to operate the capsule imprinter in Table 5.48. And indicate if they make contact with the product.

**Table 5.47. Component Material**

| Component | Material |
|---|---|
| Feed disk | *Enamel coated stainless steel* |
| Feed guard, feed deflector | *Plexiglass* |
| Offset roll sleeve | *Silicon rubber* |
| Guide fingers, stabilizer transfer roll stripper, lower reservoir, upper reservoir, side wipe (2), knurl nut | *Stainless steel* |
| Printing ink | *Food grade* |

**Table 5.48. Lubricants**

| Where Used | Type | Manufacturer | Product Contact (Yes/No) |
|---|---|---|---|
| *Gears* | *SAE Oil* | *Any Oil Co.* | *Yes* |
| Is there is a preventive maintenance procedure on file? (Yes/No) | | | *Yes* |

### *Equipment Safety Features*

The objective of testing the safety features is to verify that the safety features on the capsule imprinter function according to manufacturer specifications. Remove the gear box cover and verify that the imprinter motor will not start when the START button is pressed. Record the results in Table 5.49.

## Operational Qualification

An OQ evaluation should establish that the equipment can operate within specified tolerances and limits (Figure 5.12). The mechanical ranges of the capsule imprinter are challenged, along with the basic capsule imprinter operations. The capsule imprinter will be validated for its operating ability, not how well it imprints capsules. The information required for the OQ evaluation is calibration of the instrument used to control the capsule imprinter, equipment control functions (switches and push buttons), and equipment operation (capsule imprinter rotation direction, capsule imprinter speed control, disc and bowl assembly pickup operation).

***Table 5.49. Safety Features Test Results***

| Test Function | Expected Results | Acceptable (Yes/No) |
|---|---|---|
| Safety interlock switch operation | When the START button is pressed, with the gear box cover removed, the imprinter motor will not start. | *Yes* |

***Figure 5.12. OQ Elements of a Capsule Imprinter***

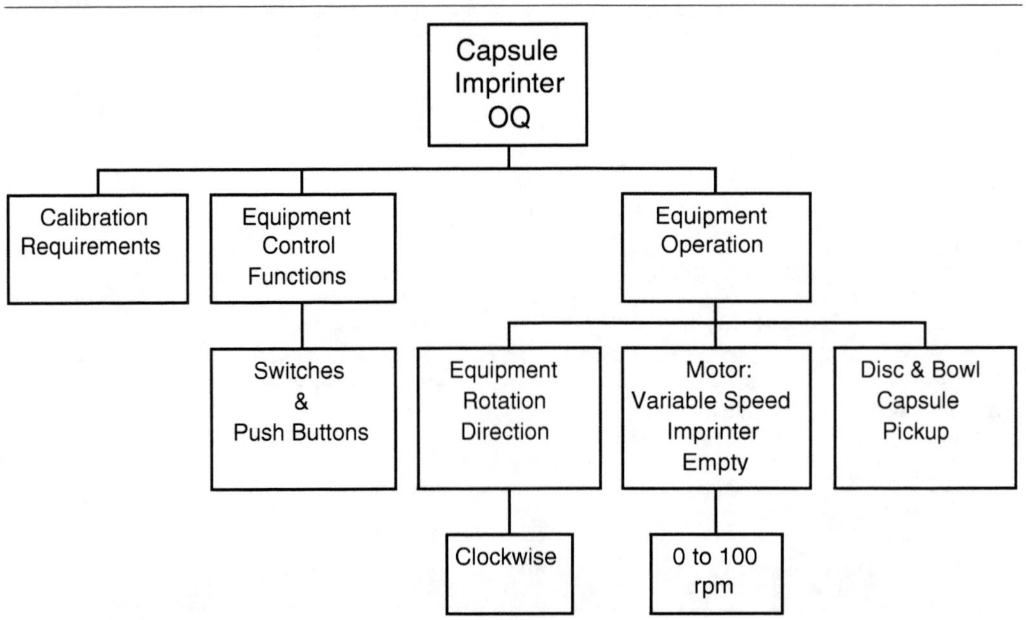

### *Calibration Requirements*

Verify that all critical instruments on the equipment have been logged into the calibration system, have calibration procedures in place, and are currently in calibration at the time of qualification testing. Record all information for the calibrated instruments used to control the capsule imprinter in Table 5.50.

### *Equipment Control Functions*

The objective of testing equipment control functions is to verify that the push buttons on the capsule imprinter operate according to manufacturer specifications. Operate each control listed in Table 5.51. and verify its proper operation. The capsule imprinter will be operated without capsules.

### *Equipment Operation: Capsule Imprinter Rotation Direction Test*

The objective of the rotation direction test is to verify that the capsule imprinter rotates in the proper direction. The capsule imprinter will be operated with the imprinter empty. Press the start push button and observe the direction of rotation of the capsule imprinter as viewed from the front and record the results in Table 5.52.

### *Capsule Imprinter Speed Control Test*

The objective of the speed control test is to document the speed control of the capsule imprinter with the imprinter empty. A tachometer will be required for this test.

**Table 5.50. Calibrated and Noncalibrated Instruments**

*There were no noncalibrated instruments on the equipment.*

| Calibrated Instrument | As-Found Conditions |
|---|---|
| Identification number | *GP-540380* |
| Type | *Air pressure gauge, analog* |
| Manufacturer | *Any Gauge Co.* |
| Model number | *442311* |
| Serial number | *Not Specified* |
| Range | *0–160 psig* |
| Scale division | *1 psig* |
| Location | *Control panel* |
| Use | *Measures the air pressure to the machine* |
| Calibration due date | *02/03/97* |
| Critical or not critical | *Critical to the operation of the imprinter* |

**Table 5.51. *Control Function Test Results***

| Test Function | Expected Results | Acceptable (Yes/No) |
|---|---|---|
| Air control knob | When the air control knob is rotated clockwise, the air pressure increases. | *Yes* |
| | When the air control knob is rotated counter clockwise, the air pressure decreases. | *Yes* |
| Air jet | When the switch is on, the air flows to all three air jets and when it is off the air stops. | *Yes* |
| | When the adjustment knob is turned clockwise, the air flow is decreased. | *Yes* |
| | When the adjustment knob is turned counter clockwise, the air flow is increased. | *Yes* |
| Ejector knob | When the switch is on, the air flows to the ejector and stops when the switch is off. | *Yes* |
| | When the knob is turned clockwise, the air flow is decreased. | *Yes* |
| | When the Knob is turned counter clockwise, the air flow is increased. | *Yes* |
| Motor switch | When the motor switch is on and the start button is press, the equipment: | |
| | stops when the stop button is pushed | *Yes* |
| | starts/stops when when the jog button is pressed and released | *Yes* |
| | motor does not start or jog when the motor switch is placed in the position. | *Yes* |

**Table 5.52. *Capsule Imprinter Rotation Direction Test Results***

| Item | Expected Results | Results | Acceptable (Yes/No) |
|---|---|---|---|
| Capsule imprinter rotation direction | Rotation should be clockwise as viewed from the front of the capsule imprinter. | *Clockwise rotation was observed.* | *Yes* |

- Measure the speed of the imprinter with a calibrated tachometer and record the results in Table 5.53. Verify that the measured speed is within ±10% of the variable speed from 0 to 100 rpm. Record the instrument used to measure the speed in Table 5.54.

  Capsule imprinter speed specification = 0 to 100 rpm

  The measured minimum rpm of 0 falls within specification.

  The measured maximum rpm of 100 falls within specification.

**Table 5.53. *Capsule Imprinter Speed Control Test Results***

| Speed | Specification rpm ± 10% | Measure Speed rpm | Acceptable (Yes/No) |
|---|---|---|---|
| Minimum | 0 | *0* | Yes |
| Maximum | 100 | *100* | Yes |

**Table 5.54 *Instrument Used***

| Test Instrument | Identification Number | Calibration Due Date |
|---|---|---|
| *Tachometer* | *64020* | *06/21/97* |

### *Disc and Bowl Assemblies Pickup Operation Test*

The objective of the pickup test is to verify that the disc and bowl assemblies operate according to the manufacturer's specifications. A stopwatch will be required for this test.

- Perform two 15 min tests, one at minimum operating speed and the second at maximum operating speed. Observe the disc for 3 min at the beginning, the middle and the end of each test. Record the number of capsules picked up and the number of times the disc slot was empty. Record the results in Table 5.55. Record the instrument used to measure the time in Table 5.56.

## Performance Qualification

Once it has been established that the equipment is properly installed and functioning within specified operating parameters, it must be shown that the capsule imprinter can operate reliably under routine, minimum and maximum operating conditions (Figure 5.13).

**Table 5.55. *Disc and Bowl Assemblies Pickup Test Results***

| Speed Setting | Measured Speed (rpm) | Beginning % Picked Up | Middle % Picked Up | End % Picked Up | Acceptable (Yes/No) |
|---|---|---|---|---|---|
| Minimum | *7.0* | *91* | *95* | *93* | *Yes* |
| Maximum | *9.5* | *93* | *90* | *92* | *Yes* |

**Table 5.56 *Instrument Used***

| Test Instrument | Identification Number | Calibration Due Date |
|---|---|---|
| *Stopwatch* | *54039* | *02/16/97* |

**Figure 5.13. PQ Elements of a Capsule Imprinter**

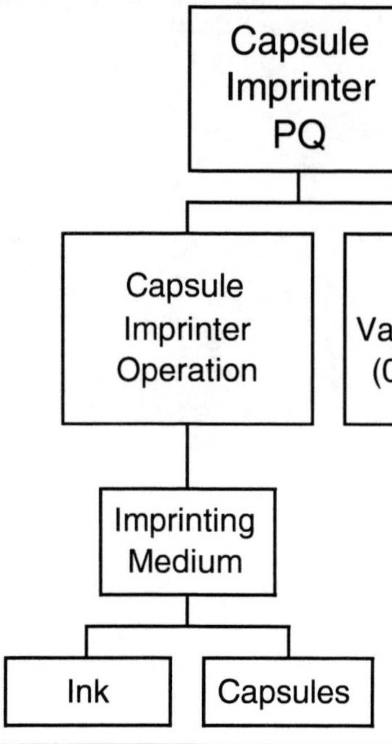


### *Capsule Imprinter Operation*

The objective of the imprinter operation test is to document the performance and speed of the capsule imprinter using capsules. A tachometer will be required according to the following procedure:

- Visually inspect 100 capsules at the beginning and the end of a production run and verify that the capsules are imprinted. AQL = 100%. Record the material used in Table 5.57. Record the results in Table 5.58.

**Table 5.57. Required Materials**

| Item | Material |
|---|---|
| Test material | *Capsules* |

**Table 5.58. Capsule Imprinter Performance Test Results**

| Production Run | Number of Capsules | Number of Capsules Imprinted | Acceptable (Yes/No) |
|---|---|---|---|
| Beginning | 100 | *100* | *Yes* |
| End | 100 | *100* | *Yes* |

- Measure the speed of the imprinter with a calibrated tachometer and record the results in Table 5.59. Verify that the measured speed is within ±10 percent of the maximum speed from 100 rpm. Record the instrument used to measure the speed in Table 5.60.

  Capsule imprinter speed specification = 0 to 100 rpm

  The measured maximum rpm of *0* falls within specification.

  The measured maximum rpm of *100* falls within specification.

## METAL DETECTOR

The metal detector is a high performance metal detector (Figure 5.14) used for detecting ferrous and nonferrous metallic contamination in a wide range of nonmetallic products. It consists of an inspection head and a power supply relay unit that may be mounted in any position. The product to be examined passes through the aperture in the detector head on an inclined chute. A penetrating electromagnetic field is generated within the aperture and metal, no matter how deeply embedded in the product, will distort the field and can be detected. Upon detection, the detect lamp goes on and the relay in the power supply relay unit operates. The relay contacts are used to close a trap door on the inclined chute.

The validation protocol for a metal detector is described in the following sections. This protocol example can be used as cGMP compliance guide, and it should be tailored to meet individual company requirements. Italicized words in this book represent examples of information that could be recorded in a validation protocol for a metal detector.

### Installation Qualification

An IQ evaluation establishes confidence that the equipment is properly installed (Figure 5.15). The installation must meet the manufacturer's specified guidelines,

**Table 5.59. *Capsule Imprinter Speed Test Results***

| Speed | Specification (rpm ± 10%) | Measure Speed (rpm) | Acceptable (Yes/No) |
|---|---|---|---|
| Minimum | 0 | *0* | *Yes* |
| Maximum | 100 | *100* | *Yes* |

**Table 5.60. *Instrument Used***

| Test Instrument | Identification Number | Calibration Due Date |
|---|---|---|
| *Tachometer* | *64020* | *06/21/97* |

**Figure 5.14. Metal Detector (courtesy of Lock Inspection Systems, Inc.)**

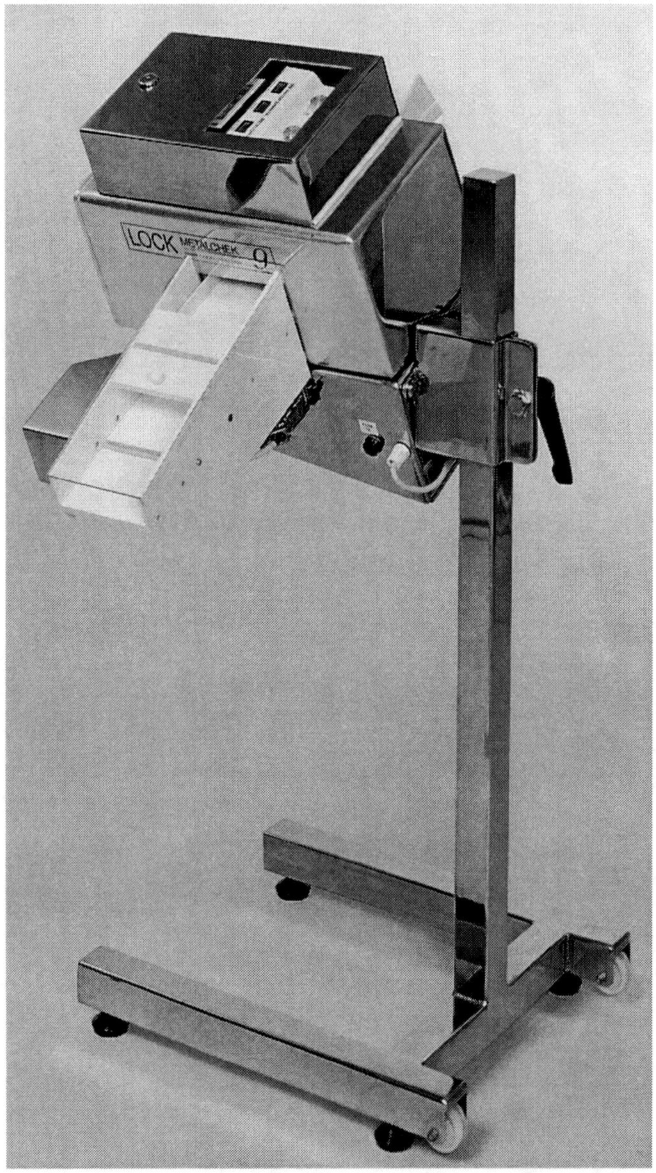


**Figure 5.15. IQ Elements of a Metal Detector**

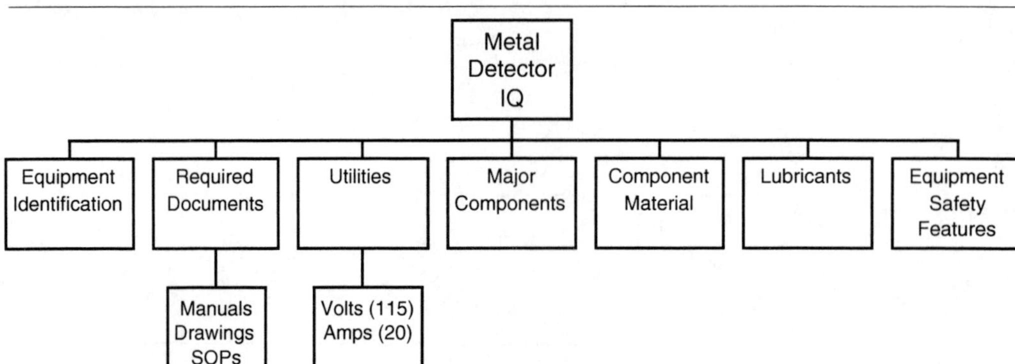

along with design changes at installation. Also, the supporting electrical utilities must meet all electrical codes. The information required for an IQ evaluation is equipment identification, required documentation, equipment utility requirements, major component specifications, component material, lubricants, and equipment safety features.

### *Equipment Identification*

Record the equipment identification numbers in Table 5.61, along with the following information: equipment manufacturers', purchase order number model number, serial number, company assigned equipment number, and the location of the equipment.

### *Required Documentation*

Record the equipment manufacturer's operation and maintenance manual and drawings in Table 5.62. Record the SOPs that cover the setup, operation, and cleaning of the metal detector in Table 5.63.

**Table 5.61. Equipment Identification**

| Required Information | As-Found Conditions |
|---|---|
| Manufacturer | *Any Metal Detector Co.* |
| Purchase order number | *54456* |
| Model number | *281* |
| Serial number | *972-44* |
| Equipment number | *1810* |
| Location | *Room 110 (This equipment is portable.)* |

**Table 5.62. Manufacturers' Manuals and Drawings**

| Number | Description | Date |
|---|---|---|
| *None* | *Operation Instructional Manual* | *None* |

**Table 5.63. Standard Operating Procedures**

| Number | Description | Date |
|---|---|---|
| *INS032* | *Inspection Equipment Setup* | *08/19/95* |
| *INS033* | *Inspection Department—Equipment Cleaning Procedure* | *08/11/95* |

### *Equipment Utility Requirements*

Compare the manufacturer's specified volts (V), and amps (A), requirements to their as-found conditions at the time of qualification testing and record the results in Table 5.64. Also record the location of the power supply source. Record the instrument used to measure the volts and amps in Table 5.65.

***Volt calculation:***

Volt specification = 115 V ± 10%

±10% of 115 = ±11.5

115 + 11.5 = 126.5

115 – 11.5 = 103.5

The measured voltage of 118 falls within ± 10%.

***Amp calculation:***

Circuit rating = 20 A

Equipment current draw = 2 A

The circuit amp rating of 20 is greater than the maximum current draw of the equipment.

### *Major Component Specifications*

The section is used to verify that the metal detector components purchased were delivered and installed. Record the major components in Table 5.66.

### *Component Material*

Record the material of each component that contacts the product in Table 5.67.

***Table 5.64. Utilities***

| Utility | Specified | Measured Results | Acceptable (Yes/No) |
|---|---|---|---|
| Volts | 115 ± 10% | *118* | *Yes* |
| Amps | 2 | 20 circuit rating | *Yes* |

Power supply source, breaker box *BB37, wire numbers: 1.*

***Table 5.65. Instrument Used***

| Test Instrument | Identification Number | Calibration Due Date |
|---|---|---|
| *Multimeter* | *ME-025* | *04/19/97* |

**Table 5.66. Major Components**

| Components | As-Found Conditions |
|---|---|
| Metal Detector | Manufacturer: *Any Metal Detector Co.*<br>Model number: *281*<br>Serial number: *91244*<br>Volts: *115*<br>Amperes: *2* |
| Power Supply | Manufacturer: *Any Power Supply Co.*<br>Part number: *45267*<br>Serial number: *25144* |

**Table 5.68. Component Material**

| Component | Material |
|---|---|
| Inclined chute | *Stainless steel* |
| Trap door | *Stainless steel* |

### *Lubricants*

Record the lubricants used to operate the metal detector in Table 5.68. and indicate if they make contact with the product.

### *Equipment Safety Features*

There are no safety features on this equipment.

## Operational Qualification

An OQ evaluation should establish that the equipment can operate within specified tolerances and limits (Figure 5.16). The mechanical ranges of the metal detector are challenged, along with the basic metal detector operations. The metal detector will be validated for its operating ability, not how well it detects metals. The information required for the OQ evaluation is calibration of the instrument

**Table 5.68. Lubricants**

| Where Used | Type | Manufacturer | Product Contact (Yes/No) |
|---|---|---|---|
| *There is no lubrication used on this equipment.* | | | |
| Is there is a preventive maintenance procedure on file? (Yes/No) | | | *Yes* |

**Figure 5.16. OQ Elements of a Metal Detector**

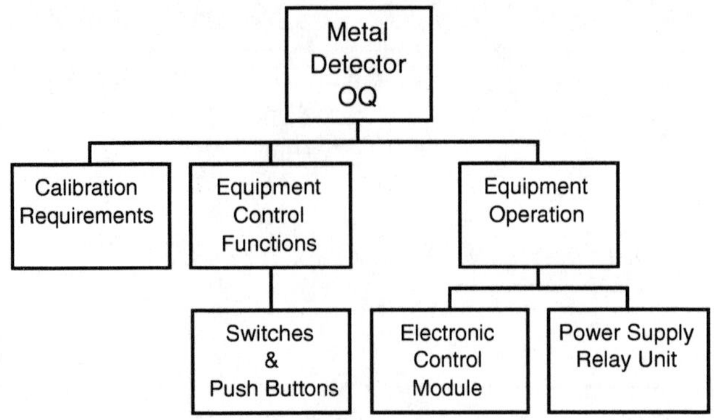


used to control the metal detector, equipment control functions, (switches and push buttons) and equipment operation (electronic control module operation, power supply relay unit operation).

### *Calibration Requirements*

Verify that all critical instruments on the equipment have been logged into the calibration system, have calibration procedures in place, and are currently in calibration at the time of qualification testing. Record all information for the calibrated instruments used to control the metal detector in Table 5.69.

### *Equipment Control Functions*

The objective of testing equipment control functions is to verify that the switches on the metal detector operate according to manufacturer specifications. The metal detector will be operated without placebo tablets. The controls that need to be tested are: ON/Off switch, sensitivity control knob, chute actuator. A metal sample will be required for the following test:

- Press the On/Off switch and verify that the equipment goes on and off. Record the results in Table 5.70.
- Press the sensitivity control knob and verify that the sensitivity increases when the knob is rotated from 0 to 10 and decreases when it is rotated from 10 to 0. Record the results in Table 5.70.
- Pass a metal sample through the unit and verify that the chute actuator closes the trap door to catch the sample. Record the results in Table 5.70.

**Table 5.69. Calibrated and Noncalibrated Instruments**

| |
|---|
| *There were no calibrated or noncalibrated instruments on the capsule polisher.* |

**Table 5.70. *Control Function Test Results***

| Test Function | Expected Results | Acceptable (Yes/No) |
|---|---|---|
| On/Off switch | When the On/Off switch is pressed, the equipment goes on and off. | *Yes* |
| Sensitivity control knob operation | When the sensitivity control knob is rotated from 0 to 10, the sensitivity increases. | *Yes* |
| | When the sensitivity control knob is rotated from 10 to 0, the sensitivity decreases. | *Yes* |
| Chute actuator operation | When a metal sample is passed through the unit, the chute actuator closes the trap door to catch the sample. | *Yes* |

### *Equipment Operation: Electronic Control Module Operation Test*

The objective of control module operation test is to verify that the electronic control module operates according to manufacturer's specification. Turn the equipment on and verify that the POWER and BALANCE/FAULT lamps go on then verify that the BALANCE/FAULT lamp goes out after two minutes. Record the time used in Table 5.71.

### *Power Supply Relay Unit Operation Test*

The objective of relay unit operation test is to verify that the Power Supply Relay Unit operates according to manufacturers' specification.

- With the equipment operating, set the product compensation control knob to '0' and the In/Out switch to Out and set the sensitivity control knob to 5. Pass a 0.5 mm test wand through the aperture and verify that the detection lamp goes on and the trap door closes. Record the amount used in Table 5.72.

## Performance Qualification

Once it has been established that the equipment is properly installed and functioning within specified operating parameters, it must be shown that the metal de-

**Table 5.71. *Electronic Control Module Operation Test Results***

| Test Function | Expected Results | Acceptable (Yes/No) |
|---|---|---|
| Electronic control module operation | Turn the equipment on and verify that the POWER and BALANCE/FAULT lamp goes on and the BALANCE/FAULT lamp goes out after two minutes. | *Yes* |

**Table 5.72. Power Supply Relay Unit Operation Test Results**

| Test Function | Expected Results | Acceptable (Yes/No) |
|---|---|---|
| Power supply relay unit operation | When the 0.5 mm test wand is passed through the aperture, with the product compensation control set to "0" and the In/Out switch is set to Out, also the sensitivity is set to five, the detection lamp goes on and the trap door closes. | *Yes* |

tector can be operated reliably under routine, minimum and maximum operating conditions (Figure 5.17).

### *Metal Detector Operation*

The objective metal detector operation test is to verify that the metal detector operates according to manufacturer's specifications. The materials required are given in Table 5.73.

- With the equipment operating, set the product compensation control to "0" and the In/Out switch to Out. Set the sensitivity control to one and pass the standard 0.5 mm diameter test wand through the aperture and record the results in Table 5.74.
- Repeat the above test for each of the material test standards for each of the sensitivity settings from 1 through 10. Record the results in Table 5.74.

**Figure 5.17. PQ Elements of a Metal Detector**

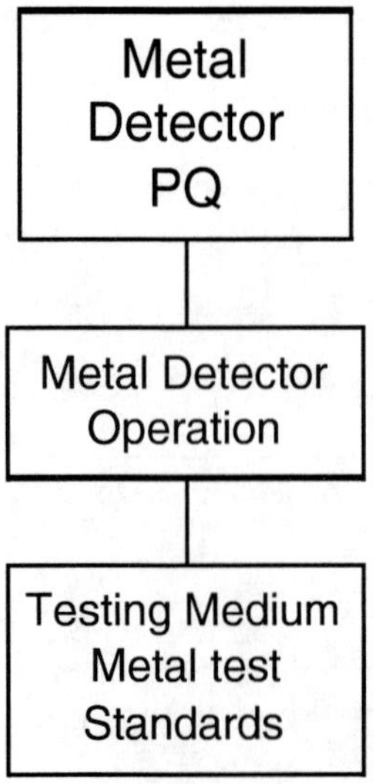

**Table 5.73. *Material Test Standards***

| Material | Sample Size-mg |
|---|---|
| Cold rolled steel | 1.064 |
| Cold rolled steel | 0.667 |
| Stainless steel | 0.369 |
| Stainless steel | 0.054 |
| Aluminum | 0.017 |
| Aluminum | 0.075 |
| Brass | 0.032 |
| Brass | 0.089 |
| Copper | 0.442 |
| Copper | 0.735 |

**Table 5.74. *Metal Detector Test Results***

N = No the metal was not detected

Y = Yes the metal was detected

| Material Test Standard (mg) | Sensitivity Settings | | | | | | | | | |
|---|---|---|---|---|---|---|---|---|---|---|
| **Weight** | 1 | 2 | 3 | 4 | 5 | 6 | 7 | 8 | 9 | 10 |
| Test wand 0.5 mm Dia. | N | N | N | N | Y | Y | Y | Y | Y | Y |
| Cold rolled steel 1.064 | Y | Y | Y | Y | Y | Y | Y | Y | Y | Y |
| Cold rolled steel 0.667 | Y | Y | Y | Y | Y | Y | Y | Y | Y | Y |
| Stainless steel 0.054 | N | N | N | N | N | Y | Y | Y | Y | Y |
| Stainless steel 0.369 | N | N | N | Y | Y | Y | Y | Y | Y | Y |
| Aluminum 0.017 | N | N | N | N | N | Y | Y | Y | Y | Y |
| Aluminum 0.075 | N | N | N | Y | Y | Y | Y | Y | Y | Y |
| Brass 0.032 | N | N | N | N | N | Y | Y | Y | Y | Y |
| Brass 0.089 | N | N | N | Y | Y | Y | Y | Y | Y | Y |
| Copper 0.442 | N | N | N | N | Y | Y | Y | Y | Y | Y |
| Copper 0.735 | N | N | N | Y | Y | Y | Y | Y | Y | Y |

All of the above tests were acceptable from setting six and above.

# Chapter 6

# Validation of Packaging Equipment

This chapter will cover the qualification testing of equipment used to perform operations in the packaging department (Figure 6.1).

***Figure 6.1. Packaging Equipment***

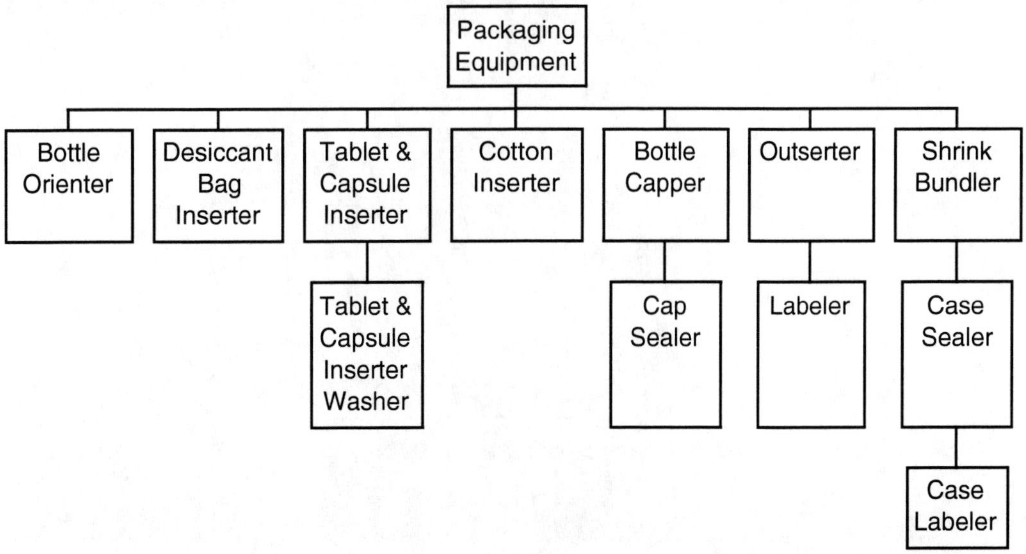

## BOTTLE ORIENTER

The validation of a bottle orienter (Figure 6.2) will be described here. Two motors drive the bottle orienter and operates at variable speeds. An elevator to the preorienter delivers bottles into a hopper. In the pre-orienter, the bottles are turned around and oriented into a single file formation. Then they are directed down a drop chute. After sliding down the drop chute the bottles enter the main orienter where rotating fingers grab one bottle at a time and by use of a sensor, orient the bottle so that a blast of compressed air cleans the bottle through the opening. The bottle is then turned right-side up and set on a conveyor to be filled with tablets or capsules.

The validation protocol for a bottle orienter is described in the following sections. This protocol example can be used as cGMP compliance guide, and it should be tailored to meet individual company requirements. Italicized words in the tables represent examples of information that could be recorded in a validation protocol for a bottle orienter.

### Installation Qualification

An IQ evaluation establishes confidence that the equipment is properly installed (Figure 6.3). The installation must meet the manufacturer's specified guidelines, along with design changes at installation. Also, the supporting electrical utilities

**Figure 6.2. Bottle Unscrambler (courtesy of Omega Design Corp.)**

***Figure 6.3. IQ Elements of a Bottle Orienter***

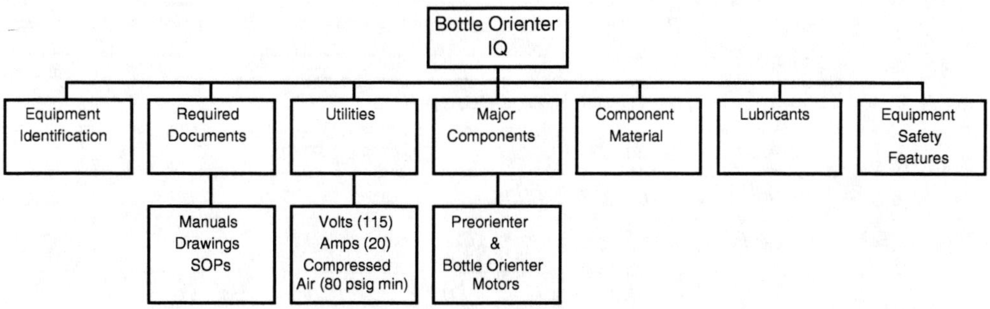


must meet all electrical codes. The information required for an IQ evaluation is equipment identification, required documentation, equipment utility requirements, major component specifications, component material, lubricants, and equipment safety features.

## *Equipment Identification*

Record the equipment identification numbers in Table 6.1, along with the following information: equipment manufacturer's purchase order number, model number, serial number, company assigned equipment number, and the location of the equipment.

## *Required Documentation*

Record the equipment manufacturer's operation and maintenance manual and drawings in Table 6.2. Record the standard operating procedures that cover the setup, operation, and cleaning of the bottle orienter in Table 6.3.

***Table 6.1. Equipment Identification***

| **Required Information** | **As-Found Conditions** |
| --- | --- |
| Manufacturer | *Any Bottle Orienter Co.* |
| Purchase order number | *001259* |
| Model number | *51-2PR D3* |
| Serial number | *3132-68* |
| Equipment number | *1540* |
| Location | *Packaging line 5* |

**Table 6.2. Manufacturers' Manuals and Drawings**

| Number | Description | Date |
|---|---|---|
| *None* | *Bottle Orienter Installation, Operation, and Maintenance Manual* | *None* |

**Table 6.3. Standard Operating Procedures**

| Number | Description | Release Date |
|---|---|---|
| *PKG001* | *Packaging Operations* | *09/22/94* |
| *PKG002* | *Packaging Equipment Cleaning Procedure* | *11/15/94* |

### *Equipment Utility Requirements*

Compare the manufacturer's specified volts (V), amps (A), and compressed air pressure requirements to their as-found conditions at the time of qualification testing and record the results in Table 6.4. Also record the location of the power supply source. Record the instruments used to measure the volts, amps, and compressed air in Table 6.5.

***Volt calculation:***

Bottle orienter motor volt specification = 115 V ± 10%

±10% of 115 = ±11.5

115 + 11.5 = 126.5

115 – 11.5 = 103.5

The measured volts of 118 falls within ± 10%.

***Amp calculation:***

Amp specification = 20 A ± 10%.

Preorienter motor = 1.5 A

Bottle orienter motor = 3.5 A

Equipment current draw = preorienter motor A + bottle orienter motor A = 1.5 + 3.5 = 5.0 A

The circuit amp rating of 20 is greater than the maximum current draw of the equipment.

***Compressed air calculation:***

Compressed air specification = 80 psig minimum

The measured air pressure of 116 psig is greater than 80 psig.

**Table 6.4. *Utilities***

| Utility | Specified | Measured Results | Acceptable (Yes/No) |
|---|---|---|---|
| Volts | 115 ± 10% | *118* | *Yes* |
| Amps | Preorienter motor = 1.5<br>Bottle orienter motor = 3.5 | 20 circuit rating | *Yes* |
| Compressed air | 80 psig minimum | *116 psig* | *Yes* |

*Power supply source, breaker box BB57, wire number: L5*

**Table 6.5. *Instruments Used***

| Test Instrument | Identification Number | Calibration Due Date |
|---|---|---|
| *Multimeter* | *ME-025* | *04/19/97* |
| *Air pressure gauge* | *P-102* | *02/10/97* |

## *Major Component Specifications*

The component specifications section of the protocol is used to verify that the bottle orienter components purchased were delivered and installed. Record the major components in Table 6.6.

**Table 6.6. *Major Components***

| Components | As-Found Conditions |
|---|---|
| Pre-orienter motor | Manufacturer: *Any Motor Co.*<br>Model number: *92174*<br>Serial number: *51981*<br>Volts: *115*<br>Amperes: *1.5*<br>Phases: *1*<br>Cycles: *60 Hz*<br>hp: *1/8*<br>rpm: *45* |
| Bottle orienter motor | Manufacturer: *Any Motor Co.*<br>Model number: *TA339MP*<br>Serial number: *50226*<br>Volts: *115*<br>Amperes: *3.5*<br>Phases: *1*<br>Cycles: *60 Hz*<br>hp: *1/3*<br>rpm: *1,725* |

### *Component Material*

Record the material of each component that contacts the product in Table 6.7.

### *Lubricants*

Record the lubricants used to operate the bottle orienter in Table 6.8 and indicate if they make contact with the product.

### *Equipment Safety Features*

The objective of testing equipment safety features is to verify that the safety features on the bottle orienter function according to manufacturer specifications. Verify the operation of the equipment safety features given in Table 6.9.

## Operational Qualification

An OQ evaluation should establish that the equipment can operate within specified tolerances and limits (Figure 6.4). The mechanical ranges of the bottle orienter are challenged, along with the basic equipment operations that the operator will use. Information required for the evaluation is: calibration of the instruments used to control the bottle orienter, equipment control functions (switches and push buttons), and equipment operation (preorienter and bottle orienter motor rotation direction, preorienter operation).

**Table 6.7. Component Material**

| Component | Material |
|---|---|
| *None of the components make contact with the product.* | |

**Table 6.8. Lubricants**

| Where Used | Type | Manufacturer | Product Contact (Yes/No) |
|---|---|---|---|
| *Gear reducer* | *Oil, Lubriplate APG 90* | *Any Oil Co.* | *No* |
| *Bearings* | *Grease, Lubriplate 1242* | *Any Oil Co.* | *No* |
| Is there a preventive maintenance procedure on file? (Yes/No) | | | *Yes* |

**Table 6.9. Safety Features Test Results**

| Test Function | Expected Results | Acceptable (Yes/No) |
|---|---|---|
| Master lockout push-button operation | When the lockout push button is pressed, the equipment cannot be started. | *Yes* |
| Door guard sensor | When the door is opened, the sensor is tripped and the equipment stops operating. | *Yes* |

**Figure 6.4. OQ Elements of a Bottle Orienter**

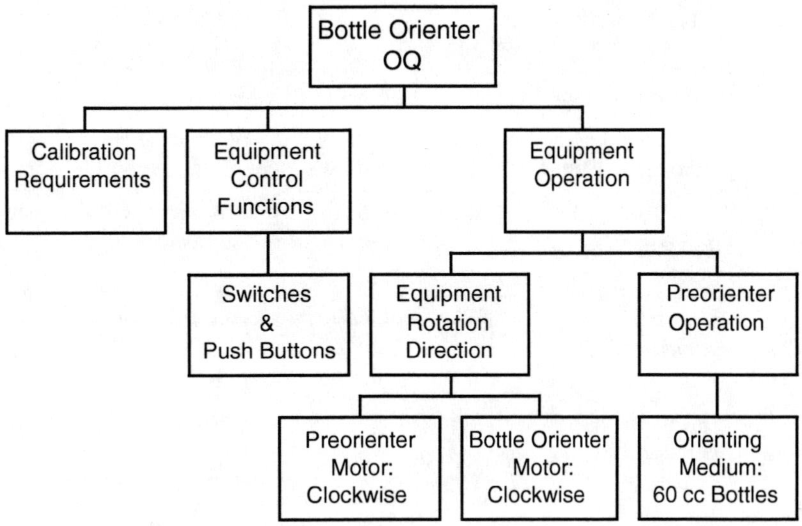


### *Calibration Requirements*

Verify that all critical instruments on the equipment have been logged into the calibration system, have calibration procedures in place, and are in calibration at the time of qualification testing. Record information for the calibrated instruments used to control the bottle orienter in Table 6.10.

### *Equipment Control Functions*

The objective of testing equipment control functions is to verify that the switches and push buttons on the bottle orienter operate per manufacturer's specifications. Operate each control listed in Table 6.11 and verify its proper operation. Record the results in Table 6.11. The tests will be performed with the bottle orienter empty.

### *Equipment Operation: Preorienter and Bottle Orienter Motor Rotation Direction Test*

The objective of the rotation direction test is to verify that the preorienter and bottle orienter rotate in the proper direction. The following tests are performed with the bottle orienter empty.

**Table 6.10. Calibrated and Noncalibrated Instruments**

| |
|---|
| *There were no calibrated or noncalibrated instruments on the bottle orienter.* |

**Table 6.11. Equipment Control Function Test Results**

| Test Function | Expected Results | Acceptable (Yes/No) |
|---|---|---|
| Start/stop push-button operation | When the stop/start push button is pressed, the equipment starts and stops. | *Yes* |
| Preorienter drive knob controller operation | When the pre-orienter drive knob is rotated, the speed of the pre-orienter increases and decreases. | *Yes* |
| Orienter main drive knob controller operation | When the main drive knob is rotated, the speed of the orienter increases and decreases. | *Yes* |
| Chute sensor operation | When bottles are not detected, the orienter stops. | *Yes* |
| | When bottles are detected, the orienter restarts. | *Yes* |
| Conveyor sensor operation | When a bottle jam is detected, the orienter stops. | *Yes* |
| | When bottles jam is removed, the orienter restarts. | *Yes* |
| Preorienter sensor operation | When bottles are required in the hopper, the elevator goes on. | *Yes* |
| | When bottles are not required for the hopper, the elevator goes off. | *Yes* |
| Container sensor operation | When the mouth of the bottle is detected, nothing happens. | *Yes* |
| | When the base of the bottle is detected, the rotating pocket trips an actuator that insures the proper position of the rotating pocket. | *Yes* |
| Air jets set handle operation | When the air jets set handle is turned, the air pressure to the jets increases and decreases. | *Yes* |
| Air cleaner handle operation | When the air cleaner handle is turned, the air pressure increases and decreases. | *Yes* |

- Press the preorienter start push button and observe the direction of rotation of the pre-orienter as viewed from behind the motor and record the results in Table 6.12.
- Press the bottle orienter start push button and observe the direction of rotation of the preorienter and bottle orienter as viewed from behind the motor, and record the results in Table 6.12.

### *Equipment Operation: Preorienter Operation Test*

The objective of the preorienter operation test is to verify that it operates according to manufacturer's specification. 60 cc bottles are required. With the machine operating verify that a cluster of bottles are oriented into a single line formation. Record the results in Table 6.13.

**Table 6.12. Preorienter and Bottle Orienter Motor Rotation Direction**

| Item | Expected Results | Results | Acceptable (Yes/No) |
|---|---|---|---|
| Preorienter rotation direction | Rotation should be clockwise as viewed from behind the motor. | *Clockwise rotation was observed.* | *Yes* |
| Bottle orienter rotation direction | Rotation should be clockwise as viewed from behind the motor. | *Clockwise rotation was observed.* | *Yes* |

**Table 6.13. Preorienter Test Results**

| Test Function | Expected Results | Acceptable (Yes/No) |
|---|---|---|
| Single line formation | *Single file formation was observed.* | *Yes* |

## Performance Qualification

Once it has been established that the equipment is properly installed and functioning within specified operating parameters, it must be shown that the bottle orienter can operate reliably under routine, minimum, and maximum operating conditions (Figure 6.5).

**Figure 6.5. PQ Elements of a Bottle Orienter**

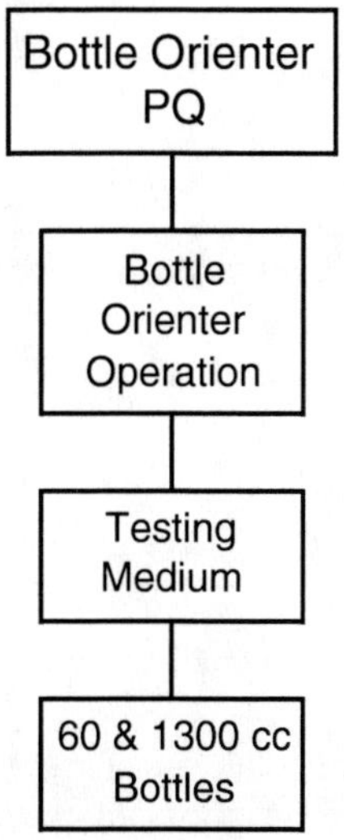

### *Bottle Orienter Operation*

The objective of testing the operation of the bottle orienter is to document its performance using the small and large size bottles. This test is for informational purposes only. 60 and 1,300 cc bottles are required. Observe 100 60 cc bottles and 100 1,300 cc bottles as they pass through the orienter and record the equipment settings and the containers per minute in Table 6.14.

## DESICCANT BAG INSERTER

The validation of a desiccant bag inserter (Figure 6.6) will be described here. Two motors drive the desiccant inserter and operates at variable speeds. The inserter is used to dispense desiccant bags into bottles prior to being filled with product. The inserter dispenses from a continuous web of desiccant bags. It is equipped with sensors designed to locate a seam on the bag web, cut the web into individual bags and dispense them into the product bottles.

The validation protocol for a desiccant bag inserter is described in the following section. This protocol example can be used as a cGMP compliance guide and it should be tailored to meet individual company requirements. Italicized words in the tables represent examples of information that could be recorded in a validation protocol for a desiccant bag inserter.

## Installation Qualification

An IQ evaluation establishes confidence that the equipment is properly installed (Figure 6.7). The installation must meet the manufacturer's specified guidelines, along with design changes at installation. Also, the supporting electrical utilities must meet all electrical codes. The information required for an IQ evaluation is equipment identification, required documentation, equipment utility requirements, major component specifications, component material, lubricants, and equipment safety features.

***Table 6.14. Bottle Orienter Performance Test Results***

| Test Function | Bottle Size (60 cc) | Bottle Size (1,300 cc) | Acceptable (Yes/No) |
|---|---|---|---|
| Preorienter speed setting | *3.7 rpm* | *4.0 rpm* | *Yes* |
| Main drive speed setting | *3.4 rpm* | *2.25 rpm* | *Yes* |
| Elevator speed setting | *6.0 rpm* | *9.0 rpm* | *Yes* |
| Containers per minute | *140** | *80** | *Yes* |

*There are not any established specification for these speeds. The next time this equipment is validated, then the specifications can be compared.

**Figure 6.6. Desiccant Bag Dispenser (courtesy of Austin-Gordon Design Corp.)**

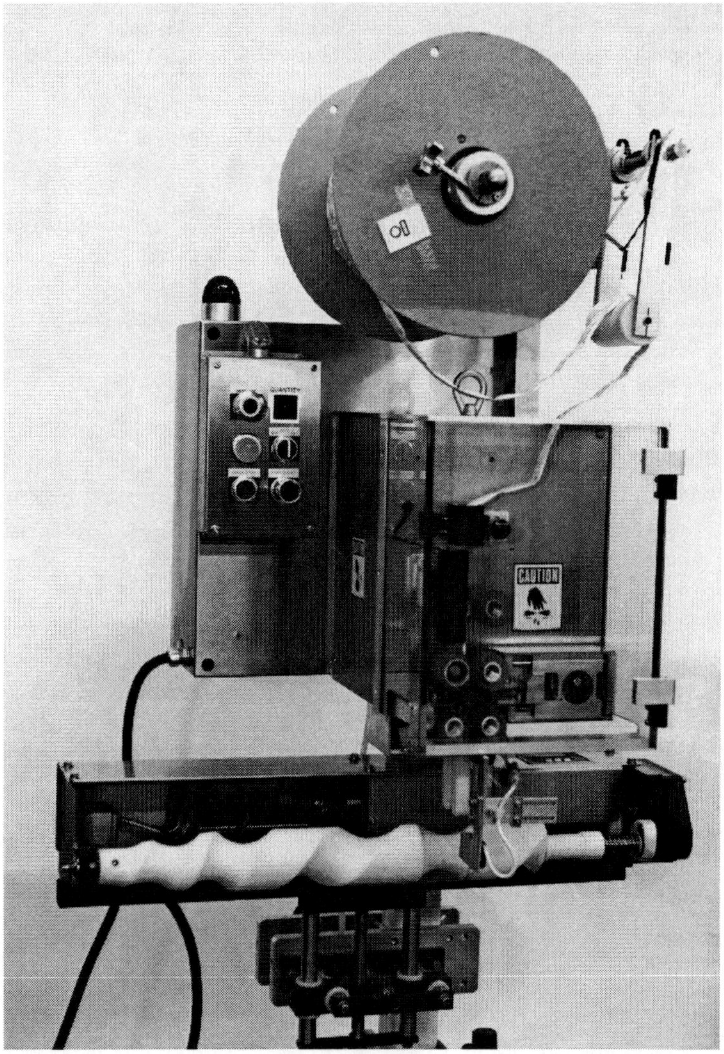


**Figure 6.7. IQ Elements of a Desiccant Bag Inserter**

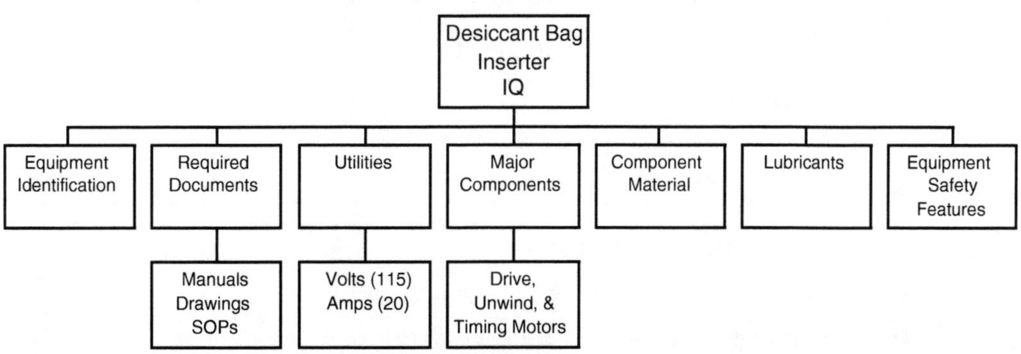

### *Equipment Identification*

Record the equipment identification numbers in Table 6.15, along with the following information: equipment manufacturer's purchase order number, model number, serial number, company assigned equipment number, and the location of the equipment.

### *Required Documentation*

Record the equipment manufacturer's operation and maintenance manual and drawings in Table 6.16. Record the SOPs that cover the setup, operation, and cleaning of the desiccant bag inserter in Table 6.17.

### *Equipment Utility Requirements*

Compare the manufacturer's specified volts (V) and amps (A) requirements to their as-found conditions at the time of qualification testing and record the results in Table 6.18. Also, record the location of the power supply source. Record the instruments used to measure the volts and amps in Table 6.19.

**Table 6.15. Equipment Identification**

| Required Information | As-Found Conditions |
|---|---|
| Manufacturer | *Any Desiccant Bag Inserter Co.* |
| Purchase order number | *562391* |
| Model number | *202-DP* |
| Serial number | *1-05384* |
| Equipment number | *0382* |
| Location | *Packaging line 2* |

**Table 6.16. Manufacturers' Manuals and Drawings**

| Number | Description | Date |
|---|---|---|
| *None* | *Desiccant Bag Inserter Installation, Operation, and Maintenance Manual* | *None* |

**Table 6.17. Standard Operating Procedures**

| Number | Description | Release Date |
|---|---|---|
| *PKG003* | *Desiccant Bag Inserter Setup, Operation, and Cleaning* | *07/14/95* |

**Table 6.18. *Utilities***

| Utility | Specified | Measured Results | Acceptable (Yes/No) |
|---|---|---|---|
| Volts | 115 ± 10% | *117* | *Yes* |
| Amps | Drive motor = 1.0<br>Unwind motor = 0.75<br>Timing screw motor = 1.8 | 20 circuit rating | *Yes* |

Power supply source, breaker box *BB22, wire number: L7*

**Table 6.19. *Instrument Used***

| Test Instrument | Identification Number | Calibration Due Date |
|---|---|---|
| *Multimeter* | *ME-025* | *04/19/97* |

***Volt calculation:***

Desiccant bag inserter volt specification = 115 V ± 10%.

±10% of 115 = ±11.5

115 + 11.5 = 126.5

115 – 11.5 = 103.5

The measured volts of 117 falls within ± 10%.

***Amp calculation:***

Amp specification = 20 A ±10%

Drive motor = 1.0 A

Unwind motor = 0.75 A

Timing screw motor = 1.8 A

Equipment current draw = drive motor A + unwind motor A + timing screw motor A = 1.0 + 0.75 + 1.8 = 3.55 A.

The circuit amp rating of 20 is greater than the maximum current draw of the equipment.

## *Major Component Specifications*

The component specifications section of the protocol verifies that the desiccant bag inserter components purchased were delivered and installed. Record the major components in Table 6.20.

**Table 6.20. Major Components**

| Components | As-Found Conditions |
|---|---|
| Drive motor | Manufacturer: *Any Motor Co.*<br>Model number: *MPEB3D24*<br>Serial number: *124961*<br>Volts: *115*<br>Amperes: *1.0*<br>Phases: *1*<br>Cycles: *60 Hz*<br>hp: *1/8*<br>rpm: *1,800* |
| Unwind motor | Manufacturer: *Any Motor Co.*<br>Model number: *Not specified*<br>Serial number: *44731*<br>Volts: *115*<br>Amperes: *0.75*<br>Phases: *1*<br>Cycles: *60 Hz*<br>hp: *1/8*<br>rpm: *43* |
| Timing screw motor | Manufacturer: *Any Motor Co.*<br>Model number: *Not specified*<br>Serial number: *4748810*<br>Volts: *115*<br>Amperes: *1.8*<br>Phases: *1*<br>Cycles: *60 Hz*<br>hp: *1/4*<br>rpm: *165* |

### *Component Material*

Record the material of each component that contacts the product in Table 6.21.

### *Lubricants*

Record the lubricants used to operate the inserter in Table 6.22 and indicate if they make contact with the product.

**Table 6.21 Component Material**

| Component | Material |
|---|---|
| *None of the components make contact with the product.* | |

**Table 6.22. *Lubricants***

| Where Used | Type | Manufacturer | Product Contact (Yes/No) |
|---|---|---|---|
| *Gears* | *Lubriplate Gear Shield Extra Heavy Grease* | *Any Oil Co.* | *No* |
| Is there a preventive maintenance procedure on file? (Yes/No) | | | *Yes* |

### *Equipment Safety Features*

The objective of testing the safety features is to verify that the safety features on the desiccant bag inserter function according to manufacturer's specifications. Verify the operation of the equipment safety features in Table 6.23 and record the results.

## Operational Qualification

An OQ evaluation should establish that the equipment can operate within specified tolerances and limits (Figure 6.8). The mechanical ranges of the desiccant bag inserter are being challenged, along with the basic equipment operations. Information required for the OQ evaluation is calibration of the instruments used to control the desiccant bag inserter, equipment control functions (switches and push buttons), and equipment operation (drive, unwind & timing screw, motor rotation direction, and speed, timing screw operation).

### *Calibration Requirements*

Verify that all critical instruments on the equipment have been logged into the calibration system, have calibration procedures in place, and are in calibration at the time of qualification testing. Record all information for the calibrated instruments used to control the desiccant bag inserter in Table 6.24.

**Table 6.23. *Safety Features Test Results***

| Test Function | Expected Results | Acceptable (Yes/No) |
|---|---|---|
| Alarm light | When there is a fail safe condition, the alarm light goes on. | *Yes* |
| Alarm buzzer | When there is a fail safe condition, the alarm buzzer sounds. | *Yes* |
| Safety guard door | When the door is opened, the equipment stops operating. | *Yes* |

**Figure 6.8. OQ Elements of a Desiccant Bag Inserter**

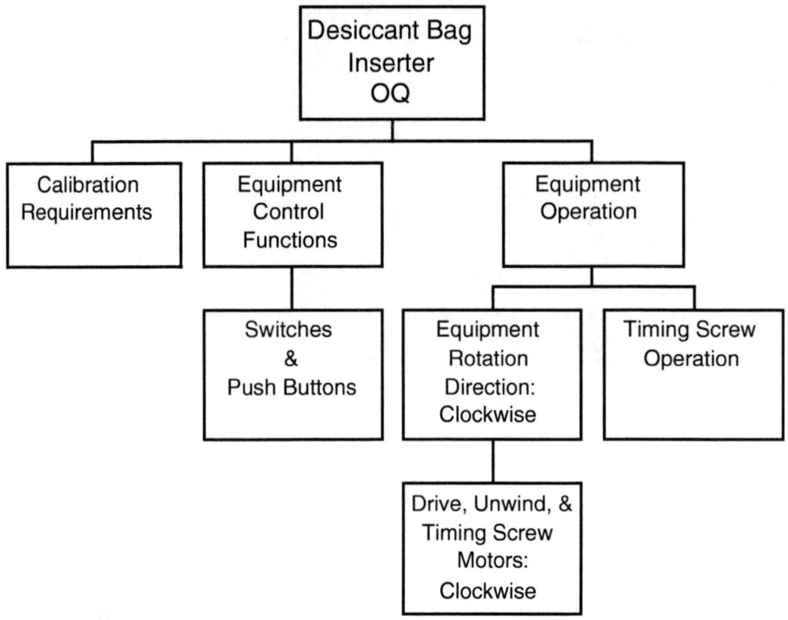


**Table 6.24. Calibrated and Noncalibrated Instruments**

*There were no calibrated or noncalibrated instruments on the desiccant bag inserter.*

### *Equipment Control Functions*

The objective of testing equipment control functions is to verify that the switches and push buttons on the desiccant bag inserter operate per manufacturer's specifications. The tests will be performed with the desiccant bag inserter empty. Operate each control listed in Table 6.25 and verify its proper operation.

### *Equipment Operation: Drive, Unwind, and Timing Screw Motor Rotation Direction*

The objective of the rotation direction test is to verify that the drive, unwind, and timing screw motor rotate in the proper direction. Press the start push button and observe the direction of rotation of the drive, unwind, and timing screw motor as viewed from behind the motors, and record the results in Table 6.26.

### *Equipment Operation: Timing Screw Operation*

The objective of the timing screw test is to verify that it operates according to manufacturer's specification. 60 cc bottles will be required. With the machine operating, visually observe 100 bottles as they pass through the timing screw and verify that they are uniformly spaced on the conveyor. Record the results in Table 6.27.

**Table 6.25. Equipment Control Function Test Results**

| Test Function | Expected Results | Acceptable (Yes/No) |
|---|---|---|
| Start push button operation | When the start push button is pressed, the inserter starts. | *Yes* |
| Stop push button | When the stop push button is pressed, the inserter stops. | *Yes* |
| Emergency stop push button | When the emergency stop push button is pressed, the inserter stops. | *Yes* |
| Reset push button | When the reset push button is pressed, the alarm condition is cleared. | *Yes* |
| Manual dispense push button | When the manual dispense push button is pressed, one desiccant bag is dispensed. | *Yes* |
| Single . . . double switch | When the single . . . double switch is set to single, one desiccant bag is dispensed. | *Yes* |
| | When the single . . . double switch is set to double, two desiccant bags are dispensed. | *Yes* |
| Time delay dial | When the time delay dial is turned, the position where the cut is made on the desiccant bag web is changed. | *Yes* |
| Desiccant bag registration sensor | When the desiccant bag registration sensor is blocked, the desiccant bag web stops and the knives activate. | *Yes* |
| Desiccant bag verification sensor | When the desiccant bag verification sensor is blocked, the desiccant bag left the outfeed area. | *Yes* |
| Bottle detection sensor | When a bottle is detected, when the unit is set to single, a desiccant bag is dispensed. | *Yes* |
| | When a bottle is detected, when the unit is set to double, two desiccant bags are dispensed. | *Yes* |
| Upper/lower limit switch | When the upper limit switch is activated, the desiccant bag web feeds more bags. | *Yes* |
| | When the lower limit switch is activated, the desiccant bag web feeds more bags. | *Yes* |

## Performance Qualification

Once it has been established that the equipment is properly installed and functioning within specified operating parameters, it must be shown that the desiccant bag inserter can operate reliably under routine, minimum, and maximum operating conditions (Figure 6.9).

**Table 6.26. Drive, Unwind, and Timing Screw Motor Rotation Direction Test Results**

| Item | Expected Results | Results | Acceptable (Yes/No) |
|---|---|---|---|
| Drive motor rotation direction | Rotation should be clockwise as viewed from behind the motor. | *Clockwise rotation was observed.* | *Yes* |
| Unwind motor rotation direction | Rotation should be clockwise as viewed from behind the motor. | *Clockwise rotation was observed.* | *Yes* |
| Timing screw motor rotation direction | Rotation should be clockwise as viewed from behind the motor. | *Clockwise rotation was observed.* | *Yes* |

**Table 6.27. Timing Screw Test Results**

| Test Function | Bottle Formation | Acceptable (Yes/No) |
|---|---|---|
| Bottle spacing | *Bottles were evenly spaced on the conveyor.* | *Yes* |

**Figure 6.9. PQ Elements of a Desiccant Bag Inserter**

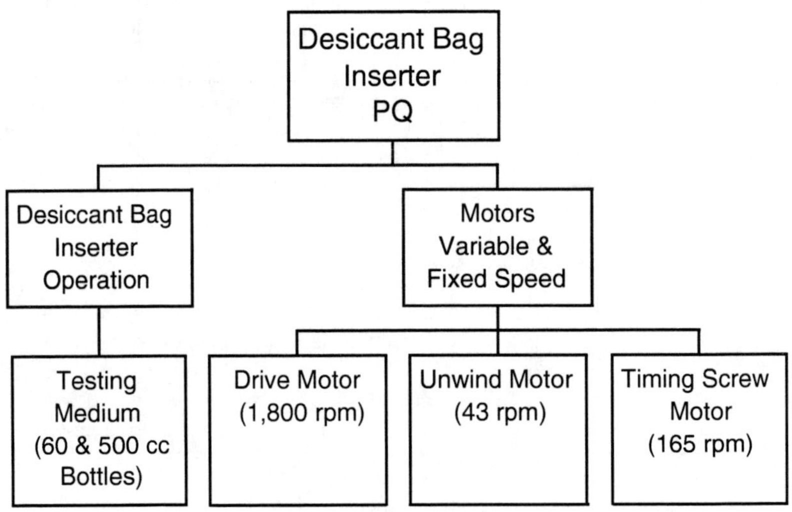


### *Desiccant Bag Inserter Operation*

The objective of the desiccant bag inserter test is to document the performance of the desiccant bag inserter using various combinations of desiccant bags and bottles. Also, the fixed and variable speeds of the motors will be measured and recorded. 60 cc bottles, 500 cc bottles, desiccant bags, and a tachometer will be required for the following tests:

- With the equipment operating, observe 100 bottles as they pass through the desiccant bag inserter and verify that the proper amount of bags are dispensed into each bottle. Repeat this test for all combinations of bags and bottles. Record the results in Table 6.28.

- Measure the speed of the drive motor with a calibrated tachometer. Verify that the measured speed is within ±10 percent of the maximum speed of 1,800 rpm. Record the results in Table 6.29. Record the instrument used to measure the speed in Table 6.30.

  Drive motor speed specification = 1,800 rpm ± 10% maximum

  ±10% of 1,800 = ±180

  1,800 + 180 = 1,980

  1,800 – 180 = 1,620

  The measured rpm of 1,840 falls within ±10%.

- Measure the speed of the unwind motor with a calibrated tachometer. Verify that the measured speed is within ±10 percent of the fixed speed of 43 rpm. Record the results in Table 6.29. Record the instrument used to measure the speed in Table 6.30.

  Unwind motor speed specification = 43 rpm ± 10%

  ±10% of 43 = ±4.3

  43 + 4.3 = 47.3

  43 – 4.3 = 38.7

  The measured rpm of 42 falls within ±10%.

- Measure the speed of the timing screw motor with a calibrated tachometer. Verify that the measured speed is within ±10 percent of the fixed speed of 165 rpm. Record the results in Table 6.29. Record the instrument used to measure the speed in Table 6.30.

  Timing screw motor speed specification = 165 rpm ± 10%

  ±10% of 165 = ±16.5

  165 + 16.5 = 181.5

  165 – 16.5 = 148.5

  The measured rpm of 167 falls within ±10%.

**Table 6.28. Desiccant Bag Inserter Performance Test Results**

**TEST 1**

| Bottle Opening 33 mm | Bottle Size 60 cc | Desiccant #7812 | 1 Desiccant Bag |
|---|---|---|---|

| Test Number | Test Description | Result (Yes/No) |
|---|---|---|
| 1 | Did a cycle start each time a bottle passed in front of the bottle sensor for each bottle? | *Yes* |
| 2 | Did the bag registration sensor signal the knives to cut on the seamed area of the web for each bottle? | *Yes* |
| 3 | Did the bag verification sensor detect the pouch dropping into each bottle? | *Yes* |
| 4 | Was the desiccant bag dispensed correctly into each bottle? | *Yes* |

**TEST 2**

| Bottle Opening 33 mm | Bottle Size 60 cc | Desiccant #7812 | 2 Desiccant Bags |
|---|---|---|---|

| Test Number | Test Description | Result (Yes/No) |
|---|---|---|
| 1 | Did a cycle start each time a bottle passed in front of the bottle sensor for each bottle? | *Yes* |
| 2 | Did the bag registration sensor signal the knives to cut on the seamed area of the web for each bottle? | *Yes* |
| 3 | Did the bag verification sensor detect the pouch dropping into each bottle? | *Yes* |
| 4 | Were the desiccant bags dispensed correctly into each bottle? | *Yes* |

**TEST 3**

| Bottle Opening 53 mm | Bottle Size 500 cc | Desiccant #7812 | 1 Desiccant Bag |
|---|---|---|---|

| Test Number | Test Description | Result (Yes/No) |
|---|---|---|
| 1 | Did a cycle start each time a bottle passed in front of the bottle sensor for each bottle? | *Yes* |
| 2 | Did the bag registration sensor signal the knives to cut on the seamed area of the web for each bottle? | *Yes* |
| 3 | Did the bag verification sensor detect the pouch dropping into each bottle? | *Yes* |
| 4 | Was the desiccant bag dispensed correctly into each bottle? | *Yes* |

*Continued on next page.*

*Continued from previous page.*

TEST 4

| Bottle Opening 53 mm | Bottle Size 500 cc | Desiccant #7812 | 2 Desiccant Bags |
|---|---|---|---|

| Test Number | Test Description | Result (Yes/No) |
|---|---|---|
| 1 | Did a cycle start each time a bottle passed in front of the bottle sensor for each bottle? | *Yes* |
| 2 | Did the bag registration sensor signal the knives to cut on the seamed area of the web for each bottle? | *Yes* |
| 3 | Did the bag verification sensor detect the pouch dropping into each bottle? | *Yes* |
| 4 | Were the desiccant bags dispensed correctly into each bottle? | *Yes* |

**Table 6.29. *Drive, Unwind, and Timing Screw Motor Speed Test Results***

| Motor | Speed Specification rpm ± 10% | Measured Speed rpm | Acceptable (Yes/No) |
|---|---|---|---|
| Drive | 1,800 | 1,840 | *Yes* |
| Unwind | 43 | 42 | *Yes* |
| Timing screw | 165 | 167 | *Yes* |

**Table 6.30. *Instrument Used***

| Test Instrument | Identification Number | Calibration Due Date |
|---|---|---|
| *Tachometer* | *64020* | *06/21/97* |

## Tablet and Capsule Inserter

The validation of a tablet and capsule inserter (Figure 6.10) will be described here. The inserter is used to fill bottles with a specified number of tablets and capsules. The tablets or capsules are placed in a feeder system which removes dust and chips before they reach the slats. Two vibrators are used to move the tablets or capsules into slots in the slats. The number of slats and the number of cycles to be run determines the number of tablets or capsules that will be placed in a single bottle through a funnel. Six bottles can be filled simultaneously. The bottles are brought under the funnels by a continuously running conveyor and positioned by a timing screw drive with scrolls.

The validation protocol for a tablet and capsule inserter is described in the following sections. This protocol example can be used as a cGMP compliance guide, and it should be tailored to meet individual company requirements. Italicized words in the tables represent examples of information that could be recorded in a validation protocol for a tablet and capsule inserter.

**Figure 6.10. Slat Counter (courtesy of Modular King Packaging Systems, Inc.)**

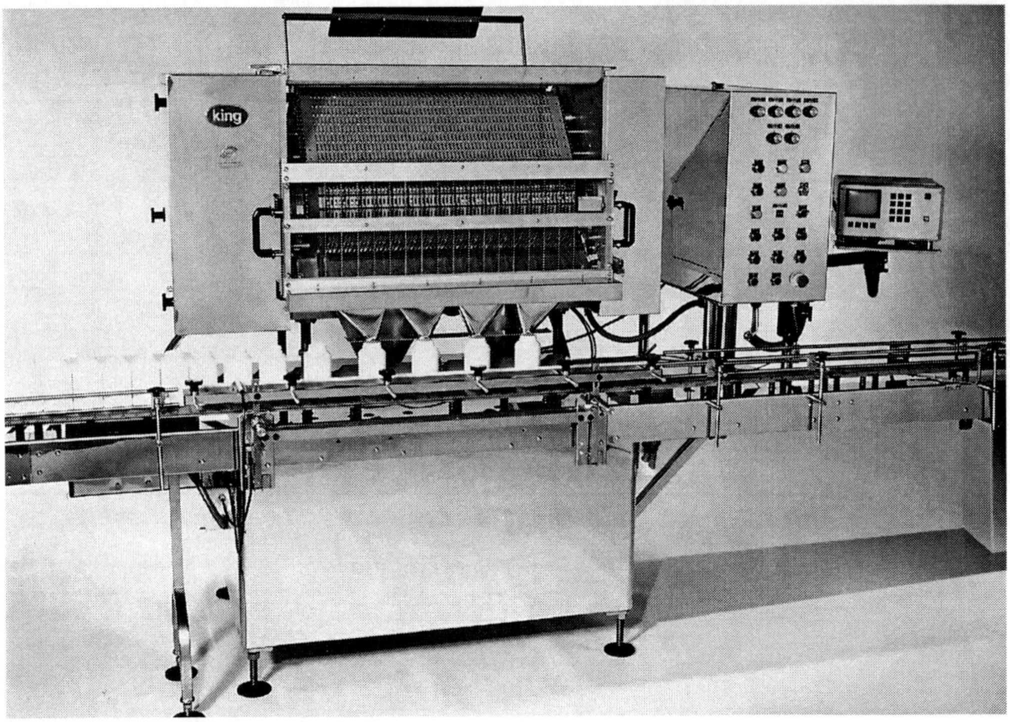


## Installation Qualification

An IQ evaluation establishes confidence that the equipment is properly installed (Figure 6.11). The installation must meet the manufacturer's specified guidelines, along with design changes at installation. Also, the supporting electrical utilities must meet all electrical codes. The information required for an IQ evaluation is equipment identification, required documentation, equipment utility requirements, major component specifications, component material, lubricants, and equipment safety features.

### *Equipment Identification*

Record the equipment identification numbers in Table 6.31, along with the following information: equipment manufacturer's purchase order number, model number, serial number, company assigned equipment number, and the location of the equipment.

### *Required Documentation*

Record the equipment manufacturer's operation and maintenance manual and drawings in Table 6.32. Record the SOPs that cover the setup, operation, and cleaning of the tablet and capsule inserter in Table 6.33.

**Figure 6.11. IQ Elements of a Tablet and Capsule Inserter**

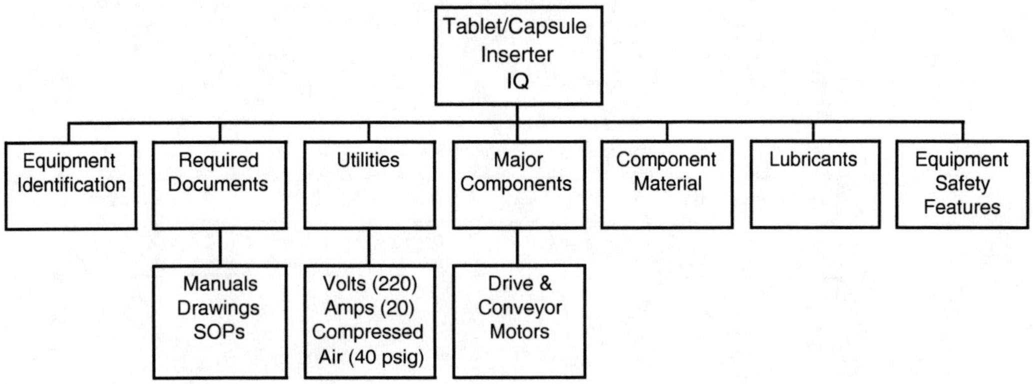


**Table 6.31. Equipment Identification**

| Required Information | As-Found Conditions |
|---|---|
| Manufacturer | *Any Tablet and Capsule Inserter Co.* |
| Purchase order number | *441870* |
| Model number | *220705* |
| Serial number | *372* |
| Equipment number | *8160* |
| Location | *Packaging line 6* |

**Table 6.32. Manufacturers' Manuals and Drawings**

| Number | Description | Date |
|---|---|---|
| *None* | *Tablet and Capsule Inserter Installation, Operation, and Maintenance Manual* | *11/15/94* |
| *285-HCS* | *Tablet and Capsule Inserter with 3-Phase Conveyor Drawing* | *12/20/94* |
| *2/287 MCS* | *Tablet and Capsule Inserter General Assembly Drawing* | *04/05/87* |

**Table 6.33. Standard Operating Procedures**

| Number | Description | Release Date |
|---|---|---|
| *PKG041* | *Tablet and Capsule Inserter Setup and Operation* | *07/14/95* |
| *PKG042* | *Tablet and Capsule Inserter Cleaning Procedure* | *09/22/94* |

### *Equipment Utility Requirements*

Compare the manufacturer's specified volts (V), amps (A), and compressed air requirements to their as-found conditions at the time of qualification testing and record the results in Table 6.34. Also record the location of the power supply

**Table 6.34. Utilities**

| Utility | Specified | Measured Results | Acceptable (Yes/No) |
|---|---|---|---|
| Volts | 220 ± 10% | *A-B 220; A-C 219; B-C 223* | *Yes* |
| Amps | Drive motor = 1.1<br>Conveyor motor = 3 | 20 circuit rating | *Yes* |
| Compressed air | Air pressure ≥ 40 psig | *58 psig* | *Yes* |
| Power supply source, breaker box *BB7A, wire numbers: 15, 17, 19* | | | |

source. Record the instruments used to measure the volts, amps, and compressed air in Table 6.35.

***Volt calculation:***

Tablet and capsule inserter volt specification = 220 V ± 10%

±10% of 220 = ±22

220 + 22 = 242

220 – 22 = 198

The measured volts of 220/219/223 fall within ±10%.

***Amp calculation:***

Amp specification = 20 A ±10%

Drive motor = 1.1 A

Conveyor motor = 3 A

Equipment current draw = drive motor A + conveyor motor A = 1.1 + 3 = 4.1 A

The circuit amp rating of 20 is greater than the maximum current draw of the equipment.

**Table 6.35. Instruments Used**

| Test Instrument | Identification Number | Calibration Due Date |
|---|---|---|
| *Multimeter* | *ME-025* | *04/19/97* |
| *Air pressure gauge* | *P-102* | *02/10/97* |

***Compressed air calculation:***

Compressed air specification ≥ 40 psig

The measured air pressure of 58 psig is greater than 40 psig.

### *Major Component Specifications*

The component specifications section of the protocol is used to verify that the tablet and capsule inserter components purchased were delivered and installed. Record the major components in Table 6.36.

### *Component Material*

Record the material of each component that contacts the product in Table 6.37.

### *Lubricants*

Record the lubricants used to operate the inserter in Table 6.38 and indicate if they make contact with the product.

***Table 6.36. Major Components***

| Components | As-Found Conditions |
|---|---|
| Drive motor | Manufacturer: *Any Motor Co.*<br>Model number: *Not specified*<br>Serial number: *Not specified*<br>Volts: *220*<br>Amperes: *1.1*<br>Phases: *3*<br>Cycles: *60 Hz*<br>hp: *1/4*<br>rpm: *1,700* |
| Conveyor motor | Manufacturer: *Any Motor Co.*<br>Model number: *9974E*<br>Serial number: *1/957781CS*<br>Volts: *220*<br>Amperes: *3*<br>Phases: *3*<br>Cycles: *60 Hz*<br>hp: *1*<br>rpm: *2,000* |

***Table 6.37. Component Material***

| Component | Material |
|---|---|
| Hopper | *Stainless steel* |
| Tablet and capsule slats | *Plastic* |

**Table 6.38. *Lubricants***

| Where Used | Type | Manufacturer | Product Contact (Yes/No) |
|---|---|---|---|
| *Chains* | *APG 90 Oil* | *Any Oil Co.* | *No* |
| *Gearbox* | *Lubriplate Grease* | *Any Oil Co.* | *No* |
| Is there a preventive maintenance procedure on file? (Yes/No) | | | *Yes* |

### *Equipment Safety Features*

The objective of testing safety features is to verify that the safety features on the tablet and capsule inserter function according to manufacturer specifications. Verify the operation of the equipment safety features in Table 6.39 and record the results.

## Operational Qualification

An OQ evaluation should establish that the equipment can operate within specified tolerances and limits (Figure 6.12). The mechanical ranges of the tablet and capsule inserter are challenged, along with the basic equipment operations that the operator will use. Information required for the OQ evaluation is calibration of the instruments used to control the tablet and capsule inserter, equipment control functions (switches and push buttons), and equipment operation (drive and conveyor motor rotation direction, and speed test).

**Table 6.39. *Safety Features Test Results***

| Test Function | Expected Results | Acceptable (Yes/No) |
|---|---|---|
| Power lift up | When the power lift up push button is pressed and held, the upward movement of the head is stopped at the limit of movement, automatically. | *Yes* |
| Power lift down | When the power lift down push button is pressed and held, the downward movement of the head is stopped at the bottom limit of movement, automatically. | *Yes* |
| Emergency stop | When the emergency stop push button is pressed, all mechanical operations stop. | *Yes* |
| Foot pedal | When the foot pedal is pressed, the filling operation pauses until the pedal is released. | *Yes* |

***Figure 6.12. OQ Elements of a Tablet and Capsule Inserter***

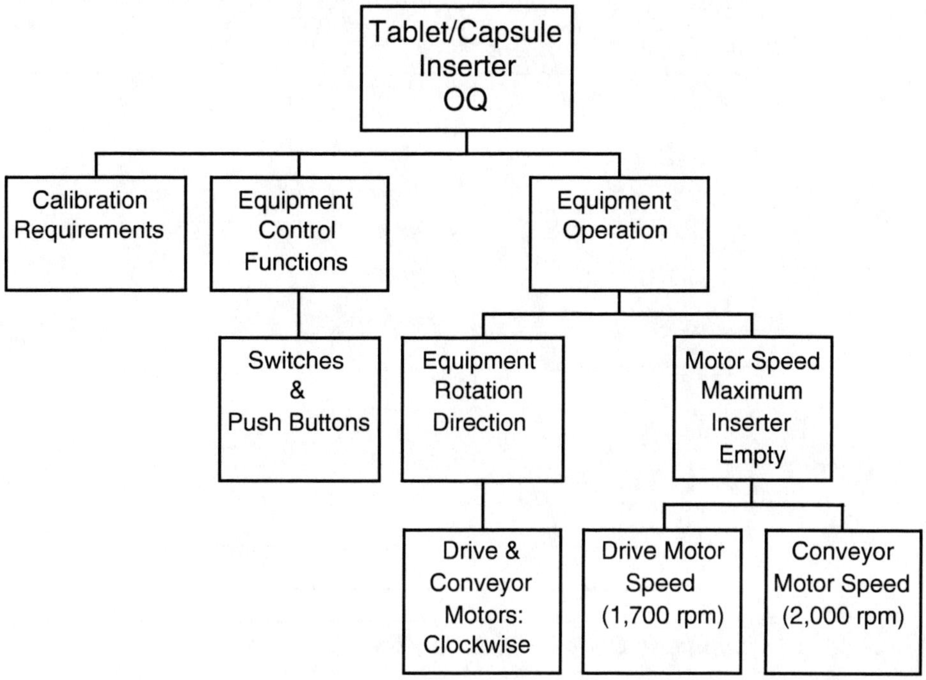


### *Calibration Requirements*

Verify that all critical instruments on the equipment have been logged into the calibration system, have calibration procedures in place, and are in calibration at the time of qualification testing. Record all information for the calibrated instruments used to control the tablet and capsule inserter in Table 6.40.

### *Equipment Control Functions*

The objective of testing equipment control functions is to verify that the switches and push buttons on the tablet and capsule inserter operate per manufacturer's specifications. Operate each control listed in Table 6.41. and verify its proper operation. Record the results in Table 6.41. The tests will be performed with the tablet and capsule inserter empty.

### *Equipment Operation: Drive and Conveyor Motor Rotation Direction Test*

The objective of the rotation direction test is to verify that the drive and conveyor motors rotate in the proper direction. The tests will be performed with the tablet and capsule inserter empty of tablets or capsules. A tachometer is required. Press the start push button and observe the direction of rotation of the drive and the conveyor motor as viewed from behind the motors and record the results in Table 6.42.

**Table 6.40. *Calibrated and Noncalibrated Instruments***

*There were no calibrated instruments on the tablet and capsule inserter.*

| Noncalibrated Instrument | As-Found Conditions |
|---|---|
| Identification number | *Slat/100 GP* |
| Type | *Air pressure gauge, analog* |
| Manufacturer | *Any Gauge Co.* |
| Model number | *Not specified* |
| Serial number | *Not specified* |
| Range | *0–140 psig* |
| Scale division | *10 psig* |
| Location | *On the inlet line* |
| Use | *Regulates the air pressure to the equipment* |
| Calibration due date | *Not required* |
| Critical or not critical | *Not critical to the operation of the inserter* |

**Table 6.41. *Equipment Control Function Test Results***

| Test Function | Expected Results | Acceptable (Yes/No) |
|---|---|---|
| Electrical disconnect switch and isolator switch | When the electrical disconnect switch and isolator switch are pressed, the power indicator light goes on. | *Yes* |
| Head motor on push button | When the head motor on push button is pressed, the head motor, upper vibrator and lower vibrator are activated. | *Yes* |
| Upper vibrator dial | When the upper vibrator dial is rotated, the amplitude increases and decreases. | *Yes* |
| Lower vibrator dial | When the lower vibrator dial is rotated, the amplitude increases and decreases. | *Yes* |
| Scroll drive on/off push button | When the scroll drive on/off push button is pressed, the screw drive with scrolls start and stop. | *Yes* |

**Table 6.42. *Drive and Conveyor Motor Rotation Direction Test Results***

| Item | Expected Results | Results | Acceptable (Yes/No) |
|---|---|---|---|
| Drive motor rotation direction | Rotation should be clockwise as viewed from behind the motor. | *Clockwise rotation was observed.* | *Yes* |
| Conveyor motor rotation direction | Rotation should be clockwise as viewed from behind the motor. | *Clockwise rotation was observed.* | *Yes* |

### *Equipment Operation: Drive Motor and Conveyor Motor Speed*

The objective of the conveyor motor test is to document the speed of the drive motor and conveyor motor with the tablet and capsule inserter empty.

- Measure the speed of the drive motor with a calibrated tachometer. Verify that the measured speed is within ±10 percent of the maximum speed of 1,700 rpm. Record the results in Table 6.43. Record the instrument used to measure the speed in Table 6.44.

  Drive motor speed specification = 1,700 rpm ± 10% maximum

  ±10% of 1,700 = ±170

  1,700 + 170 = 1,870

  1,700 – 170 = 1,530

  The measured rpm of 1,702 falls within ±10%.

- Measure the speed of the conveyor motor with a calibrated tachometer. Verify that the measured speed is within ±10 percent of the maximum speed of 2,000 rpm. Record the results in Table 6.43. Record the instrument used to measure the speed in Table 6.44.

  Unwind motor speed specification = 2,000 rpm ± 10% maximum

  ±10% of 2,000 = ±200

  2,000 + 200 = 2,000

  2,000 – 200 = 1,800

  The measured rpm of 2,010 falls within ±10%.

## Performance Qualification

Once it has been established that the equipment is properly installed and functioning within specified operating parameters, it must be shown that the tablet and capsule inserter can operate reliably under routine, minimum, and maximum operating conditions (Figure 6.13).

***Table 6.43. Drive and Conveyor Motor Speed Test Results***

| Motor | Speed Specification (rpm ± 10%) | Measured Speed (rpm) | Acceptable (Yes/No) |
|---|---|---|---|
| Drive | 1,700 | *1,702* | *Yes* |
| Conveyor | 2,000 | *2,010* | *Yes* |

***Table 6.44. Instrument Used***

| Test Instrument | Identification Number | Calibration Due Date |
|---|---|---|
| *Tachometer* | *64020* | *06/21/97* |

**Figure 6.13. PQ Elements of a Tablet and Capsule Inserter**

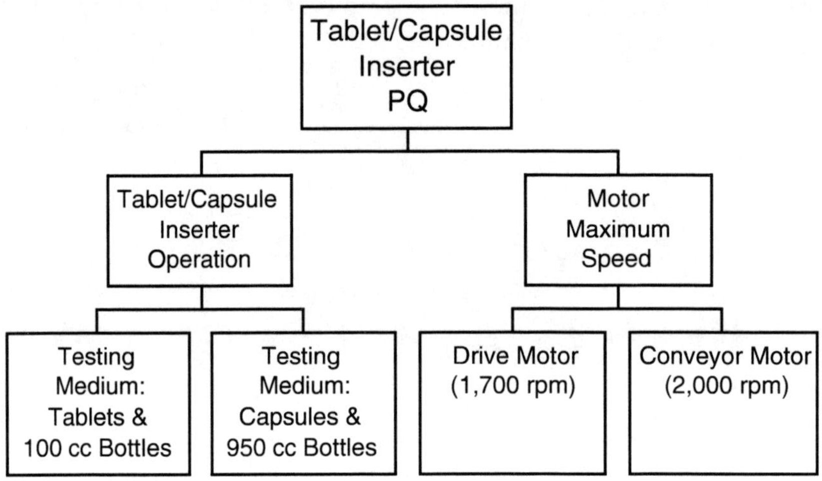


### *Tablet and Capsule Inserter Operation*

The objective of testing inserter operation is to verify that the tablet and capsule inserter is capable of inserting tablets and capsules into bottles. The speed of the motors will be measured and recorded. 100 cc bottles, 950 cc bottles, tablets, capsules, and a tachometer are required as follows:

- With the equipment operating, insert 100 tablets into 100 bottles as they pass through the tablet and capsule inserter and verify that the number of inserted tablets are within ±1 tablet of the specified quantity. Repeat this test with capsules. Record the materials used for tablets in Table 6.45 and the results in Table 6.46. Record the materials used for capsules in Table 6.47 and the results in Table 6.48.

**Table 6.45. Required Tablet Materials and Conditions**

| Item | Results |
|---|---|
| Tablet size | *0.247" × 0.502"* |
| Bottle size | *100 cc* |
| Number of bottles | *100* |
| Number of tablets | *100* |
| Equipment speed setting | *7.75 rpm* |
| Conveyor speed setting | *58 rpm* |
| Upper vibrator speed setting | *8 rpm* |
| Lower vibrator speed setting | *8 rpm* |

**Table 6.46.** ***Tablet Inserter Performance Test Results***

| Bottle No. | Tablet Qty | Bottle No. | Tablet Qty | Bottle No. | Tablet Qty | Bottle No. | Tablet Qty | Bottle No. | Tablet Qty |
|---|---|---|---|---|---|---|---|---|---|
| 1 | *100* | 11 | *100* | 21 | *100* | 31 | *100* | 41 | *100* |
| 2 | *100* | 12 | *100* | 22 | *100* | 32 | *100* | 42 | *101* |
| 3 | *100* | 13 | *100* | 23 | *100* | 33 | *100* | 43 | *100* |
| 4 | *100* | 14 | *101* | 24 | *100* | 34 | *100* | 44 | *100* |
| 5 | *100* | 15 | *100* | 25 | *100* | 35 | *100* | 45 | *100* |
| 6 | *100* | 16 | *101* | 26 | *100* | 36 | *100* | 46 | *100* |
| 7 | *100* | 17 | *100* | 27 | *100* | 37 | *100* | 47 | *100* |
| 8 | *100* | 18 | *100* | 28 | *100* | 38 | *100* | 48 | *101* |
| 9 | *100* | 19 | *100* | 29 | *100* | 39 | *100* | 49 | *99* |
| 10 | *100* | 20 | *100* | 30 | *100* | 40 | *100* | 50 | *100* |
| 51 | *99* | 61 | *100* | 71 | *100* | 81 | *100* | 91 | *100* |
| 52 | *101* | 62 | *100* | 72 | *100* | 82 | *100* | 92 | *100* |
| 53 | *100* | 63 | *100* | 73 | *100* | 83 | *101* | 93 | *99* |
| 54 | *100* | 64 | *100* | 74 | *100* | 84 | *100* | 94 | *100* |
| 55 | *101* | 65 | *100* | 75 | *101* | 85 | *100* | 95 | *100* |
| 56 | *100* | 66 | *100* | 76 | *100* | 86 | *100* | 96 | *101* |
| 57 | *100* | 67 | *99* | 77 | *100* | 87 | *100* | 97 | *100* |
| 58 | *99* | 68 | *100* | 78 | *100* | 88 | *100* | 98 | *100* |
| 59 | *100* | 69 | *100* | 79 | *100* | 89 | *100* | 99 | *100* |
| 60 | *100* | 70 | *100* | 80 | *100* | 90 | *100* | 100 | *100* |
| Were the number of inserted tablets within ± tablets of the specified quantity? (Yes/No) | | | | | | | | | *Yes* |

**Table 6.47.** ***Required Capsule Materials and Conditions***

| Item | Results |
|---|---|
| Capsule size | *Number 1* |
| Bottle size | *950 cc* |
| Number of bottles | *100* |
| Number of capsules | *100* |
| Equipment speed setting | *6 rpm* |
| Conveyor speed setting | *50 rpm* |
| Upper vibrator speed setting | *7 dial setting* |
| Lower vibrator speed setting | *6 dial setting* |

**Table 6.48. Capsule Inserter Performance Test Results**

| Bottle No. | Tablet Qty | Bottle No. | Tablet Qty | Bottle No. | Tablet Qty | Bottle No. | Tablet Qty | Bottle No. | Tablet Qty |
|---|---|---|---|---|---|---|---|---|---|
| 1 | *100* | 11 | *100* | 21 | *100* | 31 | *99* | 41 | *99* |
| 2 | *100* | 12 | *99* | 22 | *100* | 32 | *100* | 42 | *100* |
| 3 | *100* | 13 | *100* | 23 | *100* | 33 | *100* | 43 | *100* |
| 4 | *99* | 14 | *100* | 24 | *99* | 34 | *100* | 44 | *100* |
| 5 | *99* | 15 | *100* | 25 | *101* | 35 | *100* | 45 | *100* |
| 6 | *100* | 16 | *100* | 26 | *100* | 36 | *100* | 46 | *100* |
| 7 | *101* | 17 | *99* | 27 | *100* | 37 | *100* | 47 | *100* |
| 8 | *99* | 18 | *100* | 28 | *100* | 38 | *101* | 48 | *100* |
| 9 | *100* | 19 | *99* | 29 | *100* | 39 | *100* | 49 | *99* |
| 10 | *100* | 20 | *100* | 30 | *100* | 40 | *100* | 50 | *100* |
| 51 | *99* | 61 | *100* | 71 | *100* | 81 | *100* | 91 | *100* |
| 52 | *101* | 62 | *100* | 72 | *100* | 82 | *100* | 92 | *100* |
| 53 | *100* | 63 | *100* | 73 | *100* | 83 | *99* | 93 | *100* |
| 54 | *100* | 64 | *101* | 74 | *100* | 84 | *100* | 94 | *99* |
| 55 | *100* | 65 | *100* | 75 | *100* | 85 | *100* | 95 | *99* |
| 56 | *101* | 66 | *99* | 76 | *101* | 86 | *100* | 96 | *100* |
| 57 | *100* | 67 | *100* | 77 | *101* | 87 | *101* | 97 | *100* |
| 58 | *100* | 68 | *100* | 78 | *100* | 88 | *100* | 98 | *101* |
| 59 | *100* | 69 | *100* | 79 | *100* | 89 | *100* | 99 | *100* |
| 60 | *100* | 70 | *100* | 80 | *100* | 90 | *100* | 100 | *100* |

Were the number of inserted capsules within ± capsules of the specified quantity? (Yes/No) *Yes*

- Measure the speed of the drive motor with a calibrated tachometer. Verify that the measured speed is within ±10 percent of the maximum speed of 1,700 rpm. Record the results in Table 6.49. Record the instrument used to measure the speed in Table 6.50.

  Drive motor speed specification = 1,700 rpm ± 10% maximum

  ±10% of 1,700 = ±170

  1,700 + 170 = 1,870

  1,700 – 170 = 1,530

  The measured rpm of 1,694 falls within ±10%.

**Table 6.49. *Drive and Conveyor Motor Speed Test Results***

| Motor | Speed Specification (rpm ± 10%) | Measured Speed (rpm) | Acceptable (Yes/No) |
|---|---|---|---|
| Drive | 1,700 | *1,694* | *Yes* |
| Unwind | 2,000 | *1,987* | *Yes* |

**Table 6.50. *Instrument Used***

| Test Instrument | Identification Number | Calibration Due Date |
|---|---|---|
| *Tachometer* | *64020* | *06/21/97* |

- Measure the speed of the conveyor motor with a calibrated tachometer. Verify that the measured speed is within ±10 percent of the fixed speed of 2,000 rpm. Record the results in Table 6.49. Record the instrument used to measure the speed in Table 6.50.

  Conveyor motor speed specification = 2,000 rpm ± 10%

  ±10% of 2,000 = ±200

  2,000 + 200 = 2,000

  2,000 – 200 = 1,800

  The measured rpm of 1,987 falls within ±10%.

## Tablet and Capsule Inserter Washer

The validation of a tablet and capsule inserter washer (Figure 6.14) will be described here, used to dry slats from the tablet and capsule inserter. Slats from a packaging run are loaded onto a rotary rack and inserted into the washer. The washer operates through the following phases of a complete wash cycle: fill/heat, pre-wash, drain, wash, rinse, and dry.

The validation protocol for a tablet and capsule inserter washer is described in the following section. This protocol template can be used as a cGMP compliance guide, and it should be tailored to meet individual company requirements. Italicized words in the tables represent examples of information that could be recorded in a validation protocol for a tablet and capsule inserter washer.

### Installation Qualification

An IQ evaluation establishes confidence that the equipment is properly installed (Figure 6.15). The installation must meet the manufacturer's specified guidelines, along with design changes at installation. Also, the supporting electrical utilities

***Figure 6.14. Cabinet Washer (courtesy of Girton Manufacturing Co.)***

***Figure 6.15. IQ Elements of a Tablet and Capsule Inserter Washer***

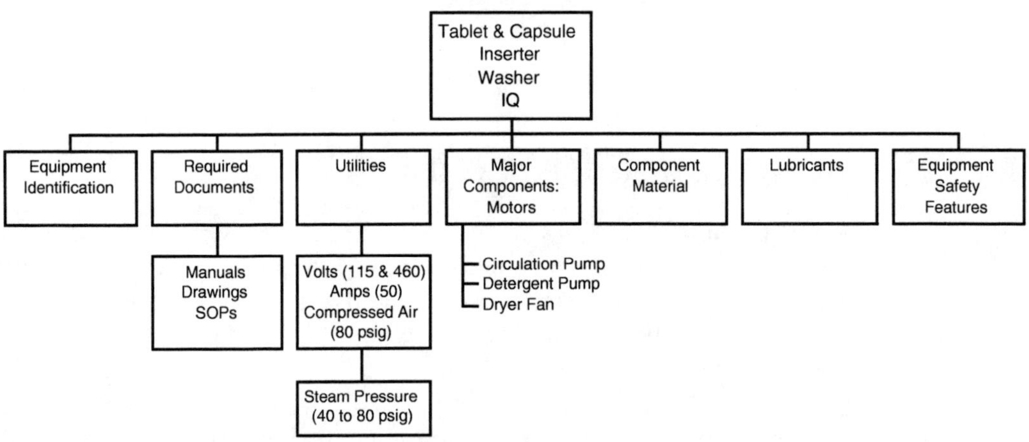


must meet all electrical codes. The information required for an IQ evaluation is equipment identification, required documentation, equipment utility requirements, major component specifications, component material, lubricants, and equipment safety features.

### *Equipment Identification*

Record the equipment identification numbers in Table 6.51, along with the following information: equipment manufacturer's purchase order number, model number, serial number, company assigned equipment number, and the location of the equipment.

### *Required Documentation*

Record the equipment manufacturer's operation and maintenance manual and drawings in Table 6.52. Record the SOPs that cover the setup, operation, and cleaning of the washer in Table 6.53.

### *Equipment Utility Requirements*

Compare the manufacturer's specified volts (V), amps (A), compressed air pressure, and steam pressure requirements to their as-found conditions at the time of qualification testing and record the results in Table 6.54. Also record the location of the power supply source. Record the instruments used to measure the volts, amps, compressed air, and steam pressure in Table 6.55.

**Table 6.51. Equipment Identification**

| Required Information | As-Found Conditions |
|---|---|
| Manufacturer | *Any Washer Co.* |
| Purchase order number | *6109900* |
| Model number | *SV-DHA497* |
| Serial number | *10422119* |
| Equipment number | *2251* |
| Location | *Room 627* |

**Table 6.52. Manufacturers' Manuals and Drawings**

| Number | Description | Date |
|---|---|---|
| *None* | *Tablet and Capsule Inserter Washer Installation, and Operation Manual* | *None* |
| *None* | *Drawings are too numerous to list.* | *None* |

**Table 6.53. Standard Operating Procedures**

| Number | Description | Release Date |
|---|---|---|
| *PKG017* | *Tablet and Capsule Inserter Washer Operating Instructions* | *01/16/95* |
| *PKG018* | *Cleaning Process Change Control* | *02/09/96* |

**Table 6.54. Utilities**

| Utility | Specified | Measured Results | Acceptable (Yes/No) |
|---|---|---|---|
| Volts | Circulation motor 115 ± 10%<br>Detergent motor = 460 ± 10% | *113*<br>*A-B 465 A-C 459 B-C 463* | *Yes* |
| Amps | Circulation pump motor = 19.8<br>Detergent pump motor = 6<br>Dryer fan motor = 9 | 50 circuit rating | *Yes* |
| Compressed air | Air pressure ≥ 80 psig | *115 psig* | *Yes* |
| Steam | Pressure 40 to 80 psig | *51 psig* | *Yes* |

Power supply source, breaker box *BB11, wire numbers: 35 and 1, 3, 5 respectively.*

**Table 6.55. Instruments Used**

| Test Instrument | Identification Number | Calibration Due Date |
|---|---|---|
| *Multimeter* | *ME-025* | *04/19/97* |
| *Air pressure gauge* | *P-102* | *02/10/97* |
| *Steam pressure gauge* | *P-102* | *02/10/97* |

***Volt calculation:***

Circulation motor volt specification = 115 V ± 10%

±10% of 115 = ±11.5

115 + 11.5 = 126.5

115 – 11.5 = 103.5

The measured volts of 113 falls within ±10%.

Detergent motor volt specification = 460 V ± 10%

±10% of 460 = ±46

460 + 46 = 482

460 – 46 = 438

The measured volts of 465/459/463 fall within ±10%.

***Amp calculation:***

Circuit rating = 50 A

Circulation pump motor = 19.8 A

Detergent pump motor = 6 A

Dryer fan motor = 9 A

Equipment current draw = (3) motors = 34.8 A

The circuit amp rating of 50 amps is greater than the maximum current draw of the equipment.

***Compressed air pressure calculation:***

Compressed air specification ≥ 80 psig

The measured compressed air pressure of 115 psig is greater than 80 psig.

***Steam pressure calculation:***

Steam pressure specification = 40 to 80 psig

The measured steam pressure of 51 psig falls within specification.

### *Major Component Specifications*

The component specifications section of the protocol is used to verify that the washer components purchased were delivered and installed. Record the major components in Table 6.56.

### *Component Material*

Record the material of each component that contacts the product in Table 6.57.

### *Lubricants*

Record the lubricants used to operate the washer in Table 6.58 and indicate if they make contact with the product.

### *Equipment Safety Features*

The objective of testing equipment safety features is to verify that the safety features on the tablet and capsule inserter washer function according to manufacturer specifications. The washer has a normally open drain valve that allows the wash tank to empty when the valve is deenergized after an alarm. Restarting the washer may cause the pump to run dry and become damaged. Therefore, it is recommended that all programs be aborted after an alarm and restarted from the beginning. Record the results in Table 6.59.

## Operational Qualification

An OQ evaluation should establish that the equipment can operate within specified tolerances and limits (Figure 6.16). The mechanical ranges of the washer being challenged, along with the basic washer operations that the operator will use. The washer will be validated for its operating ability not how well it washers slats.

**Table 6.56. Major Components**

| Components | As-Found Conditions |
|---|---|
| Washer cabinet | Manufacturer: *Any Washer Co.*<br>Material: *T-304 stainless steel*<br>Size: *45" width × 47" height × 48" deep* |
| Circulation pump motor | Manufacturer: *Any Motor Co.*<br>Model number: *ET-4CT*<br>Serial number: *F3720671U*<br>Volts: *460*<br>Amperes: *19.8*<br>Phases: *3*<br>Cycles: *60 Hz*<br>hp: *7.5*<br>rpm: *3,475* |
| Detergent pump motor | Manufacturer: *Any Motor Co.*<br>Model number: *0330367*<br>Serial number: *1R3631*<br>Volts: *115*<br>Amperes: *6*<br>Phases: *1*<br>Cycles: *60 Hz*<br>hp: *Not specified*<br>rpm: *Not specified* |
| Dryer fan motor | Manufacturer: *Any Motor Co.*<br>Model number: *52RG-57-VR*<br>Serial number: *Not specified*<br>Volts: *115*<br>Amperes: *9*<br>Phases: *1*<br>Cycles: *60 Hz*<br>hp: *1.5*<br>rpm: *Not specified* |

**Table 6.57. Component Material**

| Component | Material |
|---|---|
| *None of the components make contact with the product.* | |

**Table 6.58. Lubricants**

| Where Used | Type | Manufacturer | Product Contact (Yes/No) |
|---|---|---|---|
| *Motors* | *SAE-02 Oil* | *Any Oil Co.* | *No* |
| Is there a preventive maintenance procedure on file? (Yes/No) | | | *Yes* |

**Table 6.59. Safety Feature Test Results**

| Test Function | Expected Results | Acceptable (Yes/No) |
|---|---|---|
| Washer door open | When the washer door is opened, while a program is running, the alarm will sound and the message "door" is displayed. | *Yes* |
| | When the washer door is closed, the message "off4" is displayed. | *Yes* |
| Stop button pushed | When the stop button is pressed, the alarm sounds and the message "off1" is displayed. | *Yes* |

**Figure 6.16. OQ Elements of a Tablet and Capsule Inserter Washer**

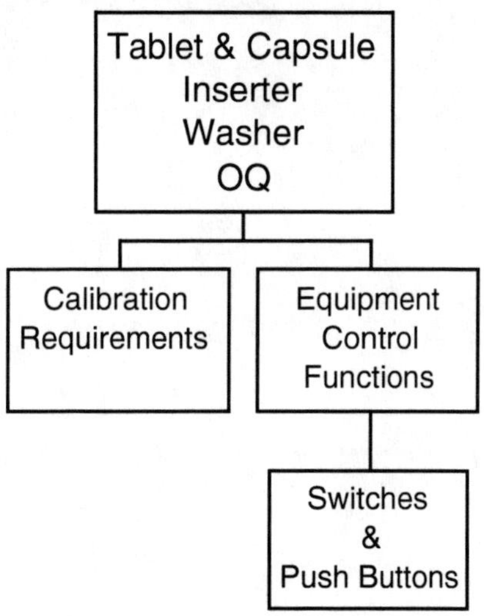


Black box software testing was performed. Black box testing evaluates the microprocessor system against the documented functional requirements or intended uses of the equipment. Information required for the OQ evaluation is calibration of the instruments used to control the washer, equipment control functions, (switches and push buttons).

### *Calibration Requirements*

Verify that all critical instruments on the equipment, have been logged into the calibration system, have calibration procedures in place, and are in calibration at the time of qualification testing. Record all information for the calibrated instruments used to control the tablet and capsule inserter washer in Table 6.60.

**Table 6.60. Calibrated and Noncalibrated Instruments**

| Calibrated Instrument | As-Found Conditions |
|---|---|
| Identification number | *AP/3083* |
| Type | *Water Pressure Gauge, Analog* |
| Manufacturer | *Any Gauge Co.* |
| Model number | *Not specified* |
| Serial number | *Not specified* |
| Range | *0–60 psig* |
| Scale division | *5 psig* |
| Location | *Mounted below the control panel* |
| Use | *Indicates the wash pump pressure* |
| Calibration due date | *04/17/97* |
| Critical or not critical | *Not critical to the operation of the washer* |
| Identification number | *AP/3028* |
| Type | *Steam Pressure Gauge, Analog* |
| Manufacturer | *Any Gauge Co.* |
| Model number | *Not specified* |
| Serial number | *Not specified* |
| Range | *0–100 psig* |
| Scale division | *1 psig* |
| Location | *Mounted on steam supply line at heat exchanger* |
| Use | *Monitors the steam pressure in the supply line* |
| Calibration due date | *04/17/97* |
| Critical or not critical | *Not critical to the operation of the washer* |
| Identification number | *ATD/3082* |
| Type | *Temperature Gauge, Digital* |
| Manufacturer | *Any Gauge Co.* |
| Model number | *Not specified* |
| Serial number | *Not specified* |
| Range | *0–999°F* |
| Scale division | *2°F* |
| Location | *Mounted on the control panel* |
| Use | *Indicates wash cycle water temperature* |
| Calibration due date | *04/17/97* |
| Critical or not critical | *Critical to the operation of the washer* |

*Continued on next page.*

*Continued from previous page.*

| Calibrated Instrument | As-Found Conditions |
|---|---|
| Identification number | *AIT/3082* |
| Type | *Timer, digital* |
| Manufacturer | *Any Gauge Co.* |
| Model number | *Not specified* |
| Serial number | *Not specified* |
| Range | *00:00–99:99 (hours: minutes)* |
| Scale division | *1 minute* |
| Location | *Mounted on the control panel* |
| Use | *Indicates and controls the cycle time* |
| Calibration due date | *04/17/97* |
| Critical or not critical | *Critical to the operation of the washer* |
| Identification number | *AIT/3082* |
| Type | *Temperature gauge, digital* |
| Manufacturer | *Any Gauge Co.* |
| Model number | *B6023RC* |
| Serial number | *Not specified* |
| Range | *50–300°F* |
| Scale division | *2°F* |
| Location | *Mounted on the heat exchanger water outlet line* |
| Use | *Indicates water temperature out from the HX* |
| Calibration due date | *04/17/97* |
| Critical or not critical | *Not critical to the operation of the washer* |
| Identification number | *Not specified* |
| Type | *Air pressure gauge, analog* |
| Manufacturer | *Any Gauge Co.* |
| Model number | *Not specified* |
| Serial number | *Not specified* |
| Range | *0–160 psig* |
| Scale division | *2 psig* |
| Location | *Mounted on the control panel* |
| Use | *Indicates regulated air pressure* |
| Calibration due date | *Calibration not required* |
| Critical or not critical | *Not critical to the operation of the washer* |

### *Equipment Control Functions*

The objective of the equipment control functions is to verify that the switches and push buttons and alarms on the tablet and capsule inserter washer operate according to manufacturer's specifications. Record the results in Table 6.61, 6.62, and 6.63. Operate each control listed in Table 6.61, 6.62, and 6.63 and verify its proper operation.

**Table 6.61. Equipment Control Function Test Results**

| Test Function | Expected Results | Acceptable (Yes/No) |
|---|---|---|
| Control on/off switch operation | When the control on/off switch is pressed, power is supplied to the control system. | *Yes* |
| Start/reset push button operation | When the start push button is pressed, with the control in standby mode the machine starts operating. | *Yes* |
| Stop push button operation | When the stop push button is pressed, the machine stops operating. | *Yes* |
| Manual drain button operation | When the manual drain button is pressed, the machine drains. | *Yes* |
| Program selector dial operation | When the program selector dial is set to "1," the equipment is programmed for one minute. | *Yes* |
| Wash time selector dial operation | When the wash, time selector dial is set to 14, the machine washes for 14 min. | *Yes* |
| Rinse time selector dial operation | When the rinse time selector dial is set to 10, the machine rinses for 10 min. | *Yes* |
| Prewash on/off switch operation | When the prewash switch is "on," the prewash is performed. | *Yes* |
| | When prewash switch is "off," the prewash step is bypassed. | *Yes* |
| Final rinse time selector dial operation | When the final rinse time selector dial is set to 1, the machine rinses for 1 min. | *Yes* |
| Hold on/off switch operation | When the hold on/off switch is "on," the machine stops. | *Yes* |
| | When the hold on/off switch is "off," the machine does not hold. | *Yes* |

**Table 6.62. *Internal Switches***

| Switch Number | Switch Position | Switch Setting | Acceptable (Yes/No) |
|---|---|---|---|
| 1 Prewash time | *D* | *40 sec* | *Yes* |
| 2 Drain time | *A* | *10 sec* | *Yes* |
| 3 Prewash & wash temperature | *0* | *130°F* | *Yes* |
| 4 Air blow-off | *A* | *30 sec* | *Yes* |
| 5 Dryer time | *9* | *45 min* | *Yes* |
| 6 Dryer temperature | *7* | *80°F* | *Yes* |
| 7 Detergent time | *B* | *10 sec* | *Yes* |
| 8 Not used | *N/A* | — | *Yes* |
| 9 Heat mode | *7* | *Hot prewash* | *Yes* |
| 10 Dryer time | *D* | *45 min* | *Yes* |

**Table 6.63. *Indicator LED Functionality***

| Indicator LED | Description | LED Illuminated | Acceptable (Yes/No) |
|---|---|---|---|
| Ready | Unit powered on and in standby mode | *Yes* | *Yes* |
| Slat cycle | Slat wash program operating | *Yes* | *Yes* |
| Standard cycle | Standard cycle program operating | *Yes* | *Yes* |
| Prewash | Pre-wash phase of cycle in operation | *Yes* | *Yes* |
| Wash | Wash phase of cycle in operation | *Yes* | *Yes* |
| 1st rinse | 1st rinse phase of cycle in operation | *Yes* | *Yes* |
| 2nd rinse | 2nd rinse phase of cycle in operation | *Yes* | *Yes* |
| DI rinse | DI rinse phase of cycle in operation | *Yes* | *Yes* |
| Pause | Pauses machine operation | *Yes* | *Yes* |
| Dryer | Dryer phase of cycle in operation | *Yes* | *Yes* |
| Cycle complete | Illuminates at the end of the cycle | *Yes* | *Yes* |
| Circ. pump | Circulation pump operating | *Yes* | *Yes* |
| Header | Header portion of washer enabled | *Yes* | *Yes* |
| Drain | Drain phase of the cycle in operation | *Yes* | *Yes* |
| Fill | Fill phase of the cycle in operation | *Yes* | *Yes* |
| Heat | Heating phase of cycle in operation | *Yes* | *Yes* |
| Slat drive | Slat carousel in operation | *Yes* | *Yes* |
| Dryer heat | Dryer heat phase of cycle in operation | *Yes* | *Yes* |
| Detergent | Detergent pump in operation | *Yes* | *Yes* |
| DI valve | DI valve open | *Yes* | *Yes* |
| Dryer fan | Dryer fan operating | *Yes* | *Yes* |
| Air blow off | Air blow off phase of the cycle in operation | *Yes* | *Yes* |
| System on | Indicates power output from the control system power supply | *Yes* | *Yes* |

## Performance Qualification

Once it has been established that the equipment is properly installed and functioning within specified operating parameters, it must be shown that the tablet and capsule inserter washer can operate reliably under routine, minimum, and maximum operating conditions (Figure 6.17).

### *Timer Test*

The objective of the timer test is to verify that the rinse timer operates according to manufacturers' specifications. A stopwatch is required. Operate the washer with the wash and rinse timers set at normal operation. Measure the times of the wash and rinse cycles and verify that the measured times are within ±10 percent of the set-point times. Record the results in Table 6.64. Record the instrument used to measure the time in Table 6.65.

***Figure 6.17. PQ Elements of a Tablet and Capsule Inserter Washer***

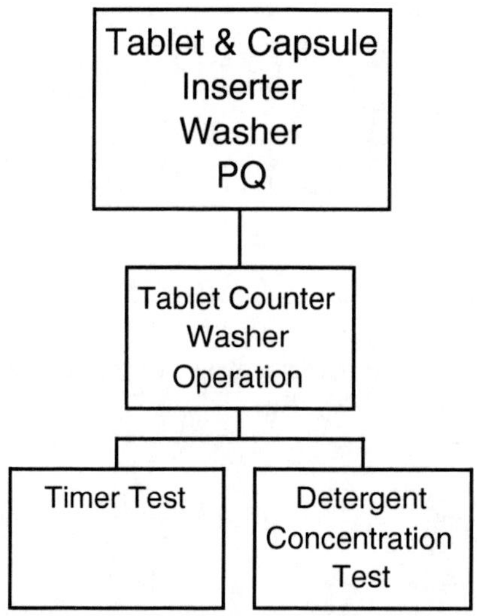


***Table 6.64. Timer Test***

| Washer Phase | Dial Setting (mm/ss) | Measured Results (mm/ss) | Acceptable (Yes/No) |
|---|---|---|---|
| Wash | *15:10* | *15:10* | *Yes* |
| Rinse | *10:00* | *10:00* | *Yes* |

***Table 6.65. Instrument Used***

| Test Instrument | Identification Number | Calibration Due Date |
|---|---|---|
| *Stopwatch* | *54039* | *02/16/97* |

### *Detergent Concentration Test*

The objective of the detergent concentration test is to document the concentration of the detergent used in the washer. This test is for informational purposes only. A stopwatch is required. The test is performed as follows:

- Remove the hose connected to the discharge side of the detergent pump. Place a container under the discharge port to collect detergent. Operate the pump by holding the prime button to collect detergent then measure the time with a calibrated stopwatch. Record the results in Table 6.66 and the instrument used to measure the time in Table 6.69.

  Detergent Pump Capacity Calculation:

  1140 mL ÷ 1,000 = 1.14 L/min

- Start the washer using the same cycle used in normal operation. Use a calibrated stopwatch to measure the time that the detergent pump operates during the wash phase. (Perform this operation three times.) Calculate the volume pumped and record the results in Table 6.67. Record the instrument used to measure the time in Table 6.69.

  Volume of Detergent Pumped Calculation:

  1.14 L/min ÷ 60 = 0.014 L/sec.

  57 sec(0.014) = 0.798 L

  55 sec(0.014) = 0.770 L

  55 sec(0.014) = 0.770 L

**Table 6.66. Pumping Capacity**

| Test Function | Expected Results | Acceptable (Yes/No) |
|---|---|---|
| Detergent pump time | *1 min* | *Yes* |
| Volume pumped | *1,140 mL* | *Yes* |
| Pumping Capacity in L/min. | *1.14 L/min* | *Yes* |

**Table 6.67. Volume Pumped**

| Test Function | Expected Results | Acceptable (Yes/No) |
|---|---|---|
| Operation 1, time pumped | *57 sec* | *Yes* |
| Calculated volume pumped | *0.798 L* | *Yes* |
| Operation 2, time pumped | *55 sec* | *Yes* |
| Calculated volume pumped | *0.770 L* | *Yes* |
| Operation 3, time pumped | *55 sec* | *Yes* |
| Calculated volume pumped | *0.770 L* | *Yes* |

**Table 6.68. *Volume Pumped***

| Test Function | Results ($mL/L\ H_2O$) | Acceptable (Yes/No) |
|---|---|---|
| Sample #1 wash water concentration | *0.0095* | *Yes* |
| Sample #2 wash water concentration | *0.0125* | *Yes* |
| Sample #3 wash water concentration | *0.0087* | *Yes* |
| Average results | *0.0101* | *Yes* |

**Table 6.69. *Instrument Used***

| Test Instrument | Identification Number | Calibration Due Date |
|---|---|---|
| *Stopwatch* | *54039* | *02/16/97* |

- At the end of the wash cycle, stop the washer and open the door. Collect a sample of the wash water and submit it to the lab for testing the concentration of potassium hydroxide. (Do this operation at three separate times.) Record the results in Table 6.68. The data was obtained from the lab.

## COTTON INSERTER

The validation of a cotton inserter (Figure 6.18) will be described here. The cotton inserter automatically stuffs strips of cotton into bottles at a speed of 100 strips per minute. The strips are fed by two pairs of rollers. One pair rotates continuously, while the other stops periodically to shear the cotton into predetermined lengths. The double ram and tube assembly stuffs cotton into the tube with the first push, rotates 180° and pushes the cotton into the bottle with the second push.

The validation protocol for a cotton inserter is described in the following sections. This protocol example can be used as a cGMP compliance guide and it should be tailored to meet individual company requirements. Italicized words in the tables represent examples of information that could be recorded in a validation protocol for a cotton inserter.

### Installation Qualification

An IQ evaluation establishes confidence that the equipment is properly installed (Figure 6.19). The installation must meet the manufacturer's specified guidelines, along with design changes at installation. Also, the supporting electrical utilities must meet all electrical codes. The information required for an IQ evaluation is equipment identification, required documentation, equipment utility requirements, major component specifications, component material, lubricants, and equipment safety features.

***Figure 6.18. King Cottoner (courtesy of Modular King Packaging Systems, Inc.)***

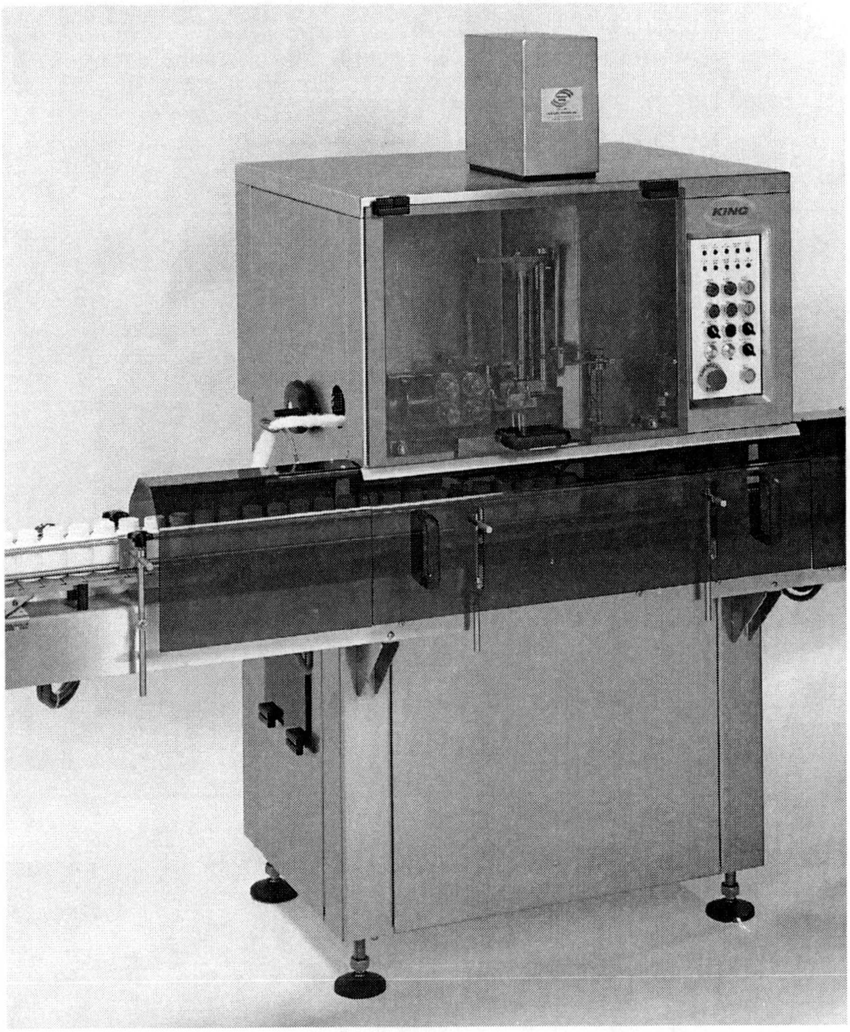


***Figure 6.19. IQ Elements of a Cotton Inserter***

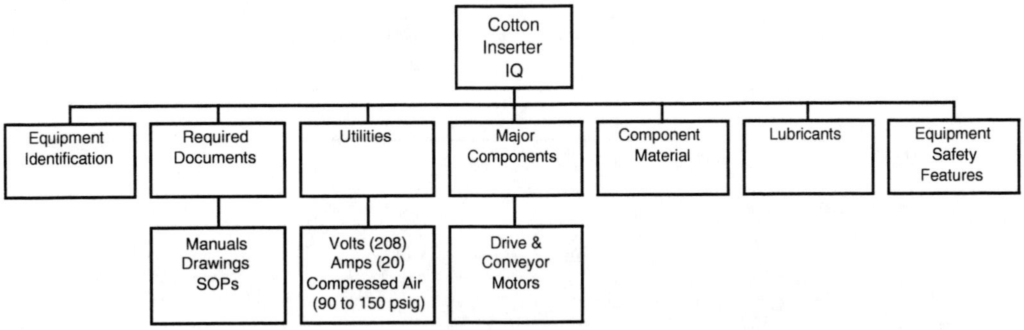

### Equipment Identification

Record the equipment identification numbers in Table 6.70, along with the following information: equipment manufacturer's purchase order number, model number, serial number, company assigned equipment number, and the location of the equipment.

### Required Documentation

Record the equipment manufacturer's operation and maintenance manual and drawings in Table 6.71. Record the standard operating procedures that cover the setup, operation, and cleaning of the cotton inserter in Table 6.72.

### Equipment Utility Requirements

Compare the manufacturer's specified volts (V), amps (A), and compressed air requirements to their as-found condition at the time of qualification testing and record the results in Table 6.73. Also record the location of the power supply source. Record the instruments used to measure the volts, amps and compressed air in Table 6.74.

**Table 6.70. Equipment Identification**

| Required Information | As-Found Conditions |
|---|---|
| Manufacturer | *Any Cotton Inserter Co.* |
| Purchase order number | *326787* |
| Model number | *LR001* |
| Serial number | *983* |
| Equipment number | *0601* |
| Location | *Packaging line 2* |

**Table 6.71. Manufacturers' Manuals and Drawings**

| Number | Description | Date |
|---|---|---|
| *LR001* | *Cotton Inserter Operators Manual* | *None* |

**Table 6.72. Standard Operating Procedures**

| Number | Description | Release Date |
|---|---|---|
| *PKG035* | *Cotton Inserter Setup and Operation* | *09/22/94* |
| *PKG036* | *Cotton Inserter Cleaning Procedure* | *11/15/94* |

**Table 6.73.** ***Utilities***

| Utility | Specified | Measured Results | Acceptable (Yes/No) |
|---|---|---|---|
| Volts | 208 ± 10% | *A-B 207; A-C 208; B-C 207* | *Yes* |
| Amps | Drive motor = 3<br>Conveyor motor = 1.9 | 20 circuit rating | *Yes* |
| Compressed air | Air pressure 90 to 150 psig | *135 psig* | *Yes* |
| Power supply source, breaker box *BB7A, wire numbers: 15, 17, 19* | | | |

**Table 6.74.** ***Instruments Used***

| Test Instrument | Identification Number | Calibration Due Date |
|---|---|---|
| *Multimeter* | *ME-025* | *04/19/97* |
| *Air pressure gauge* | *P-102* | *02/10/97* |

***Volt calculation:***

Cotton inserter volt specification = 208 V ± 10%

±10% of 208 = ±20.8

208 + 20.8 = 228.8

208 – 20.8 = 187.2

The measured volts of 207/208/207 fall within ± 10%.

***Amp calculation:***

Amp specification = 20 A ±10%

Drive motor = 3 A

Conveyor motor = 1.9 A

Equipment current draw = drive motor A + conveyor motor A = 3 + 1.9 = 4.9 A

The circuit amp rating of 20 is greater than the maximum current draw of the equipment.

***Compressed air calculation:***

Compressed air specification = 90 to 150 psig

The measured air pressure of 135 psig is within specification.

### *Major Component Specifications*

The component specifications section of the protocol is used to verify that the cotton inserter components purchased were delivered and installed. Record the major components in Table 6.75.

### *Component Material*

Record the material of each component that contacts the product in Table 6.76.

### *Lubricants*

Record the lubricants used to operate the cotton inserter in Table 6.77 and indicate if they make contact with the product.

### *Equipment Safety Features*

The objective of testing equipment safety features is to verify that the safety features on the cotton inserter function according to manufacturer specifications. Verify the operation of the equipment safety features in Table 6.78 and record the results.

**Table 6.75. Major Components**

| Components | As-Found Conditions |
|---|---|
| Drive motor | Manufacturer: *Any Motor Co.*<br>Model number: *Not specified*<br>Serial number: *99740E*<br>Volts: *208*<br>Amperes: *3*<br>Phases: *3*<br>Cycles: *60 Hz*<br>hp: *1/2*<br>rpm: *2,000* |
| Conveyor motor | Manufacturer: *Any Motor Co.*<br>Model number: *9974E*<br>Serial number: *1/957781CS*<br>Volts: *208*<br>Amperes: *1.9*<br>Phases: *3*<br>Cycles: *60 Hz*<br>hp: *1/4*<br>rpm: *1,400* |

**Table 6.76. Component Material**

| Component | Material |
|---|---|
| Insert strip | *Cotton* |

**Table 6.77. Lubricants**

| Where Used | Type | Manufacturer | Product Contact (Yes/No) |
|---|---|---|---|
| *Gearbox* | *Vitrea oil* | *Any Oil Co.* | *No* |
| *Chain drives* | *Mytilus grease* | *Any Oil Col* | *No* |
| *Bearings* | *Alvania grease* | *Any Oil Co.* | *No* |
| Is there a preventive maintenance procedure on file? (Yes/No) | | | *Yes* |

**Table 6.78. Safety Features Test Results**

| Test Function | Expected Results | Acceptable (Yes/No) |
|---|---|---|
| No-bottle micro switch | When there is a break in the bottle supply, the bottle micro switch stops the equipment. | *Yes* |
| No-cotton micro switch | When there is a break in the supply of cotton, the no-cotton micro switch stops the equipment. | *Yes* |
| Ram position switch | The ram position switch is operated when the head has stopped to ensure that the ram is clear of the container. | *Yes* |
| Safety cover door micro switch | When the safety cover door is opened, the equipment stops. | *Yes* |

## Operational Qualification

An OQ evaluation should establish that the equipment can operate within specified tolerances and limits (Figure 6.20). The mechanical ranges of the cotton inserter are challenged, along with the basic equipment operations. Information required for the OQ evaluation is calibration of the instruments used to control the cotton inserter, equipment control functions (switches and push buttons) and equipment operation (drive and conveyor motor rotation direction, and speed test).

### *Calibration Requirements*

Verify that all critical instruments on the equipment have been logged into the calibration system, have calibration procedures in place, and are in calibration at the time of qualification testing. Record all information for the calibrated instruments used to control the cotton inserter in Table 6.79.

### *Equipment Control Functions*

The objective of testing equipment functions is to verify that the switches and push buttons on the cotton inserter operate per manufacturer's specifications. Record the results in Table 6.80. The tests will be performed with the cotton inserter empty. Operate each control listed in Table 6.80 and verify its proper operation.

**Figure 6.20. OQ Elements of a Cotton Inserter**

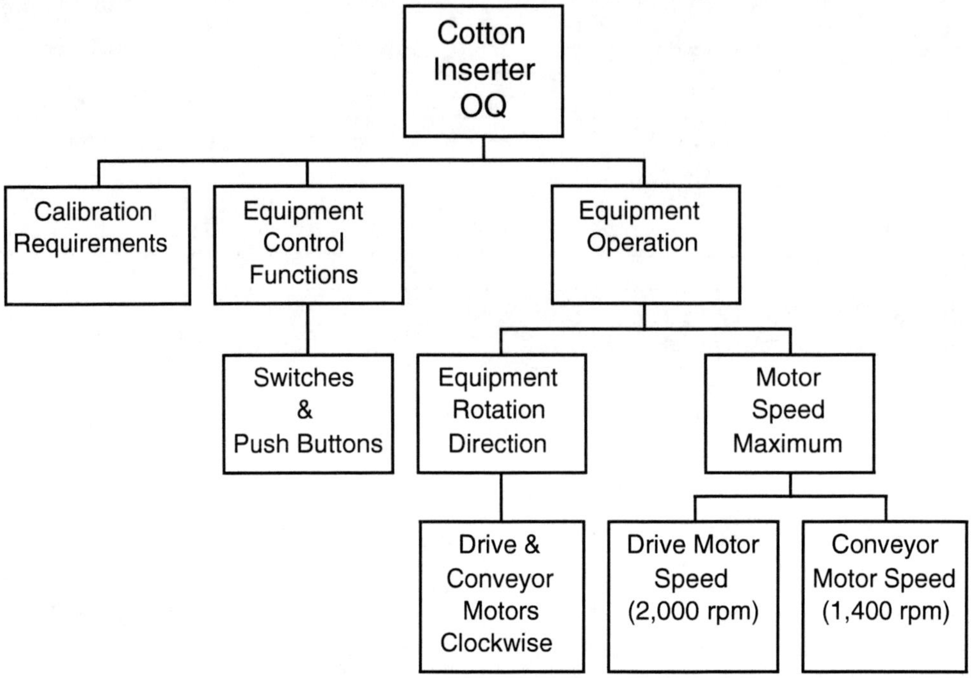


**Table 6.79. Calibrated and Noncalibrated Instruments**

*There were no calibrated instruments on the cotton inserter.*

| Noncalibrated Instrument | As-Found Conditions |
|---|---|
| Identification number | *Slat/100 GP* |
| Type | *Air pressure gauge, analog* |
| Manufacturer | *Any Gauge Co.* |
| Model number | *Not specified* |
| Serial number | *Not specified* |
| Range | *0–150 psig* |
| Scale division | *10 psig* |
| Location | *On the inlet line* |
| Use | *Records the air pressure to the equipment* |
| Calibration due date | *Not required* |
| Critical or not critical | *Not critical to the operation of the cotton inserter* |

**Table 6.80. Equipment Control Function Test Results**

| Test Function | Expected Results | Acceptable (Yes/No) |
|---|---|---|
| Power switch | When the power switch is turned, power is applied to and from the machine. | *Yes* |
| Head motor on push button | When the head motor on push button is pressed, the head motor starts and stops. | *Yes* |
| Emergency stop push button | When the emergency stop push button is pressed, the equipment stops. | *Yes* |
| Conveyor on/off push button | When conveyor on/off push button is pressed, the conveyor starts and stops. | *Yes* |
| Power lift up/down push button | When the power lift UP push button is pressed, the head raises and lowers. | *Yes* |
| Auto/hand switch | When the auto/hand switch is toggled, the equipment is switched from automatic to manual. | *Yes* |
| Jog push button | When the job push button is pressed and held, the machine can be moved manually. | *Yes* |
| Head motor control | When the head motor control is rotated, the speed of the head motor is increased and decreased. | *Yes* |
| Track vibrator | When the track vibrator control is rotated, the speed of the vibrator is increased and decreased. | *Yes* |
| Shear length crank | When the shear length crank is adjusted, the length of the cotton is varied. | *Yes* |

### *Equipment Operation: Drive and Conveyor Motor Rotation Direction*

The objective of the rotation direction test is to verify that the drive and conveyor motors rotate in the proper direction. The following tests are performed with the cotton inserter empty.

- Press the start push button and observe the direction of rotation of the drive and the conveyor motor as viewed from behind the motors, and record the results in Table 6.81.

### *Equipment Operation: Drive and Conveyor Motor Speed Test*

The objective of the speed test is to document the speed of the drive and conveyor motor with the cotton inserter empty. A tachometer will be required for the following tests:

- Measure the speed of the drive motor with a calibrated tachometer. Verify that the measured speed is within ±10 percent of the maximum speed of 2,000 rpm. Record the results in Table 6.82. Record the instrument used to measure the speed in Table 6.83.

**Table 6.81. *Drive and Conveyor Motor Rotation Direction Test Results***

| Item | Expected Results | Results | Acceptable (Yes/No) |
|---|---|---|---|
| Drive motor rotation direction | Rotation should be clockwise as viewed from behind the motor. | *Clockwise rotation was observed.* | *Yes* |
| Conveyor motor rotation direction | Rotation should be clockwise as viewed from behind the motor. | *Clockwise rotation was observed.* | *Yes* |

Drive motor speed specification = 2,000 rpm ± 10%

±10% of 2,000 = ±200

2,000 + 200 = 2,200

2,000 – 200 = 1,800

The measured rpm of 2,009 falls within ± 10%.

- Measure the speed of the conveyor motor with a calibrated tachometer. Verify that the measured speed is within ±10 percent of the maximum speed of 1,400 rpm. Record the results in Table 6.82. Record the instrument used to measure the speed in Table 6.83.

Drive motor speed specification = 1,400 rpm ± 10% maximum

±10% of 1,400 = ±140

1,400 + 140 = 1,540

1,400 – 140 = 1,260

The measured rpm of 1,403 falls within ±10%.

**Table 6.82. *Drive and Conveyor Motor Speed Test Results***

| Motor | Speed Specification (rpm ± 10%) | Measured Speed (rpm) | Acceptable (Yes/No) |
|---|---|---|---|
| Drive | 2,000 | *2,009* | *Yes* |
| Conveyor | 1,400 | *1,403* | *Yes* |

**Table 6.83. *Instrument Used***

| Test Instrument | Identification Number | Calibration Due Date |
|---|---|---|
| *Tachometer* | *64020* | *06/21/97* |

## Performance Qualification

Once it has been established that the equipment is properly installed and functioning within specified operating parameters, it must be shown that the cotton inserter can operate reliably under routine, minimum, and maximum operating conditions (Figure 6.21).

### *Cotton Inserter Operation*

The cotton inserter operation test is designed to verify that the cotton inserter is capable of inserting cotton into bottles consistently over time. Also the speed of the motors will be measured and recorded. 60 cc bottles, 1,300 cc bottles, cotton, and a tachometer will be required for the following tests:

- With the equipment operating, visually observe 100 60 cc bottles as they pass through the cotton inserter and verify that 95 percent of the bottles have cotton in them. Record the results in Table 6.84.
- With the equipment operating, visually observe 100 1,300 cc bottles as they pass through the cotton inserter and verify that 95 percent of the bottles have cotton in them. Record the results in Table 6.84.
- Measure the speed of the drive motor with a calibrated tachometer. Verify that the measured speed is within ±10 percent of the maximum speed of 2,000 rpm. Record the results in Table 6.85. Record the instrument used to measure the speed in Table 6.86.

**Figure 6.21. *PQ Elements of a Cotton Inserter***

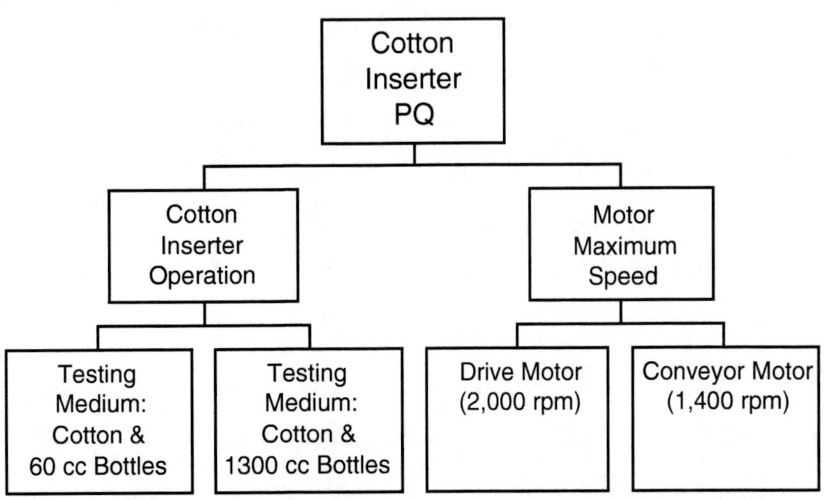

**Table 6.84. *Cotton Inserter Performance Test Results***

| Bottle Quantity | Bottle Size (cc) | # Cotton Inserted | # Cotton Not Inserted | Acceptable (Yes/No) |
|---|---|---|---|---|
| 100 | *60* | *99* | *1* | *Yes* |
| 100 | *1,300* | *98* | *2* | *Yes* |

**Table 6.85. *Drive and Conveyor Motor Speed Test Results***

| Motor | Speed Specification (rpm ± 10%) | Measured Speed (rpm) | Acceptable (Yes/No) |
|---|---|---|---|
| Drive | 2,000 | *1,987* | *Yes* |
| Conveyor | 1,400 | *1,367* | *Yes* |

**Table 6.86. *Instrument Used***

| Test Instrument | Identification Number | Calibration Due Date |
|---|---|---|
| *Tachometer* | *64020* | *06/21/97* |

Drive motor speed specification = 2,000 rpm ± 10%

±10% of 2,000 = ±200

2,000 + 200 = 2,200

2,000 – 200 = 1,800

The measured rpm of 1,987 falls within ±10%.

- Measure the speed of the conveyor motor with a calibrated tachometer. Verify that the measured speed is within ±10 percent of the maximum speed of 1,400 rpm. Record the results in Table 6.85. Record the instrument used to measure the speed in Table 6.86.

Conveyor motor speed specification = 1,400 rpm ± 10% maximum

±10% of 1,400 = ±140

1,400 + 140 = 1,540

1,400 – 140 = 1,260

The measured rpm of 1,367 falls within ±10%.

## Bottle Capper

The validation of a bottle capper (Figure 6.22) will be described here. The bottle capper accepts filled bottles from the conveyer line and seals them with threaded caps. The capper incorporates a cap feeder, photoelectric cap scanner, conveyer systems for the caps and bottles, motors to power the gripper belts and spindles, speed controllers for the gripper belts and spindles. The gripper belts move the bottles along the conveyor and caps emerge from the cap feeder bin and move down a chute onto the bottle. The spindles tighten the caps onto the bottles. The torque monitor is used to observe and monitor the torque at which the bottle capper applies caps to bottles. The monitor displays a torque measurement when a bottle passes the right rear spindle.

The validation protocol for a bottle capper is described in the following sections. This protocol example can be used as a cGMP compliance guide, and it should be tailored to meet individual company requirements. Italicized words in the tables represent examples of information that could be recorded in a validation protocol for a bottle capper.

***Figure 6.22. Kaps-All Capper (courtesy of Kaps-All Packaging Systems)***

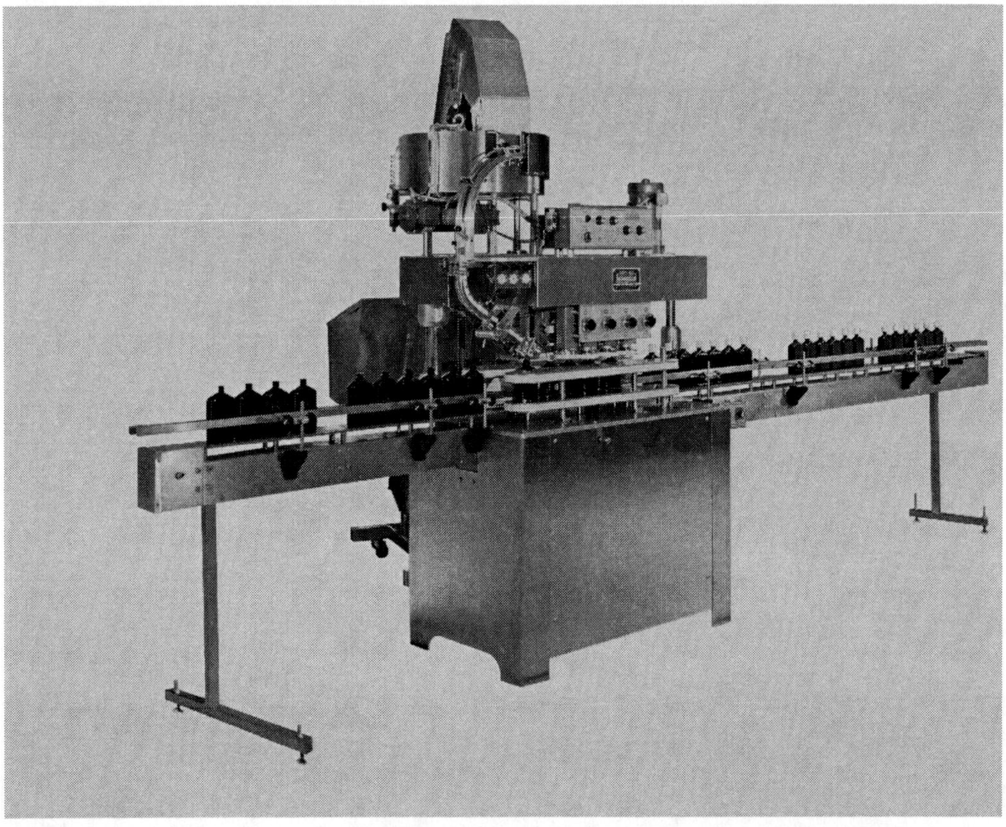

## Installation Qualification

An IQ evaluation establishes confidence that the equipment is properly installed (Figure 6.23). The installation must meet the manufacturer's specified guidelines, along with design changes at installation. Also, the supporting electrical utilities must meet all electrical codes. The information required for an IQ evaluation is equipment identification, required documentation, equipment utility requirements, major component specifications, component material, lubricants, and equipment safety features.

### *Equipment Identification*

Record the equipment identification numbers in Table 6.87, along with the following information: equipment manufacturer's purchase order number, model number, serial number, company assigned equipment number, and the location of the equipment.

### *Required Documentation*

Record the equipment manufacturer's operation and maintenance manuals and drawings in Table 6.88. Record the SOPs that cover the setup, operation, and cleaning of the bottle capper in Table 6.89.

### *Equipment Utility Requirements*

Compare the manufacturer's specified volts (V), amps (A), and compressed air requirements to their as-found conditions at the time of qualification testing and record the results in Table 6.90. Also record the location of the power supply source. Record the instruments used to measure the volts, amps, and compressed air in Table 6.91.

***Figure 6.23. IQ Elements of a Bottle Capper***

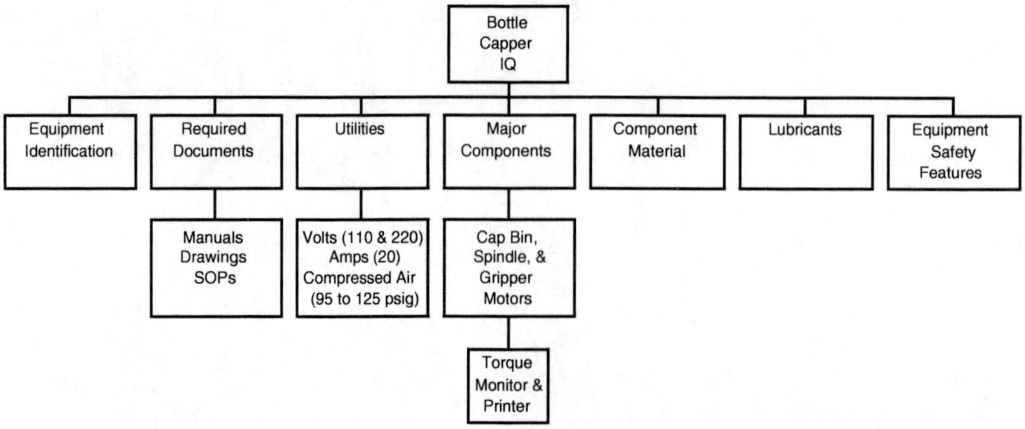

**Table 6.87. *Equipment Identification***

| Required Information | As-Found Conditions |
| --- | --- |
| Manufacturer | *Any Bottle Capper Co.* |
| Purchase order number | *551294* |
| Model number | *D* |
| Serial number | *8462* |
| Equipment number | *3380* |
| Location | *Packaging line 3* |

**Table 6.88. *Manufacturers' Manuals and Drawings***

| Number | Description | Date |
| --- | --- | --- |
| *None* | *Bottle Capper Installation, Operation, and Maintenance Manual* | *04/14/94* |
| *None* | *The Electronic Torque Monitor Manual* | *None* |
| *E514* | *Model D with RF-24 220V, 3hp, 60Hz Drawing* | *08/22/88* |
| *13-10* | *Model D Cap Tightening & Conveyor Assembly—New Grippers Drawing* | *03/87* |
| *13-4* | *Lower Cap Chute & Stabilizer Drawing* | *01/24/86* |
| *13-1* | *Adhesive Scale Locations Drawing* | *04/09/87* |
| *1-10* | *Spindle Drive Belt Installation Model Drawing* | *04/16/87* |

**Table 6.89. *Standard Operating Procedures***

| Number | Description | Release Date |
| --- | --- | --- |
| *PKG074* | *Capper/Elevator Setup, Operation, and Cleaning* | *12/20/96* |
| *PKG075* | *Torque Monitor Calibration Procedure* | *07/07/95* |

**Table 6.90. *Utilities***

| Utility | Specified | Measured Results | Acceptable (Yes/No) |
| --- | --- | --- | --- |
| Volts | 110 ± 10%<br>220 ± 10% | *111*<br>*A-B 220; A-C 219; B-C 223* | *Yes* |
| Amps | Cap bin motor = 1.7<br>Spindle motor = 2.5<br>Gripper motor = 3.7 | 20 circuit rating | *Yes* |
| Compressed air | Air pressure 95 to 125 psig | *119 psig* | *Yes* |
| Power supply source, breaker box *BB39, wire numbers: 1 and 15, 17, 19 respectively* | | | |

**Table 6.91. Instrument Used**

| Test Instrument | Identification Number | Calibration Due Date |
|---|---|---|
| *Multimeter* | *ME-025* | *04/19/97* |
| *Air pressure gauge* | *P-102* | *02/10/97* |

***Volt calculation:***

Torque monitor volt specification = 110 V ± 10%

±10% of 110 = ±11

110 + 11 = 121

110 – 11 = 109

The measured volts of 111 falls within ± 10%.

Bottle capper volt specification = 220 V ± 10%

±10% of 220 = ±22

220 + 22 = 242

220 – 22 = 198

The measured volts of 220/219/223 fall within ±10%.

***Amp calculation:***

Amp specification = 20 A

Cap bin feeder motor = 1.7 A

Spindle motor = 2.5 A

Gripper motor = 3.7 A

Equipment current draw = cap bin feeder motor A + spindle motor A + gripper motor A = 1.7 + 2.5 + 3.7 = 7.9 A

The circuit amp rating of 20 is greater than the maximum current draw of the equipment.

***Compressed air calculation:***

Compressed air specification = 95 to 125 psig

The measured air pressure of 119 psig is within specification.

### *Major Component Specifications*

The component specification section of the protocol verifies that the bottle capper components purchased were delivered and installed. Record the major components in Table 6.92.

**Table 6.92. *Major Components***

| Components | As-Found Conditions |
|---|---|
| Cap feeder bin motor | Manufacturer: *Any Motor Co.*<br>Model number: *6143 PDG*<br>Serial number: *7821W*<br>Volts: *220*<br>Amperes: *1.7*<br>Phases: *3*<br>Cycles: *60 Hz*<br>hp: *1/3*<br>rpm: *1,750* |
| Spindle motor | Manufacturer: *Any Motor Co.*<br>Model number: *6243 PDC*<br>Serial number: *88 9W*<br>Volts: *220*<br>Amperes: *2.5*<br>Phases: *3*<br>Cycles: *60 Hz*<br>hp: *1/2*<br>rpm: *1,750* |
| Gripper motor | Manufacturer: *Any Motor Co.*<br>Model number: *6343 PDC*<br>Serial number: *98 9W*<br>Volts: *220*<br>Amperes: *3.7*<br>Phases: *3*<br>Cycles: *60 Hz*<br>hp: *3/4*<br>rpm: *1,750* |
| Torque monitor | Manufacturer: *Any Torque Monitor Co.*<br>Model number: *088-ME*<br>Serial number: *9298*<br>Operating range: *0-100 KAUTS* |
| Printer | Manufacturer: *Any Printer Co.*<br>Model number: *42-PMI*<br>Serial number: *999022*<br>Operating range: *0–100 KAUTS* |

### *Component Material*

Record the material of each component that contacts the product in Table 6.93.

### *Lubricants*

Record the lubricants used to operate the bottle capper in Table 6.94 and indicate if they make contact with the product.

**Table 6.93. Component Material**

| Component | Material |
|---|---|
| *None of the components make contact with the product.* | |

**Table 6.94. Lubricants**

| Where Used | Type | Manufacturer | Product Contact (Yes/No) |
|---|---|---|---|
| *Chains* | *Mobilgear Oil* | *Any Oil Co.* | *No* |
| *Elevating Screws* | *Lubriplate 630-2 Grease* | *Any Oil Co.* | *No* |
| Is there a preventive maintenance procedure on file? (Yes/No) | | | *Yes* |

### *Equipment Safety Features*

There are no safety features on this equipment.

## Operational Qualification

An OQ evaluation should establish that the equipment can operate within specified tolerances and limits (Figure 6.24). The mechanical ranges of the bottle capper are challenged, along with the basic equipment operations. Information required for the OQ evaluation is calibration of the instruments used to control the bottle capper, equipment control functions (switches and push buttons), and equipment operation (cap bin, spindle, and gripper motor rotation direction and speed tests).

**Figure 6.24. OQ Elements of a Bottle Capper**

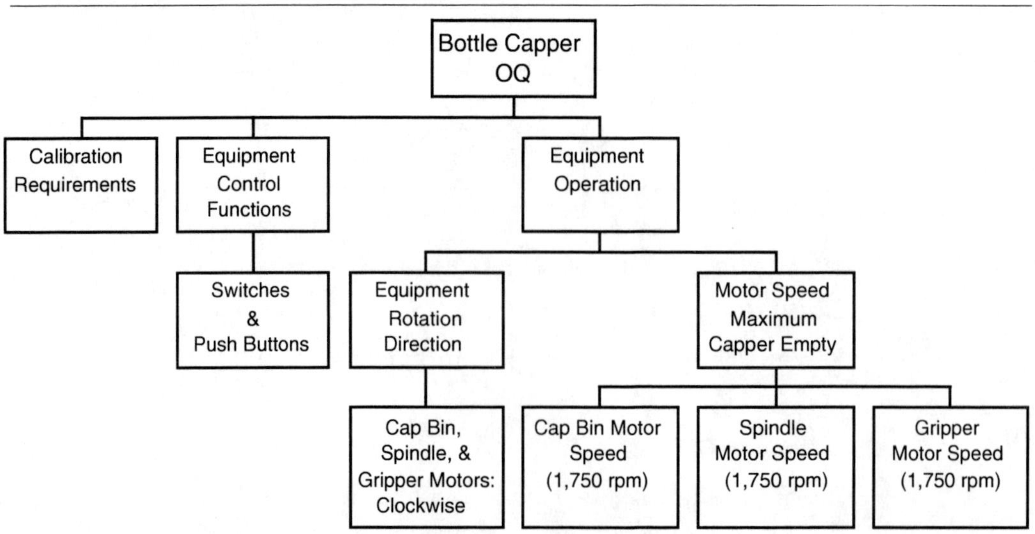

### *Calibration Requirements*

Verify that all critical instruments on the equipment have been logged into the calibration system, have calibration procedures in place, and are in calibration at the time of qualification testing. Record all information for the calibrated instruments used to control the bottle capper in Table 6.95.

### *Equipment Control Functions*

The objective of equipment control functions is to verify that the switches and push buttons on the bottle capper operate per manufacturer's specifications. Record the results in Table 6.96. The tests will be performed with the bottle capper empty. Operate each control listed in Table 6.96 and verify its proper operation.

***Table 6.95. Calibrated and Noncalibrated Instruments***

| Calibrated Instrument | As-Found Conditions |
|---|---|
| Identification number | *GP-00-130* |
| Type | *Air pressure gauge, analog* |
| Manufacturer | *Any Gauge Co.* |
| Model number | *80-595* |
| Serial number | *850-GP* |
| Range | *0–200 psig* |
| Scale division | *5 psig* |
| Location | *On top of the torque monitor* |
| Use | *Regulates the air pressure to the equipment* |
| Calibration due date | *10/04/97* |
| Critical or not critical | *Not critical to the operation of the bottle capper* |
| Identification number | *ROT-00-100* |
| Type | *Torque monitor, digital* |
| Manufacturer | *Any Gauge Co.* |
| Model number | *088-EM* |
| Serial number | *8298* |
| Range | *0–100 KATU* |
| Scale division | *0.1 KATU* |
| Location | *On top of the bottle capper* |
| Use | *Displays the application torque for each bottle* |
| Calibration due date | *04/04/98* |
| Critical or not critical | *Critical to the operation of the bottle capper* |

*Continued on next page.*

*Continued from previous page.*

| Calibrated Instrument | As-Found Conditions |
|---|---|
| Identification number | *100-GP* |
| Type | *Air pressure gauge, analog* |
| Manufacturer | *Any Gauge Co.* |
| Model number | *Not specified* |
| Serial number | *Not specified* |
| Range | *0–60 psig* |
| Scale division | *1 psig* |
| Location | *On left side of bottle capper* |
| Use | *Controls the torque pressure of the spindles* |
| Calibration due date | *11/11/97* |
| Critical or not critical | *Critical to the operation of the bottle capper* |
| Identification number | *200-GP* |
| Type | *Air pressure gauge, analog* |
| Manufacturer | *Any Gauge Co.* |
| Model number | *Not specified* |
| Serial number | *Not specified* |
| Range | *0–60 psig* |
| Scale division | *1 psig* |
| Location | *On left side of bottle capper* |
| Use | *Controls the torque pressure of the spindles* |
| Calibration due date | *11/11/97* |
| Critical or not critical | *Critical to the operation of the bottle capper* |

### *Equipment Operation: Cap Bin, Spindle, and Gripper Motor Rotation Direction Test*

The objective of the rotation direction test is to verify that the cap bin, spindle, and gripper motors rotate in the proper direction. The tests will be performed with the bottle capper empty. Press the start push button and observe the direction of rotation of the cap bin, spindle, and gripper motors as viewed from behind the motors and record the results in Table 6.97.

### *Equipment Operation: Cap Bin, Spindle, and Gripper Motor Speed Test*

The objective of the speed test is to document the speed of the cap bin, spindle, and gripper motors with the bottle capper empty. Measure the speed of the cap bin, spindle, and gripper motors with a calibrated tachometer. Verify that the measured speeds are within ±10 percent of the maximum speed of 1,750 rpm. Record the results in Table 6.98. Record the instrument used to measure the speeds in Table 6.99.

Cap bin, spindle, and gripper motor speed specification = 1,750 rpm ± 10% maximum

±10% of 1,750 = ±175

1,750 + 175 = 1,925

1,750 – 175 = 1,575

Cap bin motor measured rpm of 1,740 falls within ±10%.

Spindle motor measured rpm of 1,725 falls within ±10%.

Gripper motor measured rpm of 1,760 falls within ±10%.

***Table 6.96. Equipment Control Function Test Results***

| Test Function | Expected Results | Acceptable (Yes/No) |
|---|---|---|
| **Bottle Capper Control Panel** | | |
| Start push button | When the start push button is pressed, power is supplied to the unit. | *Yes* |
| Stop switch | When the stop switch is pressed, power supplied to the unit is stopped. | *Yes* |
| Cap feeder on/off switch | When the cap feeder on/off switch is pressed, power is supplied to/from the cap feeder. | *Yes* |
| Cap feeder speed control | When the cap feeder speed control knob is rotated, the cap feeder speed increases and decreases. | *Yes* |
| Cap bin on/off switch | When the cap bin on/off switch is pressed, power is supplied to/from the cap feeder bin. | *Yes* |
| Cap bin speed control | When the cap bin speed control knob is rotated, the spindle speed increases and decreases. | *Yes* |
| Spindle speed control | When the spindle speed control knob is rotated, the spindle speed increases and decreases. | *Yes* |
| Conveyer/gripper speed control | When the conveyer/gripper speed control knob is rotated, the conveyer and grippers speed increases. | *Yes* |
| Torque control 1 (left) | When the cap torque control 1 (left) knob is rotated, the torque applied to the caps increases and decreases. | *Yes* |
| Torque control 2 (right) | When the cap torque control 2 (right) knob is rotated, the torque applied to the caps increases and decreases. | *Yes* |

*Continued on next page.*

*Continued from previous page.*

| Test Function | Expected Results | Acceptable (Yes/No) |
|---|---|---|
| **Torque Monitor Control Panel** | | |
| Power on/off switch (control panel) | When the switch is pressed, power is supplied to and from the torque monitor. | *Yes* |
| Power on/off switch (printer | When the switch is pressed, power is supplied to and from the printer. | *Yes* |
| Set-up switch (Control panel) | When the switch is pressed, the program scrolls through the set-up procedure. | *Yes* |
| Valve-up switch (Control panel) | When the switch is pressed, values are increased during setup. | *Yes* |
| Valve-down switch (Control panel) | When the switch is pressed, values are decreased during setup. | *Yes* |
| Printer on/off switch (control panel) | When the switch is pressed, every container's torque is printed in KATUs. | *Yes* |
| Zero switch (control panel) | When the switch is pressed, the display zeros. | *Yes* |
| Shift switch (control panel) | When the switch is pressed, the auxiliary functions are provided. | *Yes* |
| Print stats switch (control panel) | When the switch is pressed, the statistical analysis is printed out. | *Yes* |

**Table 6.97. Cap Bin, Spindle, and Gripper Motor Rotation Direction Test Results**

| Item | Expected Results | Results | Acceptable (Yes/No) |
|---|---|---|---|
| Cap bin motor rotation direction | Rotation should be clockwise as viewed from behind the motor. | *Clockwise rotation was observed.* | *Yes* |
| Spindle motor rotation direction | Rotation should be clockwise as viewed from behind the motor. | *Clockwise rotation was observed.* | *Yes* |
| Gripper motor rotation direction | Rotation should be clockwise as viewed from behind the motor. | *Clockwise rotation was observed.* | *Yes* |

**Table 6.98. Cap Bin, Spindle, and Gripper Motor Speed Test Results**

| Motor | Speed Specification (rpm ± 10%) | Measured Speed (rpm) | Acceptable (Yes/No) |
|---|---|---|---|
| Cap bin | 1,750 | *1,740* | *Yes* |
| Spindle | 1,750 | *1,725* | *Yes* |
| Gripper | 1,750 | *1,760* | *Yes* |

**Table 6.99. Instrument Used**

| Test Instrument | Identification Number | Calibration Due Date |
|---|---|---|
| *Tachometer* | *64020* | *06/21/97* |

## Performance Qualification

Once it has been established that the equipment is properly installed and functioning within specified operating parameters, it must be shown that the bottle capper can operate reliably under routine, minimum, and maximum operating conditions (Figure 6.25).

### *Bottle Capper Operation*

The bottle capper operation test is designed to verify that the bottle capper is capable of installing caps on the small, and large size bottles continuously over time. 60 cc bottles, 1,300 cc bottles, and a tachometer is required as follows:

- Visually observe 100 60 cc bottles as they pass through the bottle capper and verify that 100 percent of the caps are installed on the bottles. If the caps are

**Figure 6.25. PQ Elements of a Bottle Capper**

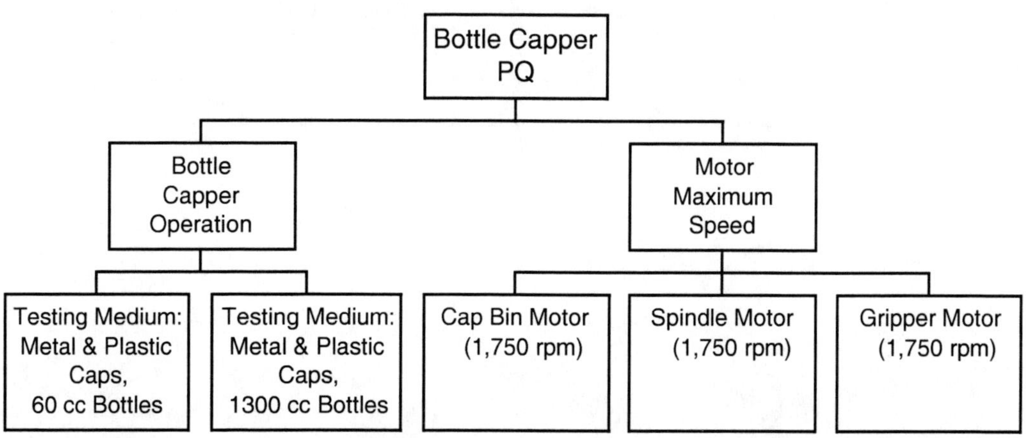

not installed properly they will be repassed through until they are. Record the results in Table 6.100.

- Remove the caps from the 100 bottles and record the removal torque value in Table 6.101. For information only.
- Visually observe 100 1,300 cc bottles as they pass through the bottle capper and verify that 100 percent of the caps are installed on the bottles. If the caps are not installed properly they will be repassed through until they are. Record the results in Table 6.100.
- Remove the caps from the 100 bottles and record the removal torque value in Table 6.102. For information purposes only.
- Measure the speed of the cap bin, spindle, and gripper motor with a calibrated tachometer. Verify that the measured speed is within ±10 percent of the maximum speed of 1,750 rpm. Record the results in Table 6.103. Record the instrument used to measure the speed in Table 6.104.

Cap bin, spindle, and gripper motor speed specification =
1,750 rpm ± 10% maximum

±10% of 1,750 = ±175

1,750 + 175 = 1,925

1,750 – 175 = 1,575

Cap bin motor measured rpm of 1,739 falls within ±10%.

Spindle motor measured rpm of 1,720 falls within ±10%.

Gripper motor measured rpm of 1,759 falls within ±10%.

***Table 6.100. Bottle Capper Performance Test Results***

| Bottle Qty | Bottle Size | Number of Caps Installed | Number of Caps Not Installed | AQL = 100% Acceptable (Yes/No) |
|---|---|---|---|---|
| 100 | 60 cc | *100* | *0* | *Yes* |
| 100 | 1,300 cc | *100* | *0* | *Yes* |

**Table 6.101. Reference Removal Torque Values for the 60 cc Size Bottles**

**Torque Values (in-lb)**

| | | | | | | | | | | | | | |
|---|---|---|---|---|---|---|---|---|---|---|---|---|---|
| 1 | *11* | 16 | *13* | 31 | *10* | 46 | *10* | 61 | *10* | 76 | *9* | 91 | *9* |
| 2 | *7* | 17 | *8* | 32 | *11* | 47 | *8* | 62 | *8* | 77 | *9* | 92 | *12* |
| 3 | *10* | 18 | *7* | 33 | *7* | 48 | *12* | 63 | *11* | 78 | *12* | 93 | *10* |
| 4 | *4* | 19 | *13* | 34 | *11* | 49 | *11* | 64 | *10* | 79 | *11* | 94 | *11* |
| 5 | *12* | 20 | *14* | 35 | *7* | 50 | *8* | 65 | *11* | 80 | *13* | 95 | *11* |
| 6 | *10* | 21 | *8* | 36 | *10* | 51 | *12* | 66 | *11* | 81 | *6* | 96 | *10* |
| 7 | *11* | 22 | *10* | 37 | *8* | 52 | *9* | 67 | *9* | 82 | *9* | 97 | *10* |
| 8 | *6* | 23 | *6* | 38 | *9* | 53 | *12* | 68 | *8* | 83 | *10* | 98 | *8* |
| 9 | *12* | 24 | *10* | 39 | *14* | 54 | *7* | 69 | *7* | 84 | *12* | 99 | *6* |
| 10 | *9* | 25 | *12* | 40 | *9* | 55 | *8* | 70 | *10* | 85 | *9* | 100 | *9* |
| 11 | *10* | 26 | *12* | 41 | *10* | 56 | *11* | 71 | *11* | 86 | *12* | | |
| 12 | *6* | 27 | *9* | 42 | *11* | 57 | *9* | 72 | *10* | 87 | *9* | | |
| 13 | *12* | 28 | *11* | 43 | *9* | 58 | *9* | 73 | *11* | 88 | *11* | | |
| 14 | *10* | 29 | *14* | 44 | *9* | 59 | *9* | 74 | *9* | 89 | *10* | | |
| 15 | *9* | 30 | *9* | 45 | *8* | 60 | *7* | 75 | *9* | 90 | *11* | | |

Average Removal Torque = *9.74 in-lb*

**Table 6.102. Reference Removal Torque Values for the 1,300 cc Size Bottles**

**Torque Values (in-lb)**

| | | | | | | | | | | | | | |
|---|---|---|---|---|---|---|---|---|---|---|---|---|---|
| 1 | *11* | 16 | *14* | 31 | *11* | 46 | *9* | 61 | *9* | 76 | *15* | 91 | *12* |
| 2 | *6* | 17 | *15* | 32 | *15* | 47 | *13* | 62 | *11* | 77 | *17* | 92 | *14* |
| 3 | *10* | 18 | *20* | 33 | *11* | 48 | *14* | 63 | *19* | 78 | *8* | 93 | *12* |
| 4 | *6* | 19 | *16* | 34 | *18* | 49 | *20* | 64 | *15* | 79 | *11* | 94 | *10* |
| 5 | *12* | 20 | *18* | 35 | *14* | 50 | *15* | 65 | *8* | 80 | *16* | 95 | *8* |
| 6 | *10* | 21 | *12* | 36 | *14* | 51 | *10* | 66 | *10* | 81 | *19* | 96 | *17* |
| 7 | *9* | 22 | *14* | 37 | *20* | 52 | *17* | 67 | *11* | 82 | *11* | 97 | *14* |
| 8 | *13* | 23 | *16* | 38 | *12* | 53 | *13* | 68 | *12* | 83 | *16* | 98 | *7* |
| 9 | *14* | 24 | *6* | 39 | *16* | 54 | *6* | 69 | *12* | 84 | *12* | 99 | *10* |
| 10 | *10* | 25 | *8* | 40 | *7* | 55 | *20* | 70 | *11* | 85 | *20* | 100 | *9* |
| 11 | *10* | 26 | *20* | 41 | *14* | 56 | *6* | 71 | *15* | 86 | *12* | | |
| 12 | *12* | 27 | *22* | 42 | *10* | 57 | *10* | 72 | *9* | 87 | *14* | | |
| 13 | *10* | 28 | *8* | 43 | *15* | 58 | *13* | 73 | *15* | 88 | *7* | | |
| 14 | *13* | 29 | *10* | 44 | *14* | 59 | *8* | 74 | *13* | 89 | *9* | | |
| 15 | *14* | 30 | *13* | 45 | *10* | 60 | *19* | 75 | *17* | 90 | *11* | | |

Average Removal Torque = *9.74 in-lb*

**Table 6.103. Cap Bin, Spindle, and Gripper Motor Speed Test Results**

| Motor | Speed Specification (rpm ± 10%) | Measured Speed (rpm) | Acceptable (Yes/No) |
|---|---|---|---|
| Cap bin | 1,750 | *1,739* | *Yes* |
| Spindle | 1,750 | *1,720* | *Yes* |
| Gripper | 1,750 | *1,759* | *Yes* |

**Table 6.104. Instrument Used**

| Test Instrument | Identification Number | Calibration Due Date |
|---|---|---|
| *Tachometer* | *64020* | *06/21/97* |

## Cap Sealer

The validation of a cap sealer (Figure 6.26) will be described here. The cap sealer bonds caps to bottles by using inductivity generated heat to seal a plastic coating inside the cap to the rim of the bottle. Capped bottles pass under the induction coil, which generates an electromagnetic field. The electromagnetic field heats the edge of the aluminum foil disk inside the cap. The aluminum disk melts the edges of the plastic coating, sealing the plastic to the rim of the bottle. The cap sealer has two sealing heads, one for plastic caps and one for metal caps.

The validation protocol for a cap sealer is described in the following sections. This protocol example can be used as a cGMP compliance guide and it should be tailored to meet individual company requirements. Italicized words in the tables represent examples of information that could be recorded in a validation protocol for a cap sealer.

## Installation Qualification

An IQ evaluation establishes confidence that the equipment is properly installed (Figure 6.27). The installation must meet the manufacturer's specified guidelines,

***Figure 6.26. Cap Sealer (courtesy of Lepel Corporation)***

**Figure 6.27. IQ Elements of a Cap Sealer**

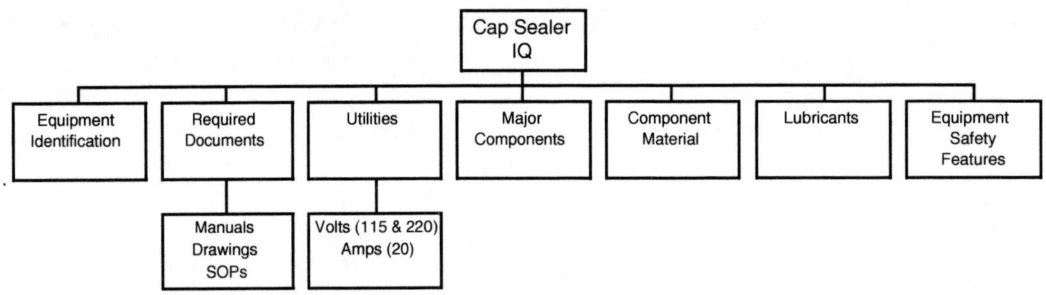


along with design changes at installation. Also, the supporting electrical utilities must meet all electrical codes. The information required for an IQ evaluation is: equipment identification, required documentation, equipment utility requirements, major component specifications, component material, lubricants, and equipment safety features.

### *Equipment Identification*

Record the equipment identification numbers in Table 6.105, along with the following information: equipment manufacturer's purchase order number, model number, serial number, company assigned equipment number, and the location of the equipment.

### *Required Documentation*

Record the equipment manufacturer's operation and maintenance manuals and drawings in Table 6.106. Record the standard operating procedures that cover the setup, operation, and cleaning of the cap sealer in Table 6.107.

**Table 6.105. Equipment Identification**

| Required Information | As-Found Conditions |
|---|---|
| Manufacturer | *Any Cap Sealer Co.* |
| Purchase order number | *641985* |
| Model number | *002RT* |
| Serial number | *8139* |
| Equipment number | *0850* |
| Location | *Packaging line 4* |

**Table 6.106. Manufacturers' Manuals and Drawings**

| Number | Description | Date |
|---|---|---|
| *058* | *Cap Sealer Installation, Operation, and Maintenance Manual* | *12/01/94* |
| *73023191* | *Schematic Diagram Cap Sealer 052 Drawing* | *03/17/80* |

**Table 6.107. Standard Operating Procedures**

| Number | Description | Release Date |
|---|---|---|
| *PKG041* | *Cap Sealer Setup, Operation, and Cleaning* | *07/14/95* |

### *Equipment Utility Requirements*

Compare the manufacturer's specified volts (V) and amps (A) requirements to their as-found condition at the time of qualification testing and record the results in Table 6.108. Also, record the location of the power supply source. Record the instruments used to measure the volts and amps in Table 6.109.

***Water cooling recirculator volt calculation:***

Cap sealer volt specification = 115 V ± 10%

±10% of 115 = ±11.5

115 + 11.5 = 126.5

115 – 11.5 = 103.5

The measured volts of 117 fall within ± 10%.

**Table 6.108. Utilities**

| Utility | Specified | Measured Results | Acceptable (Yes/No) |
|---|---|---|---|
| Volts | Power supply = 115 ± 10%<br>Recirculator 220 ± 10% | *117*<br>*A-B 214; A-C 215; B-C 210* | *Yes* |
| Amps | Power supply = 6<br>Recirculator = 12 | 20 circuit rating | *Yes* |

Power supply source, breaker box *BB25A, wire numbers: 5 and 15, 17, 19 respectively*

**Table 6.109. Instrument Used**

| Test Instrument | Identification Number | Calibration Due Date |
|---|---|---|
| *Multimeter* | *ME-025* | *04/19/97* |

***Power supply volt calculation:***

Cap sealer volt specification = 220 V ± 10%

±10% of 220 = ±22

220 + 22 = 242

220 – 22 = 198

The measured volts of 214/215/210 fall within ±10%.

***Amp calculation:***

Amp specification = 20 A ± 10%

Power supply = 6

Water cooling recirculator = 12

Equipment current draw = power supply A + recirculator A = 6 + 12 = 18 A

The circuit amp rating of 20 is greater than the maximum current draw of the equipment.

### *Major Component Specifications*

The component specifications section of the protocol verifies that the cap sealer components purchased were delivered and installed. Record the major components in Table 6.110.

### *Component Material*

Record the material of each component that contacts the product in Table 6.111.

***Table 6.110. Major Components***

| Components | As-Found Conditions |
|---|---|
| Power supply | Manufacturer: *Any Power Supply Co.*<br>Model number: *0002-RT*<br>Serial number: *8139*<br>Volts: *220*<br>Amperes: *6* |
| Water cooling recirculator | Manufacturer: *Any Recirculator Co.*<br>Model number: *GAWR-2.5*<br>Serial number: *8009*<br>Volts: *115*<br>Amperes: *12* |

**Table 6.111. *Component Material***

| Component | Material |
|---|---|
| *None of the components make contact with the product.* | |

### *Lubricants*

Record the lubricants used to operate the cap sealer in Table 6.112 and indicate if they make contact with the product.

### *Equipment Safety Features*

The objective of testing the safety features is to verify that the safety features on the cap sealer function according to manufacturer specifications. Verify the operation of the following equipment safety features in Table 6.113 and record the results.

## Operational Qualification

An OQ evaluation should establish that the equipment can operate within specified tolerances and limits (Figure 6.28). The mechanical ranges of the cap sealer are challenged, along with the basic equipment operations that the operator will use. Black box software testing was performed. Black box testing evaluates the microprocessor system against the documented functional requirements or intended uses of the equipment. Information required for the OQ evaluation is calibration of the instruments used to control the cap sealer, equipment control functions (switches and push buttons).

**Table 6.112. *Lubricants***

| Where Used | Type | Manufacturer | Product Contact (Yes/No) |
|---|---|---|---|
| *Height adjuster* | *Lubriplate 630-2 Grease* | *Any Oil Co.* | *No* |

Is there a preventive maintenance procedure on file? (Yes/No) *Yes*

**Table 6.113. *Safety Features Tests Results***

| Test Function | Acceptable Expected Results | (Yes/No) |
|---|---|---|
| Water supply cutoff | When the water supply is cutoff from the cap sealer head, the cap sealer shuts off. | *Yes* |

**Figure 6.28. OQ Elements of a Cap Sealer**

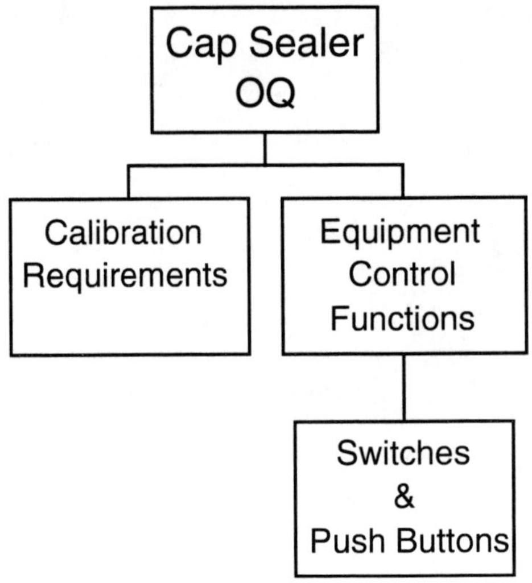


### *Calibration Requirements*

Verify that all critical instruments on the equipment have been logged into the calibration system, have calibration procedures in place, and are in calibration at the time of qualification testing. Record all information for the calibrated instruments used to control the cap sealer in Table 6.114.

**Table 6.114. Calibrated and Noncalibrated Instruments**

*There were no calibrated instruments on the blender.*

| Noncalibrated Instrument | As-Found Conditions |
|---|---|
| Identification number | *0580* |
| Type | *Voltage output meter, digital* |
| Manufacturer | *Any Gauge Co.* |
| Model number | *0002RT* |
| Serial number | *8139* |
| Range | *0–100%* |
| Scale division | *1%* |
| Location | *On the front panel of the generator* |
| Use | *Monitors the generator voltage output* |
| Calibration due date | *Not required* |
| Critical or not critical | *Not critical to the operation of the cap sealer* |

### *Equipment Control Functions*

The objective of testing equipment control functions is to verify that the switches and push buttons on the cap sealer operate according to manufacturer's specifications. Operate each control listed in Table 6.115 and verify its proper operation. Record the results in Table 6.115. The tests will be performed with the cap sealer empty.

## Performance Qualification

Once it has been established that the equipment is properly installed and functioning within specified operating parameters, it must be shown that the cap sealer can operate reliably under routine, minimum, and maximum operating conditions (Figure 6.29).

### *Cap Sealer Operation*

The cap sealer operation test is designed to verify that the cap sealer is capable of consistently sealing bottles with small (33 mm) and large (53 mm) openings.

The procedure to test cap sealer operation is as follows. The materials and instruments required for this test are bottles, caps, and a tachometer. Record the results in Table 6.116. Record the instrument used to measure the speed of the conveyor in Table 6.118.

- Observe 100 metal capped bottles (33 mm opening) as they pass through the cap sealer and verify that 100 percent of the bottles are sealed.

**Table 6.115. *Equipment Control Function Test Results***

| Test Function | Expected Results | Acceptable (Yes/No) |
|---|---|---|
| **Power Supply** | | |
| Start push-button operation | When start push button is pressed, the generator starts and the yellow light on the cap sealer head goes on. | *Yes* |
| Stop push-button operation | When stop push button is pressed, the generator stops. | *Yes* |
| Voltage output potentiometer operation | When the voltage output potentiometer is rotated, the voltage increases and decreases. | *Yes* |
| **Water Cooling Recirculator** | | |
| Start switch operation | When start switch is pressed, the water cooling recirculator starts operating. | *Yes* |
| Stop switch operation | When stop switch is pressed, the water cooling recirculator stops operating. | *Yes* |

**Figure 6.29. PQ Elements of a Cap Sealer**

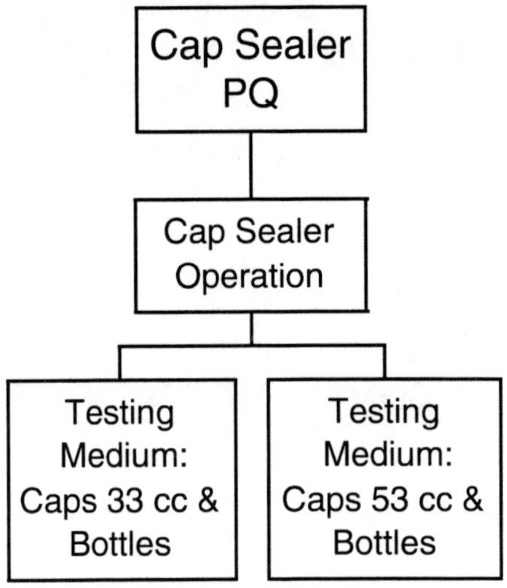


**Table 6.116. Cap Sealer Performance Test Results**

| Test Number | Number of Bottles | Bottle Opening | Cap Type | % Sealed | Acceptable (Yes/No) |
|---|---|---|---|---|---|
| 1 | *100* | *33* | *Metal* | *100* | *Yes* |
| 2 | *100* | *53* | *Metal* | *100* | *Yes* |
| 3 | *100* | *33* | *Plastic* | *100** | *Yes* |
| 4 | *100* | *53* | *Plastic* | *100** | *Yes* |

*These tests were performed by a special setup of caps and bottles by validation.

- Observe 100 metal capped bottles (53 mm opening) as they pass through the cap sealer and verify that 100 percent of the bottles are sealed.
- Observe 100 plastic capped bottles (33 mm opening) as they pass through the cap sealer and verify that 100 percent of the bottles are sealed.
- Observe 100 plastic capped bottles (53 mm opening) as they pass through the cap sealer and verify that 100 percent of the bottles are sealed.
- Measure the distance from the bottom of the cap sealer head to the top of the bottle for each of the above tests. Record the results in Table 6.117. This test is for information purposes only.
- Measure the speed of the conveyor with a calibrated tachometer, for each of the above tests. Record the results in Table 6.117. This test is for information purposes only.

**Table 6.117. Reference Dimensions (Information only)**

| Test Number | Measured Distance (in.) | Conveyor Speed (in./sec.) |
|---|---|---|
| 1 | *0.50* | *4.67* |
| 2 | *0.87* | *3.81* |
| 3 | *0.55* | *3.64* |
| 4 | *0.75* | *4.93* |

**Table 6.118. Instrument Used**

| Test Instrument | Identification Number | Calibration Due Date |
|---|---|---|
| *Tachometer* | *64020* | *06/21/97* |

# OUTSERTER

The validation of an outserter (Figure 6.30) will be described here. The outserter is used to attach a variety of accordion folded literature (hereafter known as an outsert) to bottles. A bottle sensor signals the dispenser drive motor to start rotating and the fail safe timer starts. The feed and pinch rollers grip and pull the adhesive web around the peel plate, causing the folded outsert to peel away from the adhesive transfer tape. The outsert is placed on the side of the bottle. The web continues to feed until the outsert sensor signals the dispenser drive motor to stop rotation, and the fail safe timer resets. The outsert solenoid activates to move the adhesive web into contact with the next outsert to be dispensed.

The validation protocol for an outserter is described in the following sections. This protocol example can be used as a cGMP compliance guide and it should be tailored to meet individual company requirements. Italicized words in the tables represent examples of information that could be recorded in a validation protocol for an outserter.

## Installation Qualification

An IQ evaluation establishes confidence that the equipment is properly installed (Figure 6.31). The installation must meet the manufacturer's specified guidelines, along with design changes at installation. Also, the supporting electrical utilities must meet all electrical codes. The information required for an IQ evaluation is equipment identification, required documentation, equipment utility requirements, major component specifications, component material, lubricants, and equipment safety features.

### *Equipment Identification*

Record the equipment identification numbers in Table 6.119, along with the following information: equipment manufacturer's purchase order number, model

***Figure 6.30. Outserter (courtesy of New Jersey Machine, Inc.)***

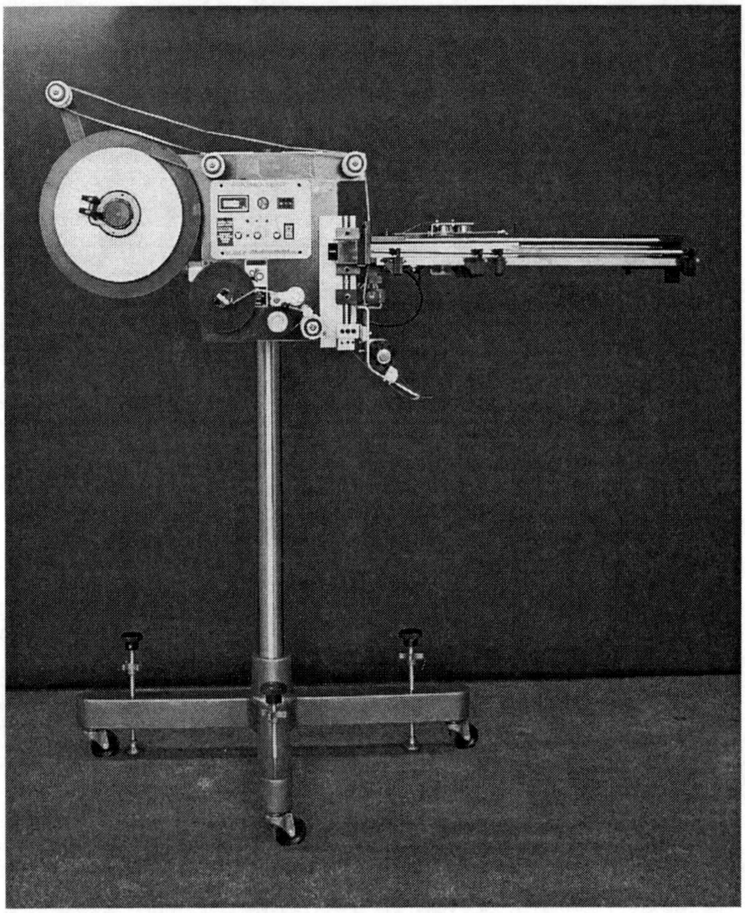

***Figure 6.31. IQ Elements of an Outserter***

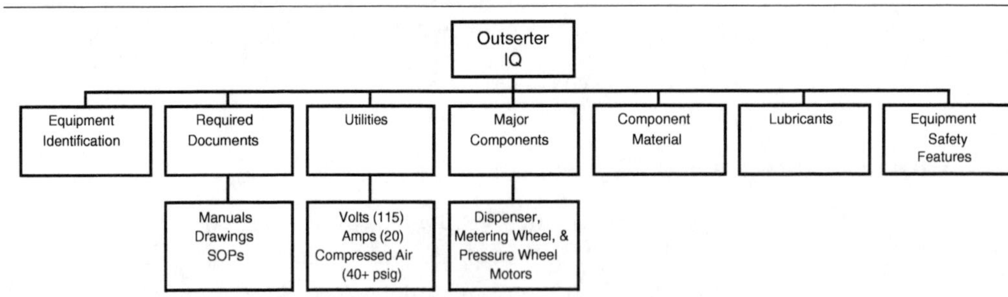

**Table 6.119. *Equipment Identification***

| Required Information | As-Found Conditions |
|---|---|
| Manufacturer | *Any Outserter Co.* |
| Purchase order number | *612954* |
| Model number | *P503* |
| Serial number | *111-LP503* |
| Equipment number | *02015* |
| Location | *Packaging line 7* |

number, serial number, company assigned equipment number, and the location of the equipment.

### *Required Documentation*

Record the equipment manufacturer's operation and maintenance manuals and drawings in Table 6.120. Record the SOPs that cover the setup, operation, and cleaning of the outserter in Table 6.121.

### *Equipment Utility Requirements*

Compare the manufacturer's specified volts (V), amps (A), and compressed air requirements to their as-found condition at the time of qualification testing and record the results in Table 6.122. Also record the location of the power supply source. Record the instruments used to measure the volts, amps, and compressed air in Table 6.123.

**Table 6.120. *Manufacturers' Manuals and Drawings***

| Number | Description | Date |
|---|---|---|
| *09ES* | *Outserter Installation, Operation, and Maintenance Manual* | *11/21/94* |

**Table 6.121. *Standard Operating Procedures***

| Number | Description | Release Date |
|---|---|---|
| *PKG124* | *Outserter Setup and Operation* | *07/14/95* |
| *PKG125* | *Outserter Cleaning Procedure* | *09/22/94* |

**Table 6.122. Utilities**

| Utility | Specified | Measured Results | Acceptable (Yes/No) |
|---|---|---|---|
| Volts | 115 ± 10% | *117* | *Yes* |
| Amps | Dispenser motor = 1<br>Metering wheel motor = 1.8<br>Pressure wheel motor = 1 | 20 circuit rating | *Yes* |
| Compressed air | ≥ 40 psig | *57 psig* | *Yes* |

Power supply source, breaker box *BB7A, wire number: 17*

**Table 6.123. Instruments Used**

| Test Instrument | Identification Number | Calibration Due Date |
|---|---|---|
| *Multimeter* | *ME-025* | *04/19/97* |
| *Air pressure gauge* | *P-102* | *02/10/97* |

***Volt calculation:***

Outserter volt specification = 115 V ± 10%

±10% of 115 = ±11.5

115 + 11.5 = 126.5

115 − 11.5 = 103.5

The measured volts of 117 falls within ± 10%.

***Amp calculation:***

Amp specification = 20 A ±10%

Dispenser motor = 1 A

Metering wheel motor = 1.8 A

Pressure wheel motor = 1 A

Equipment current draw = dispenser motor A + metering wheel motor A + pressure wheel A = 1 + 1.8 + 1 = 3.8 A

The circuit amp rating of 20 is greater than the maximum current draw of the equipment.

***Compressed air calculation:***

Compressed air specification: air pressure ≥ 40 psig

The measured air pressure of 57 psig is greater than 40 psig.

### *Major Component Specifications*

The component specifications section of the protocol verifies that the outserter components purchased were delivered and installed. Record the major components in Table 6.124.

### *Component Material*

Record the material of each component that contacts the product in Table 6.125.

### *Lubricants*

Record the lubricants used to operate the outserter in Table 6.126 and indicate if they make contact with the product.

**Table 6.124. Major Components**

| Components | As-Found Conditions |
|---|---|
| Dispenser motor | Manufacturer: *Any Motor Co.*<br>Model number: *Not Specified*<br>Serial number: *7010UEQY*<br>Volts: *115*<br>Amperes: *1*<br>Phases: *1*<br>Cycles: *60 Hz*<br>hp: *1/8*<br>rpm: *1,750* |
| Metering wheel motor | Manufacturer: *Any Motor Co.*<br>Model number: *A8271D13MC*<br>Serial number: *195773*<br>Volts: *115*<br>Amperes: *1.8*<br>Phases: *1*<br>Cycles: *60 Hz*<br>hp: *1/6*<br>rpm: *57* |
| Pressure wheel motor | Manufacturer: *Any Motor Co.*<br>Model number: *3W-MPEB5032*<br>Serial number: *472910*<br>Volts: *115*<br>Amperes: *1*<br>Phases: *1*<br>Cycles: *60 Hz*<br>hp: *1/8*<br>rpm: *84* |

**Table 6.125. Component Material**

| Component | Material |
|---|---|
| *None of the components make contact with the product.* | |

**Table 6.126. Lubricants**

| Where Used | Type | Manufacturer | Product Contact (Yes/No) |
|---|---|---|---|
| *Pinch roll bushing*<br>*Feed roll bushing*<br>*Idler roll bushing* | *Light Machine Oil* | *Any Oil Co.* | *No* |
| Is there a preventive maintenance procedure on file? (Yes/No) | | | *Yes* |

### *Equipment Safety Features*

The objective of testing equipment safety features is to verify that the safety features on the outserter function according to manufacturer specifications.

- Verify the operation of the equipment safety features in Table 6.127 and record the results.

## Operational Qualification

An OQ evaluation should establish that the equipment can operate within specified tolerances and limits (Figure 6.32). The mechanical ranges of the outserter are challenged, along with the basic equipment operations. Information required for the OQ evaluation is calibration of the instruments used to control the outserter, equipment control functions (switches and push buttons), and equipment

**Table 6.127. Safety Features Test Results**

| Test Function | Expected Results | Acceptable (Yes/No) |
|---|---|---|
| Fail safe timer | When the bottle sensor beam is broken, the fail safe timer resets. | *Yes* |
| Bottles missing | When there are no bottles present, the alarm sounds, the fail safe timer times out and the outserter goes into a fail safe (stop) condition. | *Yes* |
| Outserts missing | When there are no outserts present, the alarm sounds, the fail safe timer times out and the outserter goes into a fail safe (stop) condition. | *Yes* |
| Motor drive guard | The motor drive guard was in place. | *Yes* |

**Figure 6.32. OQ Elements of an Outserter**

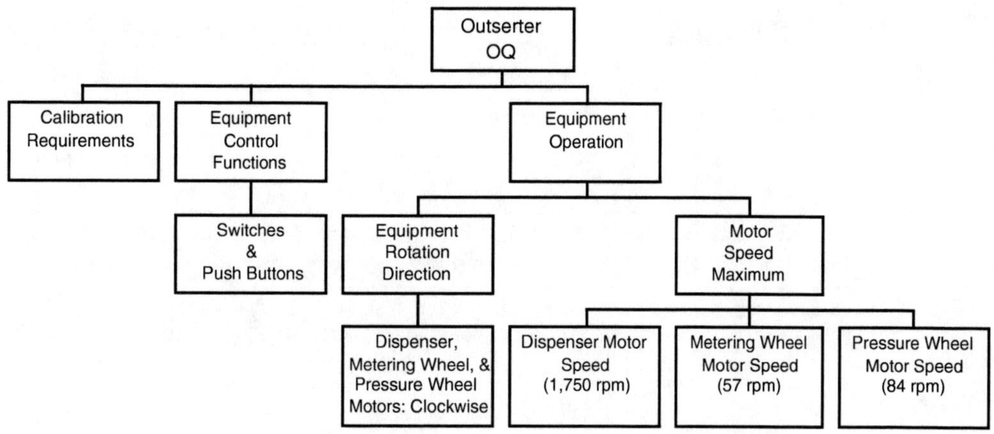


operation (dispenser, metering wheel, and pressure wheel motors rotation direction and speed tests).

### *Calibration Requirements*

Verify that all critical instruments on the equipment are logged into the calibration system, have calibration procedures in place and are in calibration at the time of qualification testing. Record all of the necessary information for the calibrated instruments used to control the outserter in Table 6.128.

### *Equipment Control Functions*

The objective of testing equipment control functions is to verify that the switches and push buttons on the outserter operate per manufacturer's specifications. Operate each control listed in Table 6.129 and verify its proper operation. Record the results in Table 6.129. The tests will be performed with the outserter empty.

### *Equipment Operation: Dispenser, Metering Wheel, and Pressure Wheel Motor Rotation Direction Test*

The objective of rotation direction test is to verify that the dispenser, metering wheel and pressure wheel motors rotate in the proper direction. The tests will be performed with the outserter empty of tablets or capsules.

Press the start push button and observe the direction of rotation of the dispenser, metering wheel and pressure wheel motors as viewed from behind the motors and record the results in Table 6.130.

**Table 6.128. Calibrated and Noncalibrated Instruments**

| |
|---|
| *There were no calibrated or noncalibrated instruments on the outserter.* |

**Table 6.129. Equipment Control Function Test Results**

| Test Function | Expected Results | Acceptable (Yes/No) |
|---|---|---|
| On/off switch (Outserter) operation | When the on/off switch is pressed, the outserter goes on and off. | *Yes* |
| Job push button operation | When the job push button is pressed, the transfer tape jogs. | *Yes* |
| Slow/fast control knob operation | When the slow/fast control knob is rotated, the dispenser motor speed increases and decreases. | *Yes* |
| Preset push button operation | When the preset push button is pressed, the outsert solenoid is activated independently. | *Yes* |
| Fail safe reset push button operation | When the fail safe reset push button is pressed, the fail safe timer resets. | *Yes* |
| Fail safe timer operation | When the fail safe timer is set to a specified time, the duration of the fail safe time is set. | *Yes* |
| Fail safe LED operation | When there is a fail safe condition, the LED goes on. | *Yes* |
| Indicator light operation | When the alarm is activated, the indicator light goes on. | *Yes* |
| Alarm switch operation | When the alarm switch is set to on, the audible alarm is activated. | *Yes* |
| On/off switch (Metering wheel) operation | When the on/off switch is pressed, the metering wheel is activated and deactivated. | *Yes* |
| On/off switch (Pressure wheel) operation | When the on/off switch is pressed, the pressure wheel is activated and deactivated. | *Yes* |
| Display counter operation | When the display counter is activated, the number of outserts that have been applied is displayed. | *Yes* |
| Speed control knob (Metering wheel) operation | When the speed control knob is rotated, the metering wheel's speed increases and decreases. | *Yes* |
| Speed control knob (Pressure wheel) operation | When the speed control knob is rotated, the pressure wheel's speed increases and decreases. | *Yes* |
| Outsert solenoid operation | When the outsert solenoid is activated, the adhesive transfer tape is moved to make contact with the outserts. | *Yes* |

### *Equipment Operation: Dispenser, Metering Wheel, and Pressure Wheel Motor Speed Test*

The objective of the speed test is to document the speed of the dispenser, metering wheel, and pressure wheel motors with the outserter empty. A tachometer is

**Table 6.130. *Dispenser, Metering Wheel, and Pressure Wheel Motor Rotation Direction Test Results***

| Item | Expected Results | Results | Acceptable (Yes/No) |
|---|---|---|---|
| Dispenser, metering wheel and pressure wheel rotation direction | Rotation should be clockwise as viewed from behind the motors. | *Clockwise rotation was observed for all three motors.* | *Yes* |

required. Perform the speed test as follows. Record the results in Table 6.131 and the instrument used to measure the speed in Table 6.132.

- Measure the speed of the dispenser motor with a calibrated tachometer. Verify that the measured speed is within ±10 percent of the fixed speed of 1,750 rpm.

  Dispenser motor speed specification = 1,750 rpm ± 10%

  ±10% of 1,750 = ±175

  1,750 + 175 = 1,925

  1,750 – 175 = 1,575

  The measured rpm of 1,762 falls within ±10%.

- Measure the speed of the metering wheel motor with a calibrated tachometer. Verify that the measured speed is within ±10 percent of the maximum speed of 57 rpm.

  Metering wheel motor speed specification = 57 rpm ± 10% maximum

  ±10% of 57 = ±5.7

  57 + 5.7 = 62.7

**Table 6.131. *Dispenser, Metering Wheel, and Pressure Wheel Motor Speed Test Results***

| Motor | Speed Specification (rpm ± 10%) | Measured Speed (rpm) | Acceptable (Yes/No) |
|---|---|---|---|
| Dispenser | 1,750 | *1,762* | *Yes* |
| Metering wheel | 57 | *62* | *Yes* |
| Pressure wheel | 84 | *86* | *Yes* |

**Table 6.132. *Instrument Used***

| Test Instrument | Identification Number | Calibration Due Date |
|---|---|---|
| *Tachometer* | *64020* | *06/21/97* |

57 – 5.7 = 51.3

The measured rpm of 62 falls within ±10%.

- Measure the speed of the pressure wheel motor with a calibrated tachometer. Verify that the measured speed is within ±10 percent of the maximum speed of 57 rpm.

  Pressure wheel motor speed specification = 84 rpm ± 10% maximum

  ±10% of 84 = ±8.4

  84 + 8.4 = 92.4

  84 – 8.4 = 75.6

  The measured rpm of 86 falls within ±10%.

## Performance Qualification

Once it has been established that the equipment is properly installed and functioning within specified operating parameters, it must be shown that the outserter can operate reliably under routine, minimum, and maximum operating conditions (Figure 6.33).

### *Outserter Operation*

The outserter operation test is designed to verify that the outserter is capable of consistently attaching outserts to small and large size bottles. The speed of the dispenser, metering wheel, and pressure wheel motors will be measured and recorded.

The materials and instruments required for this test are outserts, 60 cc bottles, 1,300 cc bottles, and a tachometer. Use the following procedure to test outserter operation:

**Figure 6.33. PQ Elements of an Outserter**

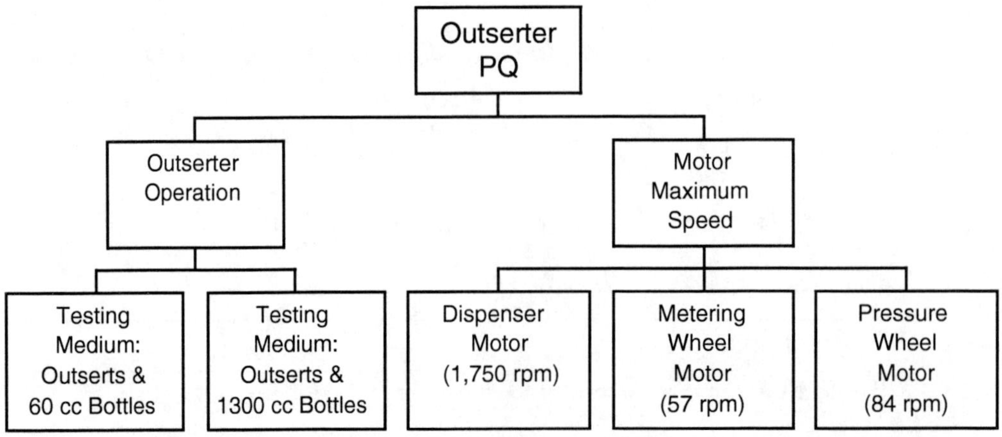

- With the equipment operating, visually observe 100 (60 cc) bottles as they pass through the outserter and verify that 95 percent of the bottles have outserts on them. Record the results in Table 6.133.
- With the equipment operating, visually observe 100 (1300 cc) bottles as they pass through the outserter and verify that 95 percent of the bottles have outserts on them. Record the results in Table 6.133.
- Measure the speed of the dispenser motor with a calibrated tachometer. Verify that the measured speed is within ±10 percent of the fixed speed of 1,750 rpm. Record the results in Table 6.134. Record the instrument used to measure the speed in Table 6.135.

  Dispenser motor speed specification = 1,750 rpm ± 10%

  ±10% of 1,750 = ±175

  1,750 + 175 = 1,925

  1,750 – 175 = 1,575

  The measured rpm of 1,760 falls within ±10%.

- Measure the speed of the metering wheel motor with a calibrated tachometer. Verify that the measured speed is within ±10 percent of the maximum speed of 57 rpm. Record the results in Table 6.134. Record the instrument used to measure the speed in Table 6.135.

**Table 6.133. Outserter Performance Test Results**

| Bottle Quantity | Bottle Size (cc) | # Outserts Attached | # Outserts Not Attached | Acceptable (Yes/No) |
|---|---|---|---|---|
| 100 | *60* | *99* | *1* | *Yes* |
| 100 | *1,300* | *98* | *2* | *Yes* |

**Table 6.134. Dispenser, Metering Wheel, and Pressure Wheel Motor Speed Test Results**

| Motor | Speed Specification (rpm ± 10%) | Measured Speed (rpm) | Acceptable (Yes/No) |
|---|---|---|---|
| Dispenser | 1,750 | *1,760* | *Yes* |
| Metering wheel | 57 | *61* | *Yes* |
| Pressure wheel | 84 | *85* | *Yes* |

**Table 6.135. Instrument Used**

| Test Instrument | Identification Number | Calibration Due Date |
|---|---|---|
| *Tachometer* | *64020* | *06/21/97* |

Metering wheel motor speed specification = 57 rpm ± 10% maximum

±10% of 57 = ±5.7

57 + 5.7 = 62.7

57 – 5.7 = 51.3

The measured rpm of 61 falls within ±10%.

- Measure the speed of the pressure wheel motor with a calibrated tachometer. Verify that the measured speed is within ±10 percent of the maximum speed of 57 rpm. Record the results in Table 6.134. Record the instrument used to measure the speed in Table 6.135.

Pressure wheel motor speed specification = 84 rpm ± 10% maximum

±10% of 84 = ±8.4

84 + 8.4 = 92.4

84 – 8.4 = 75.6

The measured rpm of 85 falls within ±10%.

## Labeler

The validation of a bottle labeler (Figure 6.34) will be described here. This machine is used to attach literature and imprinted labels to various sizes of bottles. There are several components in this configuration. There is a metering wheel that is used to space the bottles on the conveyor; an overhead stabilizing belt that stabilizes the bottles as the literature and labels are attached to the bottles; a literature dispensing unit is used to attach literature to the bottles; and a labeling unit is used to attach labels to the bottles; there is an imprinter that codes labels with a serial number and date; and a bar code reader that is used to verify the accuracy of bar code labels and a pressure belt that is used to finish rolling the labels around the bottles.

The validation protocol for a labeler is described in the following sections. This protocol example can be used as a cGMP compliance guide, and it should be tailored to meet individual company requirements. Italicized words in the tables represent examples of information that could be recorded in a validation protocol for a labeler.

### Installation Qualification

An IQ evaluation establishes confidence that the equipment is properly installed (Figure 6.35). The installation must meet the manufacturer's specified guidelines, along with design changes at installation. Also, the supporting electrical utilities must meet all electrical codes. The information required for an IQ evaluation is equipment identification, required documentation, equipment utility requirements,

***Figure 6.34. Labeler (courtesy of New Jersey Machine, Inc.)***

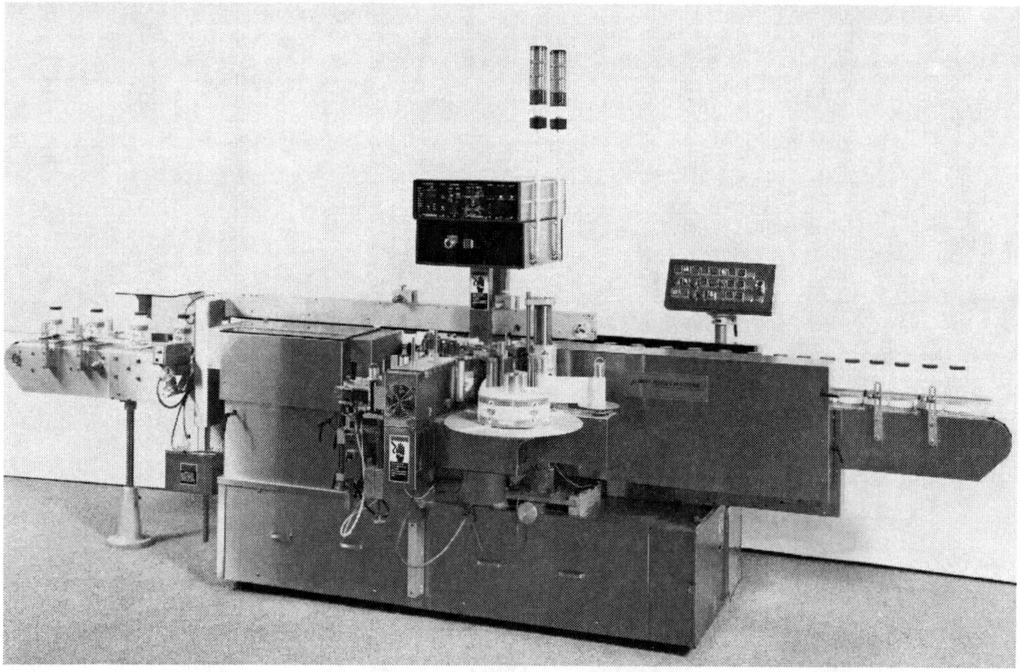

major component specifications, component material, lubricants, and equipment safety features.

### *Equipment Identification*

Record the equipment identification numbers in Table 6.136, along with the following information: equipment manufacturer's purchase order number, model

***Figure 6.35. IQ Elements of a Bottle Labeler***

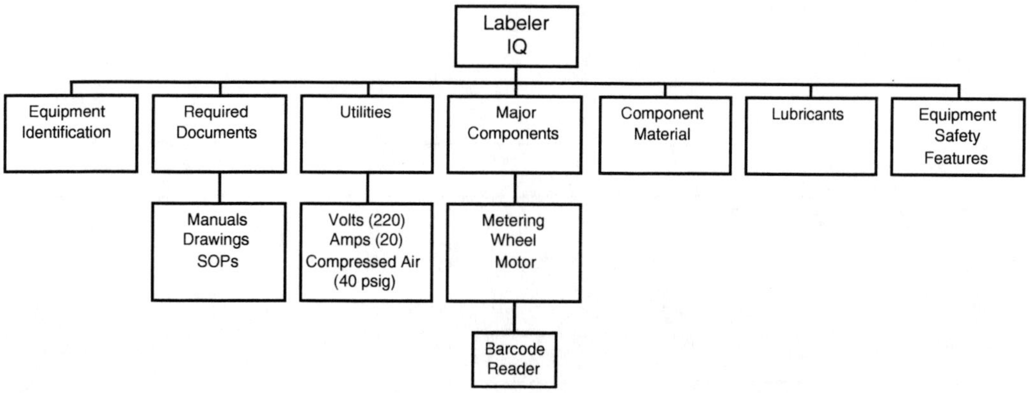

**Table 6.136. Equipment Identification**

| Required Information | As-Found Conditions |
|---|---|
| Manufacturer | *Any Labeler Co.* |
| Purchase order number | *975201* |
| Model number | *112-PBWRL433* |
| Serial number | *2700D79M* |
| Equipment number | *5313* |
| Location | *Packaging line 7* |

number, serial number, company assigned equipment number, and the location of the equipment.

### *Required Documentation*

Record the equipment manufacturer's operation and maintenance manual and drawings in Table 6.137. Record the SOPs that cover the setup, operation, and cleaning of the labeler in Table 6.138.

### *Equipment Utility Requirements*

Compare the manufacturer's specified volts (V), amps (A), and compressed air requirements to their as-found conditions at the time of qualification testing and record the results in Table 6.139. Also record the location of the power supply source. Record the instruments used to measure the volts, amps, and compressed air in Table 6.140.

**Table 6.137. Manufacturers' Manuals and Drawings**

| Number | Description | Date |
|---|---|---|
| *None* | *Instructions for Model 112-PBWRL433 Manual* | *9/92* |
| *None* | *Automatic On-Line Inspection System Software Manual* | *7.96* |
| *48-DW-433* | *Power Distribution Schematic Drawing* | *06/02/97* |
| *58-DW-334* | *Drawer Panel Layout* | *06/23/97* |

**Table 6.138. Standard Operating Procedures**

| Number | Description | Release Date |
|---|---|---|
| *PKG072* | *Labeler Setup and Operation* | *07/14/95* |
| *PKG073* | *Labeler Cleaning Procedure* | *09/22/94* |

**Table 6.139. *Utilities***

| Utility | Specified | Measured Results | Acceptable (Yes/No) |
|---|---|---|---|
| Volts | 220 ± 10% | *A-B 219; A-C 217; B-C 220* | *Yes* |
| Amps | Metering wheel motor =1 | 20 circuit rating | *Yes* |
| Compressed air | Air pressure = 40 psig | 40 psig (regulated) | *Yes* |
| Power supply source, breaker box *BB745A, wire numbers: 21, 23, 25* | | | |

**Table 6.140. *Instruments Used***

| Test Instrument | Identification Number | Calibration Due Date |
|---|---|---|
| *Multimeter* | *ME-025* | *04/19/97* |
| *Air pressure gauge* | *P-102* | *02/10/97* |

***Volt calculation:***

Labeler volt specification = 220 V ± 10%

±10% of 220 = ±22

220 + 22 = 242

220 – 22 = 198

The measured volts of 219/217/220 fall within ± 10%.

***Amp calculation:***

Amp specification = 20 A ± 10%

Metering wheel motor = 1 A
The circuit amp rating of 20 is greater than the maximum current draw of the equipment.

***Compressed air calculation:***

Compressed air specification = 40 psig

The compressed air pressure is regulated to 40 psig.

### *Major Component Specifications*

The component specifications section of the protocol is used to verify that the labeler components purchased were delivered and installed. Record the major components in Table 6.141.

**Table 6.141. Major Components**

| Components | As-Found Conditions |
|---|---|
| Metering wheel motor | Manufacturer: *Any Motor Co.*<br>Model number: *Not specified*<br>Serial number: *701DUEQY45*<br>Volts: *220*<br>Amperes: *1*<br>Phases: *3*<br>Cycles: *60 Hz*<br>hp: *1/8*<br>rpm: *57* |
| Barcode reader | Manufacturer: *Any Barcode Reader Co.*<br>Model number: *220EA28CPM*<br>Serial number: *8007G79*<br>Software: *Datalogic 5209-01*<br>*A08MCP Configuration*<br>*Program Revision 04* |

### *Component Material*

Record the material of each component that contacts the product in Table 6.142.

### *Lubricants*

Record the lubricants used to operate the labeler in Table 6.143 and indicate if they make contact with the product.

**Table 6.142. Component Material**

| Component | Material |
|---|---|
| *None of the components make contact with the product.* | |

**Table 6.143. Lubricants**

| Where Used | Type | Manufacturer | Product Contact (Yes/No) |
|---|---|---|---|
| *Open gears* | *Lithium #2 Grease* | *Any Oil Co.* | *No* |
| *Roller chain & sprocket* | *Suntec 302, 30W Oil* | *Any Oil Co.* | *No* |
| *Main drive* | *Texaco #620 Grease* | *Any Oil Co.* | *No* |
| Is there a preventive maintenance procedure on file? (Yes/No) | | | *Yes* |

### *Equipment Safety Features*

The objective of testing equipment safety features is to verify that the safety features on the labeler function according to manufacturer specifications. Verify the operation of the equipment safety features in Table 6.144 and record the results.

## Operational Qualification

An OQ evaluation should establish that the equipment can operate within specified tolerances and limits (Figure 6.36). The mechanical ranges of the labeler are challenged along with the basic equipment operations that the operator will use. Information required for the OQ evaluation is calibration of the instruments used to control the labeler, equipment control functions (switches and push buttons) and equipment operation (metering wheel rotation direction, speed test).

***Table 6.144. Safety Features Test Results***

| Test Function | Expected Results | Acceptable (Yes/No) |
|---|---|---|
| Emergency stop push button | When the emergency stop push button is pressed, the alarms screen will appear on the operator interface panel. | *Yes* |

***Figure 6.36. OQ Elements of a Bottle Labeler***

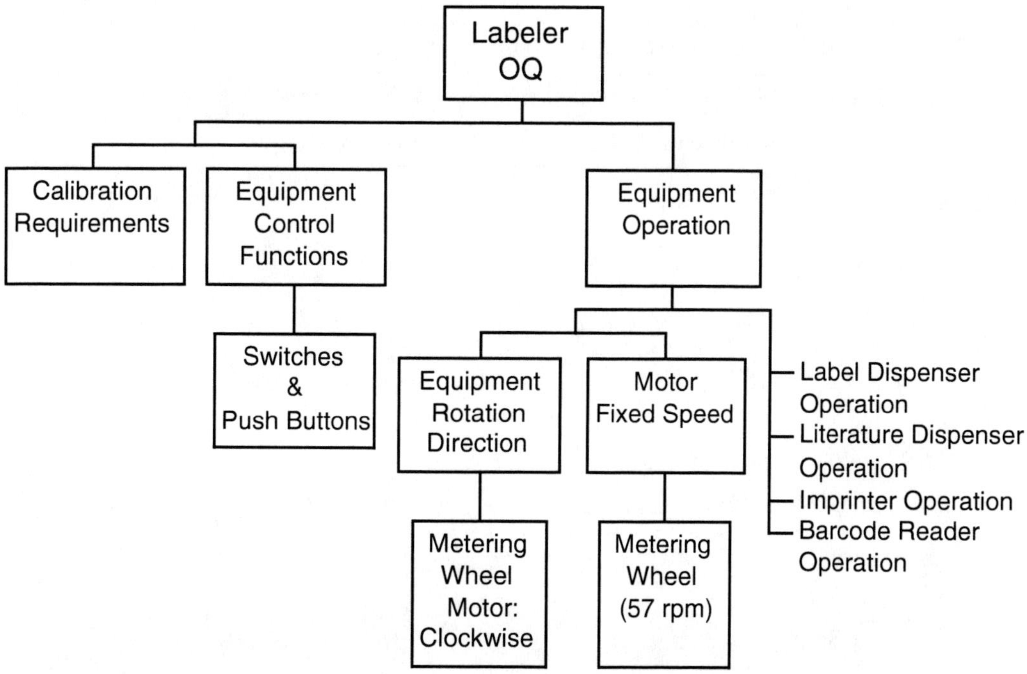

### *Calibration Requirements*

Verify that all critical instruments on the equipment have been logged into the calibration system, have calibration procedures in place, and are in calibration at the time of qualification testing. Record all information for the calibrated instruments used to control the labeler in Table 6.145.

### *Equipment Control Functions*

The objective of testing equipment control functions is to verify that the switches and push buttons on the labeler operate according to manufacturer's specifications. Record the results in Table 6.146. The tests will be performed with the labeler empty. Operate each control listed in Table 6.146 and verify its proper operation.

### *Labeler Control Panel*

The objective of testing the labeler control panel is to verify that the switches and push buttons on the labeler operate per manufacturer's specifications. The tests will be performed with the labeler empty. Operate each control listed in Table 6.147 and verify its proper operation.

### *Equipment Operation: Metering Wheel Motor Rotation Direction Test*

The objective of the rotation direction test is to verify that the metering wheel motor rotates in the proper direction. The tests will be performed with the labeler empty. Press the start push button and observe the direction of rotation of the

**Table 6.145. Calibrated and Noncalibrated Instruments**

*There were no noncalibrated instruments on the labeler.*

| Calibrated Instrument | As-Found Conditions |
|---|---|
| Identification number | *00-032-GP* |
| Type | *Air pressure gauge, analog* |
| Manufacturer | *Any Gauge Co.* |
| Model number | *80-595* |
| Serial number | *710-GP* |
| Range | *0–200 psig* |
| Scale division | *5 psig* |
| Location | *On a pole next to the labeler* |
| Use | *Regulates the air pressure to the equipment* |
| Calibration due date | *10/04/98* |
| Critical or not critical | *Not critical to the operation of the labeler* |

**Table 6.146. Equipment Control Function Test Results**

| Test Function | Expected Results | Acceptable (Yes/No) |
|---|---|---|
| Separator speed dial operation | When the separator speed dial is rotated from 0 to 10 and from 10 to 0, the separator speed increases and decreases. | *Yes* |
| Set-up on/off switch (key) operation | When the set-up on/off switch (key) is turned on and off, the machine can be set up and not set up. | *Yes* |
| Machine start push button operation | When the machine start push button is pressed, the machine starts. | *Yes* |
| Machine stop push button operation | When the machine stop push button is pressed, the machine stops. | *Yes* |
| Conveyor speed control dial operation | When the conveyor speed control dial is rotated from 0 to 10 and from 10 to 0, the conveyor speed increases and decreases. | *Yes* |
| Emergency stop push button operation | When the emergency stop push button is pressed, the conveyor stops and the alarm menu appears on the operator interface panel. | *Yes* |
| | When the emergency stop push button is pulled out the machine does not start operating. | *Yes* |

**Table 6.147. Control Function Test Results**

| Test Function | Expected Results | Acceptable (Yes/No) |
|---|---|---|
| *Because this is a programmable logic controlled piece of equipment, the control functions are too numerous to list in this book.* | | |

metering wheel motor as viewed from behind the motor and record the results in Table 6.148.

### *Equipment Operation: Metering Wheel Motor Speed Test*

The objective of the motor speed test is to document the speed of the metering wheel motor with the labeler empty. Perform the test as follows with a tachometer.

**Table 6.148. Metering Wheel Motor Rotation Direction Test Results**

| Item | Expected Results | Results | Acceptable (Yes/No) |
|---|---|---|---|
| Metering wheel motor rotation direction | Rotation should be clockwise as viewed from behind the motor. | *Clockwise rotation was observed.* | *Yes* |

- Measure the speed of the metering wheel motor with a calibrated tachometer. Verify that the measured speed is within ±10 percent of the fixed speed of 57 rpm. Record the results in Table 6.149. Record the instrument used to measure the speed in Table 6.150.

  Metering wheel motor speed specification = 57 rpm ± 10% maximum

  ±10% of 57 = ±5.7

  57 + 5.7 = 62.7

  57 − 5.7 = 51.3

  The measured rpm of 59 falls within ±10%.

### *Equipment Operation: Literature Dispenser Operation*

The objective of literature dispenser operation is to verify that the literature dispenser attaches literature to bottles consistently over time. Literature, 60 cc bottles, and 1,300 cc bottles are required for the following tests:

- Visually observe 100 60 cc bottles as they pass through the labeler and verify that 95 percent of the bottles have literature attached. Record the results in Table 6.151.
- Visually observe 100 1,300 cc bottles as they pass through the labeler and verify that 95 percent of the bottles have literature attached. Record the results in Table 6.151.

### *Equipment Operation: Label Dispenser Operation*

The objective of label dispenser operation test is to verify that the label dispenser attaches labels to bottles consistently over time. Labels, 60 cc bottles, and 1,300 cc bottles are required for the following tests:

***Table 6.149. Metering Wheel Motor Speed Test Results***

| Motor | Speed Specification (rpm ± 10%) | Measured Speed (rpm) | Acceptable (Yes/No) |
|---|---|---|---|
| Metering wheel | 57 | *59* | *Yes* |

***Table 6.150. Instrument Used***

| Test Instrument | Identification Number | Calibration Due Date |
|---|---|---|
| *Tachometer* | *64020* | *06/21/97* |

**Table 6.151. Literature Dispenser Test Results**

| Bottle Quantity | Bottle Size (cc) | # Literature Attached | # Literature Not Attached | Acceptable (Yes/No) |
|---|---|---|---|---|
| 100 | *60* | *99* | *1* | *Yes* |
| 100 | *1,300* | *98* | *2* | *Yes* |

- Visually observe 100 60 cc bottles as they pass through the labeler and verify that 95 percent of the bottles have labels attached. Record the results in Table 6.152.
- Visually observe 100 1,300 cc bottles as they pass through the labeler and verify that 95 percent of the bottles have labels attached. Record the results in Table 6.152.

### *Equipment Operation: Imprinter Operation*

The objective of imprinter operation test is to verify that the imprinter is capable of printing the serial number and date on labels consistently over time.

- Visually observe 100 connective 1 3/4" × 3 3/8" size labels as they pass through the imprinter and verify that 95 percent of the labels have a serial number and date printed on them. Record the results in Table 6.153.
- Visually observe 100 connective 2 1/2" × 6" size labels as they pass through the imprinter and verify that 95 percent of the labels have a serial number and date printed on them. Record the results in Table 6.153.

**Table 6.152. Label Dispenser Test Results**

| Bottle Quantity | Bottle Size (cc) | # Labels Attached | # Labels Not Attached | Acceptable (Yes/No) |
|---|---|---|---|---|
| 100 | *60* | *100* | *0* | *Yes* |
| 100 | *1,300* | *99* | *1* | *Yes* |

**Table 6.153. Imprinter Test Results**

| Test | Number of Labels | Label Size | Number of Labels Imprinted | Acceptable (Yes/No) |
|---|---|---|---|---|
| 1 | *100* | *1 3/4" × 3 3/8"* | *100* | *Yes* |
| 2 | *100* | *2 1/2" × 6"* | *100* | *Yes* |

### *Equipment Operation: Barcode Reader Operation*

The objective of barcode reader operation test is to verify that the imprinter is capable of detecting barcodes on labels and literature consistently over time. Literature, labels, 60 cc bottles, and 1,300 cc bottles are required for the following tests:

- Scan the barcode from a label and literature into the barcode reader controller and verify that the digital readout matches the barcode on the labels and literature.

  Small label scanned barcode number: 030781-1720-00-5

  Large label scanned barcode number: 030781-1717-12-5

  Scanned literature barcode number: 1203

- Remove the power for about 10 sec, then restore the power and verify that the scanned barcodes are retained in memory.

  Small label scanned barcode number: 030781-1720-00-5
  Acceptable? (Yes/No) *Yes*

  Large label scanned barcode number: 030781-1717-12-5
  Acceptable? (Yes/No) *Yes*

  Scanned literature barcode number: 1203
  Acceptable? (Yes/No) *Yes*

- Visually observe 100 60 cc bottles as they pass through the labeler and verify that 90 percent of the labels and literatures barcodes match the digital readouts on the control panel. Record the results in Table 6.154.

  Setup a series of 1 3/4" × 3 3/8" size labels where every third label and literature are bad. Bad being: foreign, defaced, and blank labels and literatures. Record the results in Table 6.155.

- Visually observe 100 1,300 cc bottles as they pass through the labeler and verify that 90 percent of the labels and literatures barcodes match the digital readouts on the control panel. Record the results in Table 6.155.

  Setup a series of 2 1/2" × 6" size labels where every third label and literature are bad. Bad being: foreign, defaced, and blank labels and literatures. Record the results in Table 6.155.

**Table 6.154. Barcode Reader Test Results**

| Item | Bottle Quantity | Bottle Size (cc) | # Digital Readouts Matched | # Digital Readouts Not Matched | Acceptable (Yes/No) |
|---|---|---|---|---|---|
| Label | 100 | *60* | *100* | *0* | *Yes* |
| Label | 100 | *1,300* | *99* | *1* | *Yes* |
| Literature | 100 | *60* | *98* | *2* | *Yes* |
| Literature | 100 | *1,300* | *100* | *0* | *Yes* |

***Table 6.155. Foreign, Defaced, and Blank Labels and Literatures Test Results***

| Item | Foreign | Defaced | Blank | Acceptable (Yes/No) |
|---|---|---|---|---|
| Labels (60 cc) | *4* | *25* | *4* | *Yes* |
| Labels (1,300 cc) | *6* | *22* | *5* | *Yes* |
| Literature (60 cc) | *5* | *23* | *5* | *Yes* |
| Literature (1,300 cc) | *7* | *25* | *6* | *Yes* |

## Performance Qualification

Once it has been established that the equipment is properly installed and functioning within specified operating parameters, it must be shown that the labeler can operate reliably under routine, minimum, and maximum operating conditions (Figure 6.37).

### *Labeler System Operation*

The objective of labeler system operation test is to document that the labeler is capable of attaching literature and labels to small and large bottles, marking information on the labels, and accepting and rejecting barcodes. Literature, labels, 60 cc bottles, and 1,300 cc bottles are required for the following tests:

- Visually observe 100 60 cc bottles as they pass through the labeler and verify that the literature and labels are attached to 90 percent of the bottles. Record the number of literature and labels that were attached to the containers and the number of accepted and rejected barcodes. Record the results in Table 6.156.

***Figure 6.37. PQ Elements of a Bottle Labeler***

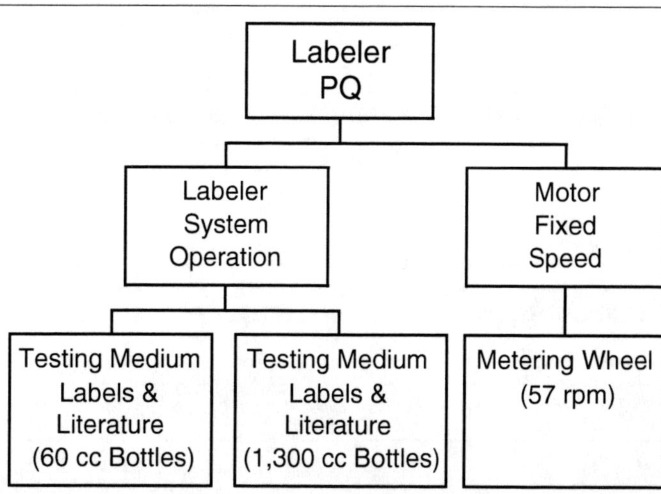

**Table 6.156. Labeler Performance Test Results**

| Test | Small Bottle Results | Large Bottle Results | Acceptable (Yes/No) |
|---|---|---|---|
| Container quantity | 100 | 100 | *Yes* |
| Container size | 60 cc | 1,300 cc | *Yes* |
| # Literature attached | 100 | 99 | *Yes* |
| # Literature not attached | 0 | 1 | *Yes* |
| # Labels attached | 98 | 100 | *Yes* |
| # Labels not attached | 2 | 0 | *Yes* |
| # of acceptable barcodes | 99 | 97 | *Yes* |
| # of rejected barcodes | 1 | 3 | *Yes* |

- Visually observe 100 1,300 cc bottles as they pass through the labeler and verify that the literature and labels are attached to 90 percent of the bottles. Record the number of literature and labels that were attached to the containers and the number of accepted and rejected barcodes. Record the results in Table 6.156.
- Measure the speed of the metering wheel motor with a calibrated tachometer. Verify that the measured speed is within ±10 percent of the fixed speed of 57 rpm. Record the results in Table 6.157. Record the instrument used to measure the speed in Table 6.158.

  Metering wheel motor speed specification = 57 rpm ± 10% maximum

  ±10% of 57 = ±5.7

  57 + 5.7 = 62.7

  57 – 5.7 = 51.3

  The measured rpm of 54 falls within ±10%.

**Table 6.157. Metering Wheel Motor Speed Test Results**

| Motor | Speed Specification (rpm) | Measured Speed (rpm) | Acceptable (Yes/No) |
|---|---|---|---|
| Metering wheel | 57 | *54* | *Yes* |

**Table 6.158. Instrument Used**

| Test Instrument | Identification Number | Calibration Due Date |
|---|---|---|
| *Tachometer* | *64020* | *06/21/97* |

## SHRINK BUNDLER

The validation of a shrink bundler (Figure 6.38) will be described here. The bundler receives bottles from the label equipment on a conveyor. The bottles are arranged into a predetermined number, typically twelve, then they are put into a rectangular configuration. The shrink bundler wraps the bundle of bottles in plastic, they then pass through the heat tunnel where the plastic is shrunk around the bundle of bottles.

The validation protocol for a shrink bundler is described in the following sections. This protocol example can be used as a cGMP compliance guide and it should be tailored to meet individual company requirements. Italicized words in the tables represent examples of information that could be recorded in a validation protocol for a shrink bundler.

### Installation Qualification

An IQ evaluation establishes confidence that the equipment is properly installed (Figure 6.39). The installation must meet the manufacturer's specified guidelines,

***Figure 6.38. Shrink Bundler (courtesy of Design Corp.)***

**Figure 6.39. *IQ Elements of a Shrink Bundler***

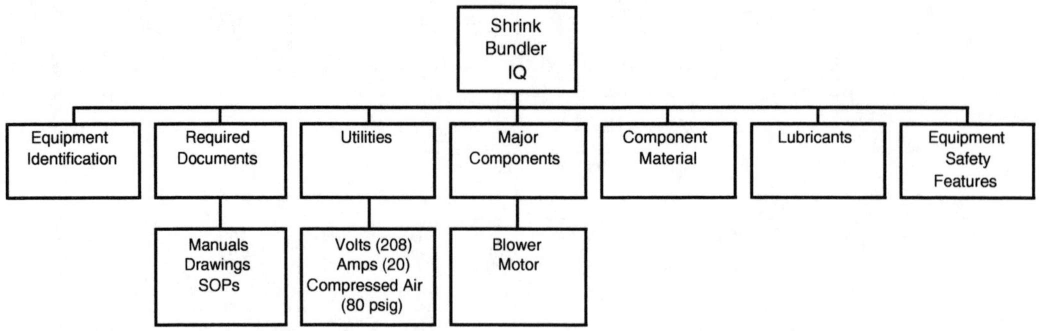


along with design changes at installation. Also, the supporting electrical utilities must meet all electrical codes. The information required for an IQ evaluation is equipment identification, required documentation, equipment utility requirements, major component specifications, component material, lubricants, and equipment safety features.

### *Equipment Identification*

Record the equipment identification numbers in Table 6.159, along with the following information: equipment manufacturer's purchase order number, model number, serial number, company assigned equipment number, and the location of the equipment.

### *Required Documentation*

Record the equipment manufacturer's operation and maintenance manual and drawings in Table 6.160. Record the SOPs that cover the setup, operation, and cleaning of the shrink bundler in Table 6.161.

**Table 6.159. *Equipment Identification***

| Required Information | As-Found Conditions |
|---|---|
| Manufacturer | *Any Shrink Bundler Co.* |
| Purchase order number | *6723810* |
| Model number | *81-LS-BSF* |
| Serial number | *93-998-61J* |
| Equipment number | *8791* |
| Location | *Packaging line 8* |

**Table 6.160. *Manufacturers' Manuals and Drawings***

| Number | Description | Date |
|---|---|---|
| *99861-J* | *Shrink Bundler Installation, Operation, and Maintenance Manual* | *None* |
| *2/287* | *Shrink Bundler General Assembly Drawing* | *04/05/92* |

**Table 6.161. *Standard Operating Procedures***

| Number | Description | Release Date |
|---|---|---|
| *PKG082* | *Shrink Bundler Setup and Operation* | *07/14/95* |
| *PKG083* | *Shrink Bundler Cleaning Procedure* | *09/22/94* |

## *Equipment Utility Requirements*

Compare the manufacturer's specified volts (V), amps (A), and compressed air requirements to their as-found conditions at the time of qualification testing and record the results in Table 6.162. Also record the location of the power supply source. Record the instruments used to measure the volts, amps, and compressed air in Table 6.163.

***Volt calculation:***

Shrink bundler volt specification = 208 V ± 10%

±10% of 208 = ±20.8

208 + 20.8 = 228.8

**Table 6.162. *Utilities***

| Utility | Specified | Measured Results | Acceptable (Yes/No) |
|---|---|---|---|
| Volts | 208 ± 10% | *A-B 220; A-C 219; B-C 223* | *Yes* |
| Amps | 0.58 | 20 circuit rating | *Yes* |
| Compressed air | Air pressure = 80 psig | 80 psig (regulated) | *Yes* |
| Power supply source, breaker box *BB23A, wire numbers: 23, 25, 27* | | | |

**Table 6.163. *Instruments Used***

| Test Instrument | Identification Number | Calibration Due Date |
|---|---|---|
| *Multimeter* | *ME-025* | *04/19/97* |
| *Air pressure gauge* | *P-102* | *02/10/97* |

208 – 20.8 = 187.2

The measured volts of 220/219/223 fall within ±10%.

***Amp calculation:***

Amp specification = 20 A ± 10%

Equipment current draw = 0.58 A

The circuit amp rating of 20 is greater than the maximum current draw of the equipment.

***Compressed air calculation:***

Compressed air specification = 80 psig

The compressed air pressure is regulated to 80 psig.

### *Major Component Specifications*

The component specifications section of the protocol verifies that the shrink bundler components purchased were delivered and installed. Record the major components in Table 6.164.

### *Component Material*

Record the material of each component that contacts the product in Table 6.165.

### *Lubricants*

Record the lubricants used to operate the shrink bundler in Table 6.166 and indicate if they make contact with the product.

***Table 6.164. Major Components***

| Components | As-Found Conditions |
|---|---|
| Blower motor | Manufacturer: *Any Motor Co.*<br>Model number: *3843-1207*<br>Serial number: *39DC*<br>Volts: *208*<br>Amperes: *0.58*<br>Phases: *3*<br>Cycles: *60 Hz*<br>hp: *1/30*<br>rpm: *3,020* |

**Table 6.165. *Component Material***

| Component | Material |
|---|---|
| *None of the components make contact with the product.* | |

**Table 6.166. *Lubricants***

| Where Used | Type | Manufacturer | Product Contact (Yes/No) |
|---|---|---|---|
| *DGO cylinder* | *Festo Gel Grease* | *Any Oil Co.* | *No* |
| *Motor* | *SAE 20 Oil* | *Any Oil Co.* | *No* |
| Is there a preventive maintenance procedure on file? (Yes/No) | | | *Yes* |

### *Equipment Safety Features*

The objective of testing equipment safety features is to verify that the safety features on the shrink bundler function according to manufacturer specifications. Verify the operation of the equipment safety features in Table 6.167 and record the results.

## Operational Qualification

An OQ evaluation should establish that the equipment can operate within specified tolerances and limits (Figure 6.40). The mechanical ranges of the shrink bundler are challenged, along with the basic equipment operations that the operator will use. Information required for the OQ evaluation is calibration of the instruments used to control the shrink bundler, equipment control functions (switches and push buttons) and equipment operation (blower motor rotation direction and speed).

**Table 6.167. *Safety Features Test Results***

| Test Function | Expected Results | Acceptable (Yes/No) |
|---|---|---|
| Auto run | During the auto run, if the heat tunnel temperature is above or below the preset range, the equipment stops. | *Yes* |
| Guard door | If any of the guard doors are removed, the equipment stops. | *Yes* |
| Bundle pusher | During the extended movement, if the bundle pusher exceeds the preset time, the equipment stops. | *Yes* |

**Figure 6.40. OQ Elements of a Shrink Bundler**

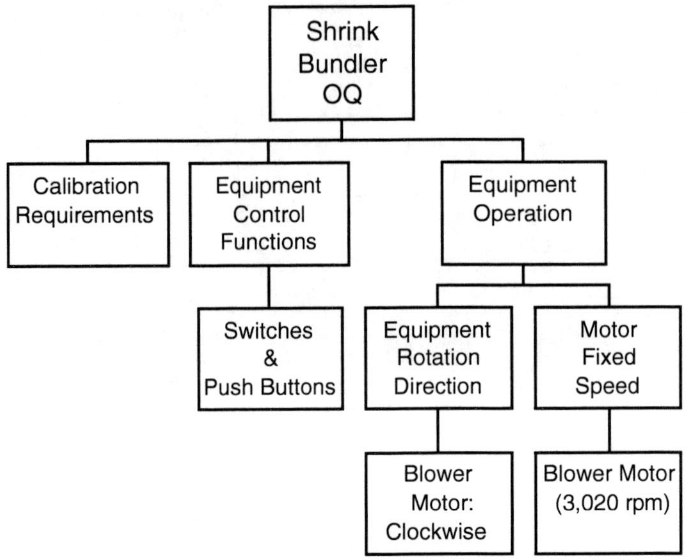


### *Calibration Requirements*

Verify that all critical instruments on the equipment have been logged into the calibration system, have calibration procedures in place, and are in calibration at the time of qualification testing. Record all information for the calibrated instruments used to control the shrink bundler in Table 6.168.

**Table 6.168. Calibrated and Noncalibrated Instruments**

*There were no noncalibrated instruments on the shrink bundler.*

| Calibrated Instrument | As-Found Conditions |
|---|---|
| Identification number | *00-033-GP* |
| Type | *Air pressure gauge, analog* |
| Manufacturer | *Any Gauge Co.* |
| Model number | *90-694* |
| Serial number | *810-GP* |
| Range | *0–200 psig* |
| Scale division | *5 psig* |
| Location | *On a pole next to the shrink bundler* |
| Use | *Regulates the air pressure to the equipment* |
| Calibration due date | *10/04/97* |
| Critical or not critical | *Not critical to the operation of the shrink bundler* |

### *Equipment Control Functions*

The objective of testing equipment control functions is to verify that the switches and push buttons on the shrink bundler operate according to manufacturer's specifications. Record the results in Table 6.169. The tests will be performed with the shrink bundler empty. Operate each control listed in Table 6.169 and verify its proper operation.

### *Equipment Operation: Blower Motor Rotation Direction Test*

The objective of the rotation direction test is to verify that the blower motor rotates in the proper direction. The following tests are performed with the shrink bundler empty. Press the start push button and observe the direction of rotation of the blower motor as viewed from behind the motor and record the results in Table 6.170.

### *Equipment Operation: Blower Motor Speed Test*

The objective of the speed test is to document the speed of the blower motor with the shrink bundler empty. A tachometer is required. Measure the speed of the blower motor with a calibrated tachometer. Verify that the measured speed is within ±10 percent of the fixed speed of 3,020 rpm. Record the results in Table 6.171. Record the instrument used to measure the speed in Table 6.172.

#### ***Blower Motor Speed Calculation:***

Blower motor speed specification = 3,020 rpm ± 10%

±10% of 3,020 = ±302

3,020 + 302 = 3,322

3,020 – 302 = 2,718

The measured rpm of 3,017 falls within ±10%.

## Performance Qualification

Once it has been established that the equipment is properly installed and functioning within specified operating parameters, it must be shown that the shrink bundler can operate reliably under routine, minimum, and maximum operating conditions (Figure 6.41).

### *Shrink Bundler Operation*

The shrink bundler operation test is designed to verify that the shrink bundler is capable of shrink wrapping various sizes of bundles of bottles consistently over time. The speed of the shrink bundler motor will be measured with the bundler full.

**Table 6.169. Equipment Control Function Test Results**

| Test Function | Expected Results | Acceptable (Yes/No) |
|---|---|---|
| Heater on/off switch | When the heater on/off switch is pressed, the shrink tunnel heaters go on and off. | *Yes* |
| System enabled push button | When the system enabled push button is pressed, the shrink bundler is activated. | *Yes* |
| Counter reset push button | When the counter reset push button is pressed, the count is reset. | *Yes* |
| Start auto run push button | When the start auto run push button is pressed, the shrink bundler is in an automatic mode. | *Yes* |
| Manual seal push button | When the manual seal push button is pressed, sheets are sealed manually. | *Yes* |
| Cycle stop push button | When the cycle stop push button is pressed, the shrink bundler stops. | *Yes* |
| Sealing bar temperature key | When the sealing bar temperature key is turned, the switching between set points is accomplished. | *Yes* |
| Sealing bar temperature ↑ key | When the sealing bar temperature ↑ key is pressed, the increment selection is accomplished. | *Yes* |
| Sealing bar temperature ↓ key | When the sealing bar temperature ↓ key is pressed, the decrement selection is accomplished. | *Yes* |
| Air supply switch | When the air supply is turned, the sealing arm is lifted. | *Yes* |
| Change recipe push button | When the change recipe push button is pressed, the recipe can be programmed. | *Yes* |
| Change part selection switch A–D | When the change part selection switch A–D is turned, the bundle configuration is chosen. | *Yes* |
| Change part selection switch E–H | When the change part selection switch E–H is turned, the bundle configuration is chosen. | *Yes* |
| Tunnel speed control knob | When the tunnel speed control knob is rotated, the speed of the conveyor increases and decreases. | *Yes* |
| Tunnel temperature control *key | When the tunnel temperature control *key is turned, select point one or two are selected. | *Yes* |
| Emergency stop push button | When the emergency stop push button is pressed, the shrink bundler stops. | *Yes* |
| Bottle sensor | When there are no bottles present, the shrink bundler stops. | *Yes* |
| Upper roll sensor | When there is no film on the upper roll, the shrink bundler stops. | *Yes* |
| Lower roll sensor | When there is no film on the lower roll, the shrink bundler stops. | *Yes* |

**Table 6.170. Blower Motor Rotation Direction Test Results**

| Item | Expected Results | Results | Acceptable (Yes/No) |
|---|---|---|---|
| Blower motor rotation direction | Rotation should be clockwise as viewed from behind the motor. | *Clockwise rotation was observed.* | *Yes* |

**Table 6.171. Blower Motor Speed Test Results**

| Motor | Speed Specification (rpm ± 10%) | Measured Speed (rpm) | Acceptable (Yes/No) |
|---|---|---|---|
| Blower | 3020 | *3017* | *Yes* |

**Table 6.172. Instrument Used**

| Test Instrument | Identification Number | Calibration Due Date |
|---|---|---|
| *Tachometer* | *64020* | *06/21/97* |

**Figure 6.41. PQ Elements of a Shrink Bundler**

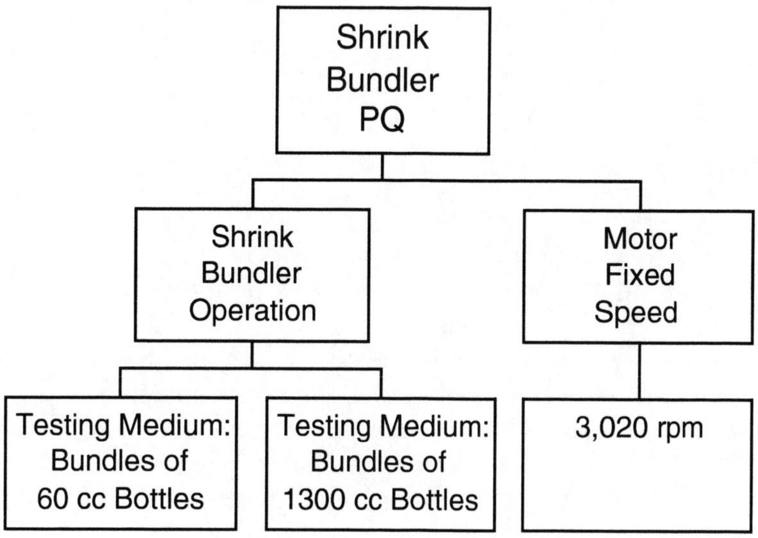


The materials and instruments required are 60 cc bottles, 1,300 cc bottles, and a tachometer. Use the following procedure to test shrink bundler:

- With the equipment operating, shrink wrap 10 bundles of 60 cc bottles and record the results in Table 6.173.
- With the equipment operating, shrink wrap 10 bundles of 1,300 cc bottles and record the results in Table 6.173.
- Measure the speed of the blower motor with a calibrated tachometer. Verify that the measured speed is within ±10 percent of the fixed speed of 3,020 rpm.

**Table 6.173. Shrink Bundler Performance Test Results**

| Bottle Size (cc) | # of Bundles | # Shrink Wrapped | # Not Shrink Wrapped | Machine Setting | Acceptable (Yes/No) |
|---|---|---|---|---|---|
| *60* | *10* | *10* | *0* | *B & E* | *Yes* |
| *1,300* | *10* | *10* | *0* | *A & E* | *Yes* |

Record the results in Table 6.174. Record the instrument used to measure the speed in Table 6.175.

Blower motor speed specification = 3,020 rpm ± 10%

±10% of 3,020 = ±302

3,020 + 302 = 3,322

3,020 – 302 = 2,718

The measured rpm of 3,014 falls within ±10%.

## Case Sealer

The validation of a case sealer (Figure 6.42) will be described here. The case sealer closes corrugated boxes by applying transparent tape to the tops and bottoms of boxes. An operator slides a filled carton into the front of the case sealer. The sealer guides the carton with gripper belts on either side. Tape is applied from both the upper and lower taping heads and is severed by upper and lower knives. Sealed cartons are ejected from the rear of the machine to proceed down a slideway.

The validation protocol for a case sealer is described in the following sections. This protocol example can be used as a cGMP compliance guide, and it should be tailored to meet individual company requirements. Italicized words in the tables represent examples of information that could be recorded in a validation protocol for a case sealer.

**Table 6.174. Blower Motor Speed Test Results**

| Motor | Speed Specification (rpm ± 10%) | Measured Speed (rpm) | Acceptable (Yes/No) |
|---|---|---|---|
| Blower | 3,020 | *3,014* | *Yes* |

**Table 6.175. Instrument Used**

| Test Instrument | Identification Number | Calibration Due Date |
|---|---|---|
| *Tachometer* | *64020* | *06/21/97* |

***Figure 6.42. Case Sealing System (courtesy of 3M Packaging System)***

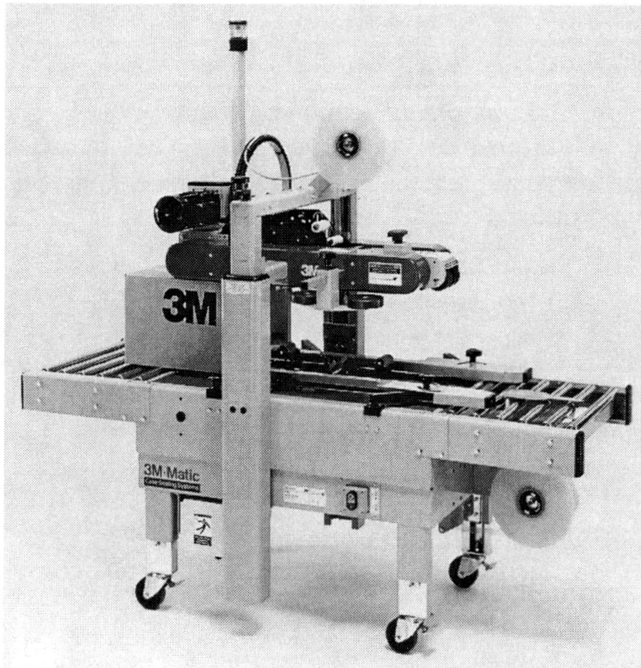


## Installation Qualification

An IQ evaluation establishes confidence that the equipment is properly installed (Figure 6.43). The installation must meet the manufacturer's specified guidelines, along with design changes at installation. Also, the supporting electrical utilities must meet all electrical codes. The information required for an IQ evaluation is equipment identification, required documentation, equipment utility requirements,

***Figure 6.43. IQ Elements of a Case Sealer***

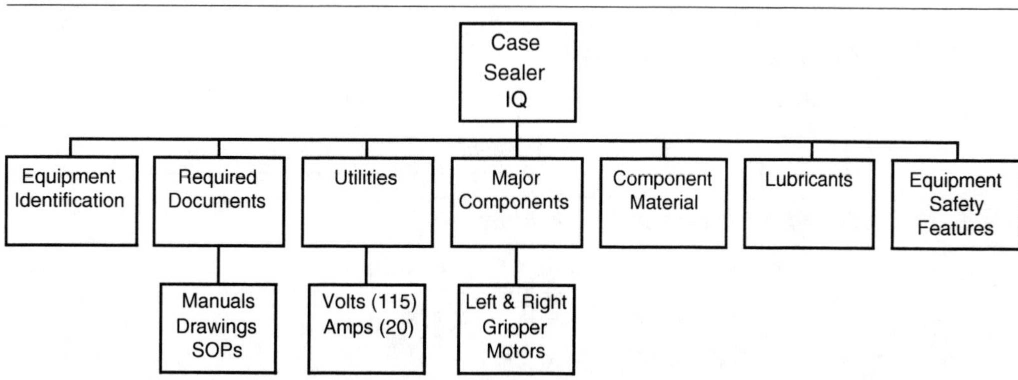

major component specifications, component material, lubricants, and equipment safety features.

### *Equipment Identification*

Record the equipment identification numbers in Table 6.176, along with the following information: equipment manufacturer's purchase order number, model number, serial number, company assigned equipment number, and the location of the equipment.

### *Required Documentation*

Record the equipment manufacturer's operation and maintenance manual and drawings in Table 6.177. Record the SOPs that cover the setup, operation, and cleaning of the case sealer in Table 6.178.

### *Equipment Utility Requirements*

Compare the manufacturer's specified volts (V) and amps (A) requirements to their as-found conditions at the time of qualification testing and record the results in Table 6.179. Also record the location of the power supply source. Record the instruments used to measure the volts and amps in Table 6.180.

**Table 6.176. Equipment Identification**

| Required Information | As-Found Conditions |
|---|---|
| Manufacturer | *Any Case Sealer Co.* |
| Purchase order number | *341876* |
| Model number | *00681* |
| Serial number | *2606* |
| Equipment number | *8990* |
| Location | *Packaging line 9* |

**Table 6.177. Manufacturers' Manuals and Drawings**

| Number | Description | Date |
|---|---|---|
| *899* | *Case Sealer Installation, Operation, and Maintenance Manual* | *None* |

**Table 6.178. Standard Operating Procedures**

| Number | Description | Release Date |
|---|---|---|
| *PKG071* | *Case Sealer Setup and Operation* | *07/14/95* |
| *PKG072* | *Case Sealer Cleaning Procedure* | *09/22/94* |

**Table 6.179. *Utilities***

| Utility | Specified | Measured Results | Acceptable (Yes/No) |
|---|---|---|---|
| Volts | 115 ± 10% | *116* | *Yes* |
| Amps | Left gripper motor = 2.4<br>Right gripper motor = 2.4 | 20 circuit rating | *Yes* |
| Power supply source, breaker box *BB53A, wire number: 51* | | | |

**Table 6.180. *Instrument Used***

| Test Instrument | Identification Number | Calibration Due Date |
|---|---|---|
| *Multimeter* | *ME-025* | *04/19/97* |

***Volt calculation:***

Case sealer volt specification = 115 V ± 10%

±10% of 115 = ±11.5

115 + 11.5 = 126.5

115 – 11.5 = 103.5

The measured volts of 116 fall within ±10%.

***Amp calculation:***

Amp specification = 20 A ± 10%

Left gripper motor = 2.4 A

Right gripper motor = 2.4 A

Equipment current draw = left gripper motor A + right gripper motor A = 2.4 + 2.4 = 4.8 A

The circuit amp rating of 20 is greater than the maximum current draw of the equipment.

### *Major Component Specifications*

The component specifications section of the protocol verifies that the case sealer components purchased were delivered and installed. Record the major components in Table 6.181.

### *Component Material*

Record the material of each component that contacts the product in Table 6.182.

**Table 6.181. Major Components**

| Components | As-Found Conditions |
|---|---|
| Left gripper motor | Manufacturer: *Any Motor Co.*<br>Model number: *3843-1207*<br>Serial number: *39DC*<br>Volts: *115*<br>Amperes: *2.4*<br>Phases: *1*<br>Cycles: *60 Hz*<br>hp: *0.12*<br>rpm: *15* |
| Right gripper motor | Manufacturer: *Any Motor Co.*<br>Model number: *3843-1207*<br>Serial number: *39DC*<br>Volts: *115*<br>Amperes: *2.4*<br>Phases: *1*<br>Cycles: *60 Hz*<br>hp: *0.12*<br>rpm: *15* |

**Table 6.182. Component Material**

| Component | Material |
|---|---|
| *None of the components make contact with the product.* | |

### *Lubricants*

Record the lubricants used to operate the case sealer in Table 6.183 and indicate if they make contact with the product.

### *Equipment Safety Features*

There are no safety features on this equipment.

## Operational Qualification

An OQ evaluation should establish that the equipment can operate within specified tolerances and limits (Figure 6.44). The mechanical ranges of the case sealer are challenged along with the basic equipment operations that the operator will use. Information required for the OQ evaluation should be calibration of the instruments used to control the case sealer, equipment control functions (switches and push buttons) and equipment operation (left and right gripper motor rotation direction and speed tests).

**Table 6.183. Lubricants**

| Where Used | Type | Manufacturer | Product Contact (Yes/No) |
|---|---|---|---|
| *Taping heads* | *SAE #30 Oil* | *Any Oil Co.* | *No* |
| Is there a preventive maintenance procedure on file? (Yes/No) | | | *Yes* |

**Figure 6.44. OQ Elements of a Case Sealer**

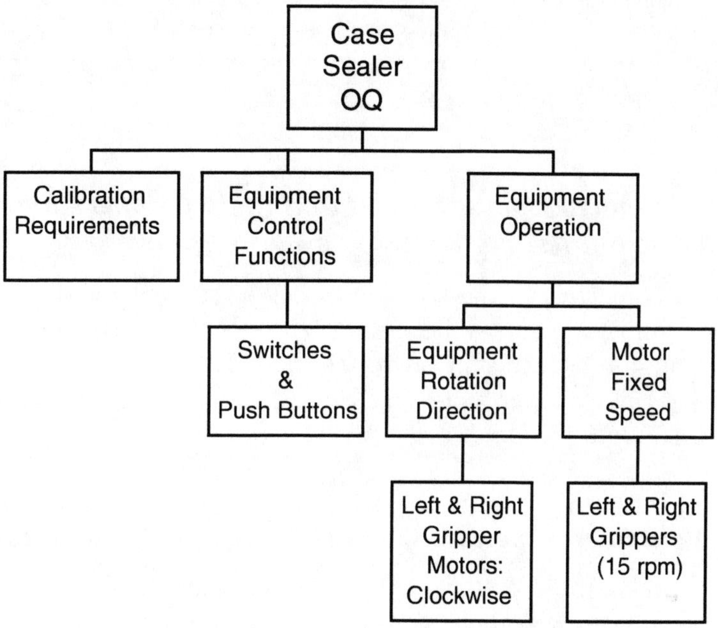


### *Calibration Requirements*

Verify that all critical instruments on the equipment have been logged into the calibration system, have calibration procedures in place, and are in calibration at the time of qualification testing. Record all information for the calibrated instruments used to control the case sealer in Table 6.184.

### *Equipment Control Functions*

The objective of testing equipment control functions is to verify that the switches and push buttons on the case sealer operate per manufacturer's specifications. Record the results in Table 6.185. The tests will be performed with the case sealer empty. Operate each control listed in Table 6.185 and verify its proper operation.

**Table 6.184. Calibrated and Noncalibrated Instruments**

*There are no calibrated or noncalibrated instruments on the case sealer.*

**Table 6.185. Equipment Control Function Test Results**

| Test Function | Expected Results | Acceptable (Yes/No) |
|---|---|---|
| Start/stop push-button operation | When the start/stop push button is pressed, the case sealer goes on and off. | *Yes* |
| Height adjustment crank operation | When the height adjustment crank is turned, the height goes up and down. | *Yes* |
| Width adjustment crank operation | When the width adjustment crank is turned, the width goes in and out. | *Yes* |

### *Equipment Operation: Left and Right Gripper Motor Rotation Direction Test*

The objective of the rotation direction test is to verify that the left and right gripper motors rotate in the proper direction. The test is performed with the case sealer empty. Press the start push button, and observe the direction of rotation of the left and right gripper motors as viewed from behind the motors and record the results in Table 6.186.

### *Equipment Operation: Left and Right Gripper Motors Speed Test*

The objective of the speed test is to document the speed of the left and right gripper motors with the case sealer empty. A tachometer is required. Measure the speed of the left and right gripper motors with a calibrated tachometer. Verify that the measured speed is within ±10 percent of the fixed speed of 15 rpm. Record the results in Table 6.187. Record the instrument used to measure the speed in Table 6.188.

Left gripper motor speed specification = 15 rpm ± 10%

±10% of 15 = ±1.5

15 + 1.5 = 16.5

15 – 1.5 = 13.5

The measured rpm of 15 falls within ±10%.

**Table 6.186. Blower Motor Rotation Direction Test Results**

| Item | Expected Results | Results | Acceptable (Yes/No) |
|---|---|---|---|
| Left and right gripper motors rotation direction | Rotation should be clockwise as viewed from behind the motors. | *Clockwise rotation was observed.* | *Yes* |

**Table 6.187. *Left and Right Gripper Motors Speed Test Results***

| Motor | Speed Specification (rpm ± 10%) | Measured Speed (rpm) | Acceptable (Yes/No) |
|---|---|---|---|
| Left gripper | 15 | *15* | *Yes* |
| Right gripper | 15 | *15* | *Yes* |

**Table 6.188. *Instrument Used***

| Test Instrument | Identification Number | Calibration Due Date |
|---|---|---|
| *Tachometer* | *64020* | *06/21/97* |

Right gripper motor speed specification = 15 rpm ± 10%

±10% of 15 = ±1.5

15 + 1.5 = 16.5

15 − 1.5 = 13.5

The measured rpm of 15 falls within ±10%.

## Performance Qualification

Once it has been established that the equipment is properly installed and functioning within specified operating parameters, it must be shown that the case sealer can operate reliably under routine, minimum, and maximum operating conditions (Figure 6.45).

### *Case Sealer Operation*

The objective of the case sealer operation test is to verify that the case sealer is capable of taping small and large size boxes consistently over time. The speed of the left and right gripper motors will be measured and recorded. This test will be performed with small and large boxes. Small boxes, large boxes, and a tachometer are required for the following tests:

- With the equipment operating, seal 100 small boxes and record the results in Table 6.189.
- With the equipment operating, seal 100 large boxes and record the results in Table 6.189.
- Measure the speed of the left and right gripper motors with a calibrated tachometer. Verify that the measured speed is within ±10 percent of the fixed speed of 15 rpm. Record the results in Table 6.190. Record the instrument used to measure the speed in Table 6.191.

**Figure 6.45. PQ Elements of a Case Sealer**

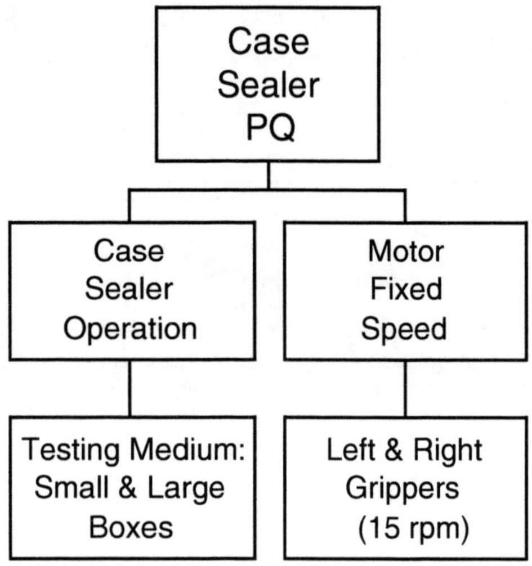


**Table 6.189. Case Sealer Performance Test Results**

| Bottle Size | # of Boxes | # Cases Sealed | # Cases Not Sealed | Acceptable (Yes/No) |
|---|---|---|---|---|
| Small | *100* | *100* | *0* | *Yes* |
| Large | *100* | *100* | *0* | *Yes* |

**Table 6.190. Left and Right Gripper Motors Speed Test Results**

| Motor | Speed Specification (rpm ± 10%) | Measured Speed (rpm) | Acceptable (Yes/No) |
|---|---|---|---|
| Left gripper | 15 | *14* | *Yes* |
| Right gripper | 15 | *15* | *Yes* |

**Table 6.191. Instrument Used**

| Test Instrument | Identification Number | Calibration Due Date |
|---|---|---|
| *Tachometer* | *64020* | *06/21/97* |

Left gripper motor speed specification = 15 rpm ± 10%

±10% of 15 = ±1.5

15 + 1.5 = 16.5

15 – 1.5 = 13.5

The measured rpm of 15 falls within ±10%.

Right gripper motor speed specification = 15 rpm ± 10%

$\pm 10\%$ of $15 = \pm 1.5$

$15 + 1.5 = 16.5$

$15 - 1.5 = 13.5$

The measured rpm of 15 falls within ±10%.

## Case Labeler

The validation of a case labeler (Figure 6.46) will be described here, that is motor driven and operates at a fixed speed. The machine prints high quality bar codes, text, and graphics on pressure sensitive labels and applies them to various sizes of cases. This equipment is controlled by a personnel computer (PC).

***Figure 6.46. Print and Apply Labeling System (courtesy of Stadia Corp.)***

The validation protocol for a case labeler is described in the following sections. This protocol example can be used as a cGMP compliance guide and it should be tailored to meet individual company requirements. Italicized words in the tables represent examples of information that could be recorded in a validation protocol for a case labeler.

## Installation Qualification

An IQ evaluation establishes confidence that the equipment is properly installed (Figure 6.47). The installation must meet the manufacturer's specified guidelines, along with design changes at installation. Also, the supporting electrical utilities must meet all electrical codes. The information required for an IQ evaluation is equipment identification, required documentation, equipment utility requirements, major component specifications, component material, lubricants, and equipment safety features.

### *Equipment Identification*

Record the equipment identification numbers in Table 6.192, along with the following information: equipment manufacturers' purchase order number, model number, serial number, company assigned equipment number, and the location of the equipment.

### *Required Documentation*

Record the equipment manufacturer's operation and maintenance manual and drawings in Table 6.193. Record the SOPs that cover the setup, operation, and cleaning of the case labeler in Table 6.194.

***Figure 6.47. IQ Elements of a Case Labeler***

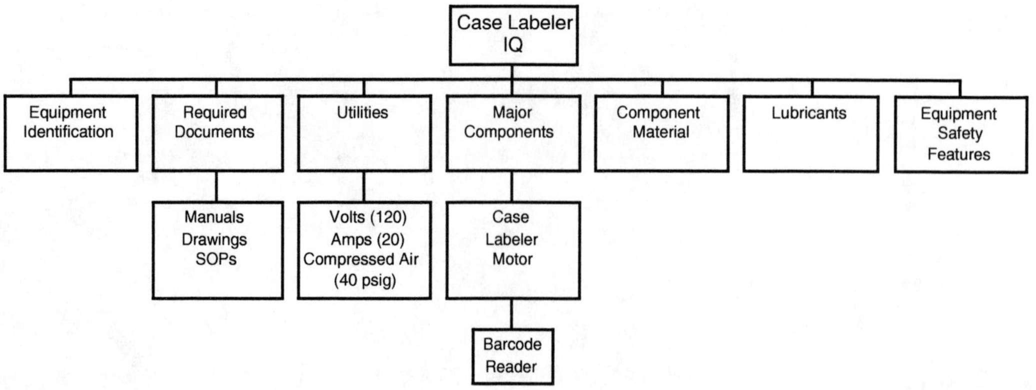

**Table 6.192. *Equipment Identification***

| Required Information | As-Found Conditions |
|---|---|
| Manufacturer | *Any Case Labeler Co.* |
| Purchase order number | *514911* |
| Model number | *0831* |
| Serial number | *L439071* |
| Equipment number | *6033* |
| Location | *Packaging line 11* |

**Table 6.193. *Manufacturers' Manuals and Drawings***

| Number | Description | Date |
|---|---|---|
| *None* | *Case Labeler Instructions Manual* | *10/96* |

**Table 6.194. *Standard Operating Procedures***

| Number | Description | Release Date |
|---|---|---|
| *PKG124* | *Case Labeler Setup and Operation* | *07/14/96* |
| *PKG125* | *Case Labeler Cleaning Procedure* | *09/22/96* |

### *Equipment Utility Requirements*

Compare the manufacturer's specified volts (V), amps (A), and compressed air requirements to their as-found conditions at the time of qualification testing and record the results in Table 6.195. Also record the location of the power supply source. Record the instruments used to measure the volts, amps, and compressed air in Table 6.196.

**Table 6.195. *Utilities***

| Utility | Specified | Measured Results | Acceptable (Yes/No) |
|---|---|---|---|
| Volts | 120 ± 10% | *125* | *Yes* |
| Amps | 10 | 20 circuit rating | *Yes* |
| Compressed air | 40 psig | 40 psig (regulated) | *Yes* |
| Power supply source, breaker box *BB10A, wire number: 5* | | | |

**Table 6.196. Instruments Used**

| Test Instrument | Identification Number | Calibration Due Date |
|---|---|---|
| *Multimeter* | *ME-025* | *04/19/97* |
| *Air pressure gauge* | *P-102* | *02/10/97* |

***Volt calculation:***

Case labeler volt specification = 120 V ± 10%

±10% of 120 = ±12

120 + 12 = 132

120 – 12 = 108

The measured volts of 125 falls within ±10%.

***Amp calculation:***

Amp specification = 20 A ± 10%

Motor = 10 A

The circuit amp rating of 20 is greater than the maximum current draw of the equipment.

***Compressed air calculation:***

Compressed air specification = 40 psig

The compressed air pressure is regulated to 40 psig.

### *Major Component Specifications*

The component specifications section of the protocol is used to verify that the case labeler components purchased were delivered and installed. Record the major components in Table 6.197.

### *Component Material*

Record the material of each component that contacts the product in Table 6.198.

### *Lubricants*

Record the lubricants used to operate the case labeler in Table 6.199 and indicate if they make contact with the product.

**Table 6.197. *Major Components***

| Components | As-Found Conditions |
|---|---|
| Motor | Manufacturer: *Any Motor Co.*<br>Model number: *15677*<br>Serial number: *472591*<br>Volts: *120*<br>Amperes: *10*<br>Phases: *1*<br>Cycles: *60 Hz*<br>hp: *1/6*<br>rpm: *104* |
| Barcode reader | Manufacturer: *Any Barcode Reader Co.*<br>Model number: *1C4DL-5572*<br>Serial number: *802414DLS* |

**Table 6.198. *Component Material***

| Component | Material |
|---|---|
| *None of the components make contact with the product.* | |

**Table 6.199. *Lubricants***

| Where Used | Type | Manufacturer | Product Contact (Yes/No) |
|---|---|---|---|
| *Gears* | *Lithium #2 Grease* | *Any Oil Co.* | *No* |
| Is there a preventive maintenance procedure on file? (Yes/No) | | | *Yes* |

### *Equipment Safety Features*

The objective of testing equipment safety features is to verify that the safety features on the case labeler function according to manufacturer specifications. Verify the operation of the equipment safety features in Table 6.200 and record the results.

**Table 6.200. *Safety Features Test Results***

| Test Function | Expected Results | Acceptable (Yes/No) |
|---|---|---|
| Emergency stop push button | When the emergency stop push button is pressed, the machine stops operating. | *Yes* |

## Operational Qualification

An OQ evaluation should establish that the equipment can operate within specified tolerances and limits (Figure 6.48). The mechanical ranges of the case labeler are being challenged along with the basic equipment operations that the operator will use. Black box software testing was performed. Black box testing evaluates the PC system against the documented functional requirements or intended uses of the equipment. Information required for the OQ evaluation is calibration of the instruments used to control the case labeler, equipment control functions (switches and push buttons) and equipment operation (metering wheel and conveyor motor rotation direction, speed test).

### *Calibration Requirements*

Verify that all critical instruments on the equipment have been logged into the calibration system, have calibration procedures in place, and are in calibration at the time of qualification testing. Record all information for the calibrated instruments used to control the case labeler in Table 6.201.

### *Equipment Control Functions*

The objective of testing equipment control functions is to verify that the switches and push buttons on the case labeler operate according to manufacturer's specifications. Record the results in Table 6.202. The tests will be performed with the case

**Figure 6.48. OQ Elements of a Case Labeler**

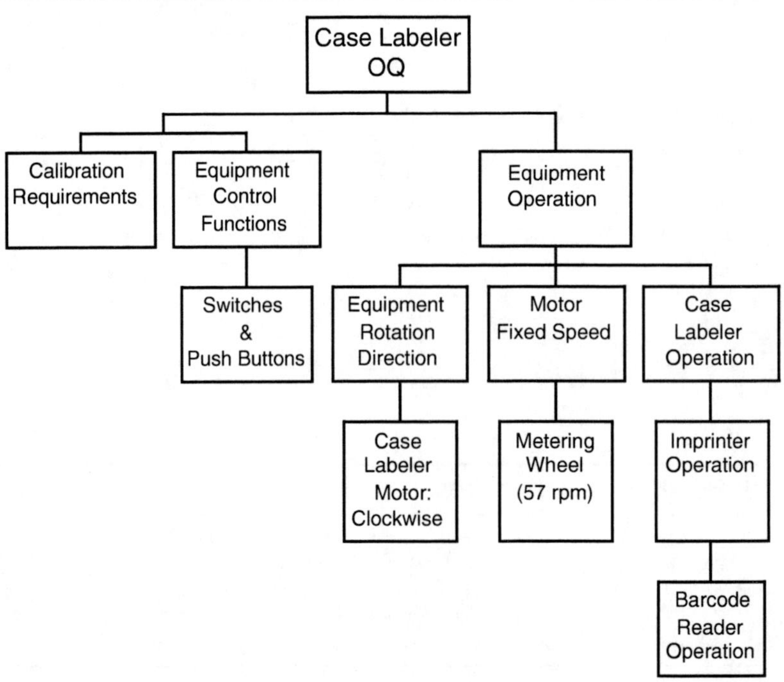

**Table 6.201. *Calibrated and Noncalibrated Instruments***

*There were no noncalibrated instruments on the case labeler.*

| Calibrated Instrument | As-Found Conditions |
|---|---|
| Identification number | *210-GP* |
| Type | *Air pressure gauge, analog* |
| Manufacturer | *Any Gauge Co.* |
| Model number | *80-595* |
| Serial number | *710-GP* |
| Range | *0–200 psig* |
| Scale division | *5 psig* |
| Location | *On a pole next to the case labeler* |
| Use | *Regulates the air pressure to the equipment* |
| Calibration due date | *10/04/98* |
| Critical or not critical | *Not critical to the operation of the case labeler* |

**Table 6.202. *Equipment Control Function Test Results***

| Test Function | Expected Results | Acceptable (Yes/No) |
|---|---|---|
| Start push button | When the start push button is pressed, the machine starts operating. | *Yes* |
| Stop push button | When the stop push button is pressed, the machine stops operating. | *Yes* |
| Power push button | When the power push button is pressed, the machine is energized. | *Yes* |
| Motor push button | When the motor push button is pressed, the motor and conveyor start. | *Yes* |
| Jog push button | When the jog push button is pressed, the motor and conveyor are jogged. | *Yes* |

labeler empty. Operate each control listed in Table 6.202 and verify its proper operation.

### *Equipment Operation: Case Labeler PC*

The objective of testing the case labeler PC is to verify that the PC that operates the case labeler operates per manufacturer's specifications. The tests will be performed with the case labeler empty. Operate each control listed in Table 6.203 and verify its proper operation.

**Table 6.203. Control Function Test Results**

| Test Function | Expected Results | Acceptable (Yes/No) |
|---|---|---|
| *Because this is a PC–controlled piece of equipment, the control functions are too numerous to list in this book.* | | |

### *Equipment Operation: Motor Rotation Direction Test*

The objective of rotation direction test is to verify that the motor rotates in the proper direction. The tests will be performed with the case labeler empty. Press the start push button and observe the direction of rotation of the motor as viewed from behind and record the results in Table 6.204.

### *Equipment Operation: Motor Speed Test*

The objective of the motor speed test is to document the speed of the motor with the case labeler empty. A tachometer is required. Measure the speed of the motor with a calibrated tachometer. Verify that the measured speed is within ±10 percent of the fixed speed of 104 rpm. Record the results in Table 6.205. Record the instrument used to measure the speed in Table 6.206.

Motor speed specification = 104 rpm ± 10% maximum

±10% of 104 = ±10.4

104 + 10.4 = 114.4

**Table 6.204. Motor Rotation Direction Test Results**

| Item | Expected Results | Results | Acceptable (Yes/No) |
|---|---|---|---|
| Motor rotation direction | Rotation should be clockwise as viewed from behind the motor. | *Clockwise rotation was observed.* | *Yes* |

**Table 6.205. Motor Speed Test Results**

| Motor | Speed Specification (rpm ± 10%) | Measured Speed (rpm) | Acceptable (Yes/No) |
|---|---|---|---|
| Labeler motor | 104 | *105* | *Yes* |

**Table 6.206. Instrument Used**

| Test Instrument | Identification Number | Calibration Due Date |
|---|---|---|
| *Tachometer* | *64020* | *06/21/97* |

104 – 10.4 = 93.6

The measured rpm of 105 falls within ±10%.

### *Equipment Operation: Case Labeler Operation*

The objective of testing the case labeler is to verify that the case labeler attaches labels to cases consistently over time. Labels, small cases, and large cases are required for the following tests:

- Visually observe 100 small cases as they pass through the case labeler and verify that 100 percent of the cases have labels attached. Record the results in Table 6.207.
- Visually observe 100 large cases as they pass through the case labeler and verify that 100 percent of the cases have labels attached. Record the results in Table 6.207.

### *Equipment Operation: Imprinter*

The objective of the imprinter operation test is to verify that the imprinter is capable of printing the serial number and date on labels consistently over time.

- Visually observe 100 connective small size labels as they pass through the imprinter and verify that 100 percent of the labels have a serial number and date printed on them. Record the results in Table 6.208.
- Visually observe 100 connective large size labels as they pass through the imprinter and verify that 100 percent of the labels have a serial number and date printed on them. Record the results in Table 6.208.

**Table 6.207. Case Labeler Test Results**

| Carton Quantity | # Labels Attached | # Labels Not Attached | Acceptable (Yes/No) |
|---|---|---|---|
| *100 small* | *99* | *1* | *Yes* |
| *100 large* | *98* | *2* | *Yes* |

**Table 6.208. Imprinter Test Results**

| Test | Number of Labels | Label Size | Number of Labels Imprinted | Acceptable (Yes/No) |
|---|---|---|---|---|
| 1 | *100* | *1 3/4" × 3 3/8"* | *100* | *Yes* |
| 2 | *100* | *2 1/2" × 6"* | *100* | *Yes* |

### *Equipment Operation: Barcode Reader Operation*

The objective of the barcode reader operation test is to verify that the imprinter is capable of detecting barcodes on labels and literature consistently over time. Labels, small cases, and large cases are required for the following tests:

- Scan the barcode from a label into the barcode reader controller and verify that the digital readout matches the barcode on the labels.

  Small label scanned barcode number: 030781-1720-00-5

  Large label scanned barcode number: 030781-1717-12-5

- Remove the power for about 10 seconds, then restore the power and verify that the scanned barcodes are retained in memory.

  Small label scanned barcode number: 030781-1720-00-5
  Acceptable? (Yes/No) *Yes*

  Large label scanned barcode number: 030781-1717-12-5
  Acceptable? (Yes/No) *Yes*

- Visually observe 100 small cases as they pass through the case labeler and verify that 90 percent of the labels barcodes match the digital readouts on the PC. Record the results in Table 6.209.
- Set up a series of 1 3/4" × 3 3/8" size labels where every third label is bad. Bad labels are: foreign, defaced, and blank labels. Record the results in Table 6.210.
- Visually observe 100 large cases as they pass through the case labeler and verify that 90 percent of the labels barcodes match the digital readouts on the PC. Record the results in Table 6.209.
- Set up a series of 2 1/2" × 6" size labels where every third label is bad. Bad labels are: foreign, defaced, and blank labels. Record the results in Table 6.210.

**Table 6.209. Barcode Reader Test Results**

| Item | Case Qty | Case Size | # Digital Readouts Matched | # Digital Readouts Not Matched | Acceptable (Yes/No) |
|---|---|---|---|---|---|
| Label | 100 | *Small* | *100* | *0* | *Yes* |
| Label | 100 | *Large* | *99* | *1* | *Yes* |

**Table 6.210. Foreign, Defaced, and Blank Labels Test Results**

| Item | Foreign | Defaced | Blank | Acceptable (Yes/No) |
|---|---|---|---|---|
| Labels—small | *4* | *25* | *4* | *Yes* |
| Labels—large | *6* | *22* | *5* | *Yes* |

## Performance Qualification

Once it has been established that the equipment is properly installed and functioning within specified operating parameters, it must be shown that the case labeler can operate reliably under routine, minimum, and maximum operating conditions (Figure 6.49).

### *Case Labeler System Operation*

The objective of the case labeler system operation test is to document that the case labeler is capable of: attaching labels to small and large cases, marking information on the labels and accepting and rejecting barcodes. Labels, small cases, and large cases are required for the following tests:

- Observe 100 small cases as they pass through the case labeler and verify that the labels are attached to 90 percent of the cases. Record the number of labels that were attached to the cases and the number of accepted and rejected barcodes. Record the results in Table 6.211.
- Visually observe 100 large cases as they pass through the case labeler and verify that the labels are attached to 90 percent of the cases. Record the number of labels that were attached to the containers and the number of accepted and rejected barcodes. Record the results in Table 6.211.
- Measure the speed of the motor with a calibrated tachometer. Verify that the measured speed is within ±10 percent of the fixed speed of 104 rpm. Record the results in Table 6.212. Record the instrument used to measure the speed in Table 6.213.

***Figure 6.49. PQ Elements of a Case Labeler***

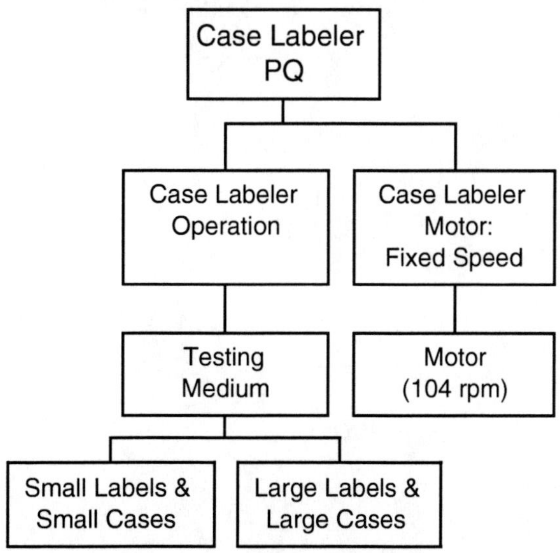

**Table 6.211. *Case Labeler Performance Test Results***

| Test | Small Case Results | Large Case Results | Acceptable (Yes/No) |
|---|---|---|---|
| Container quantity | *100* | *100* | *Yes* |
| Container size | *Small* | *Large* | *Yes* |
| Labels attached | *98* | *100* | *Yes* |
| Labels not attached | *2* | *0* | *Yes* |
| # of acceptable barcodes | *99* | *97* | *Yes* |
| # of rejected barcodes | *1* | *3* | *Yes* |

**Table 6.212. *Motor Speed Test Results***

| Motor | Speed Specification (rpm ± 10%) | Measured Speed (rpm) | Acceptable (Yes/No) |
|---|---|---|---|
| Print and apply wheel | 57 | *102* | *Yes* |

**Table 6.213. *Instrument Used***

| Test Instrument | Identification Number | Calibration Due Date |
|---|---|---|
| *Tachometer* | *64020* | *06/21/97* |

Motor speed specification = 104 rpm ± 10% maximum

±10% of 104 = ±10.4

104 + 10.4 = 114.4

104 – 10.4 = 93.6

The measured rpm of 102 falls within ±10%.

# Chapter 7

# Validation of Laboratory Equipment

This chapter will cover the qualification testing of equipment used to perform operations in a laboratory (Figure 7.1).

## Lab Mixer

The validation of a lab mixer (Figure 7.2) that is air driven will be described here. The mixer has a variable speed that is controlled by a valve. The operator establishes the desired mixing action visually. The mixer is used for mixing coating, or granulation solutions and suspensions. The mixer can be operated without the impeller being submerged in water or other representative fluid or powder.

***Figure 7.1. Laboratory Equipment***

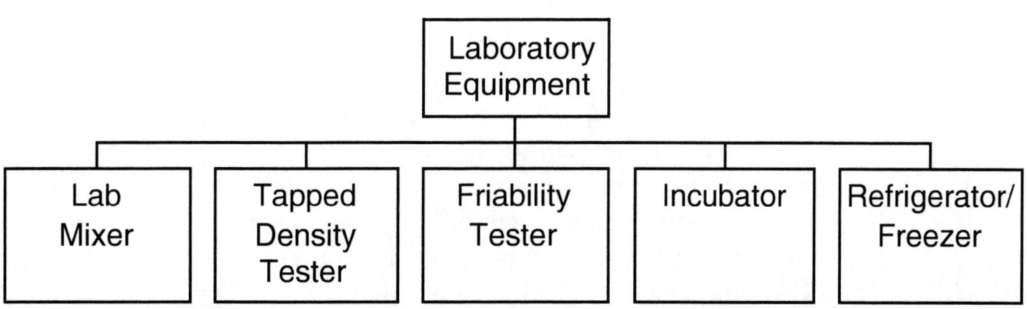

***Figure 7.2. Lab Mixer (courtesy of Lightnin)***

The validation protocol for a mixer is described in the following sections. This protocol template can be used as a cGMP compliance guide, and it should be tailored to meet individual company requirements. Italicized words in the tables represent examples of information that could be recorded in a validation protocol for a mixer.

## Installation Qualification

An IQ evaluation establishes confidence that the equipment is properly installed (Figure 7.3). The installation must meet the manufacturer's specified guidelines, along with design changes at installation. The information required for an IQ evaluation is equipment identification, required documentation, equipment utility requirements, major component specifications, materials of construction, lubricants, and equipment safety features.

**Figure 7.3. IQ Elements of a Lab Mixer**

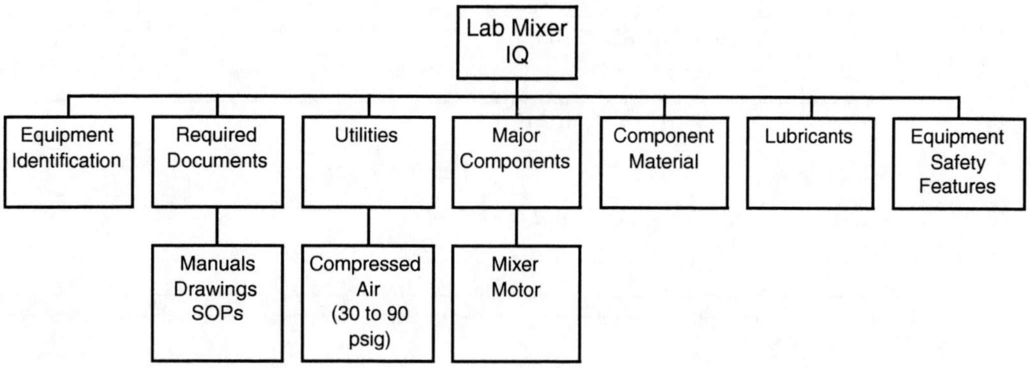


### *Equipment Identification*

Record the equipment identification numbers in Table 7.1, along with the following information: equipment manufacturer's purchase order number, model number, serial number, company assigned equipment number, and the location of the equipment.

### *Required Documentation*

Record the equipment manufacturer's operation and maintenance manual in Table 7.2. Record the SOPs that cover the setup, operation, and cleaning of the mixer in Table 7.3.

### *Equipment Utility Requirements*

Compare the manufacturer's specified compressed air pressure requirements to the as-found condition at the time of qualification testing and record the results in Table 7.4. Record the instrument used to measure the compressed air in Table 7.5.

**Table 7.1. Equipment Identification**

| Required Information | As-Found Conditions |
|---|---|
| Manufacturer | *Any Mixer Co.* |
| Purchase order number | *003481* |
| Model number | *Series 20* |
| Serial number | *800300* |
| Equipment number | *0854* |
| Location | *Room 125 (This equipment is portable.)* |

**Table 7.2. *Manufacturers' Manuals and Drawings***

| Number | Description | Date |
|---|---|---|
| *None* | *Installing and Maintaining the Lab Mixer* | *1987* |

**Table 7.3. *Standard Operating Procedures***

| Number | Description | Release Date |
|---|---|---|
| *LAB011* | *Lab Mixer Setup and Operation* | *02/08/95* |
| *LAB012* | *Lab Mixer Cleaning Procedures* | *03/08/95* |

**Table 7.4. *Utilities***

| Utility | Specified (psig) | As-Found Condition (psig) | Acceptable (Yes/No) |
|---|---|---|---|
| Compressed air | 30–90 | *89* | *Yes* |

**Table 7.5. *Instrument Used***

| Test Instrument | Identification Number | Calibration Due Date |
|---|---|---|
| *Air pressure gauge* | *PG-001* | *08/14/97* |

***Compressed air calculation:***

Compressed air specification = 30–90 psig

The measured air pressure of 89 psig falls within specification.

### *Major Component Specifications*

The component specifications section of the protocol verifies that the mixer components purchased were delivered and installed. Record the major components in Table 7.6.

### *Component Material*

Record the material of each component that contacts the product in Table 7.7.

### *Lubricants*

Record the lubricants used to operate the mixer in Table 7.8 and indicate if they make contact with the product.

### *Equipment Safety Features*

There are no safety features on this equipment. Nevertheless, there are some general safety precautions that should be followed when operating a mixer.

**Table 7.6. *Major Components***

| Components | As-Found Conditions |
|---|---|
| Mixer motor | Manufacturer: *Any Motor Co.*<br>Model number: *20*<br>Serial number: *65248*<br>Air Pressure: *30–90 psig*<br>hp: *1/30*<br>rpm: *30–1750* |
| Shaft | Part number: *63142*<br>Size: *5/16" diameter by 12" long*<br>Material: *316 stainless steel* |
| Impeller | Part number: *63143*<br>Size: *5/16" diameter by 12" long*<br>Material: *316 stainless steel* |

**Table 7.7. *Component Material***

| Component | Material |
|---|---|
| Shaft | *316 stainless steel* |
| Impeller | *316 stainless steel* |

**Table 7.8. *Lubricants***

| Where Used | Type | Manufacturer | Product Contact (Yes/No) |
|---|---|---|---|
| *Air motor* | *SAE # 10 Oil* | *Any Oil Co.* | *No* |
| Is there a preventive maintenance procedure on file? (Yes/No) | | | *Yes* |

- Never attempt to move or adjust a mixer while it is running.
- Never touch any rotating part of a mixer with bare hands, gloved hands, or with any hand-held object. Rotating parts include, but are not limited to, the mixer shaft, impeller(s), and mechanical seals.

## Operational Qualification

An OQ evaluation should establish that the equipment can operate within specified tolerances and limits (Figure 7.4). The mechanical ranges of the mixer are challenged, along with the basic mixer operations that the operator will use. The mixer will be validated for its operating ability, not how well it mixes liquids. The information required for the OQ evaluation is calibration of the instrument used to control the mixer, equipment control functions (switches and push buttons) and equipment operation (mixer speed).

**Figure 7.4. OQ Elements of a Lab Mixer**

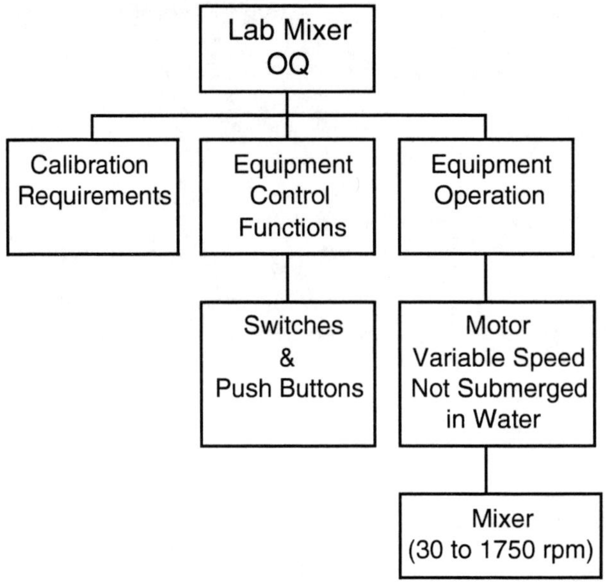


### *Calibration Requirements*

Verify that all critical instruments on the equipment, have been logged into the calibration system, have calibration procedures in place, and are currently in calibration at the time of qualification testing. Record information for the calibrated instruments used to control the mixer in Table 7.9.

### *Equipment Control Functions*

The objective of testing equipment control functions is to verify that the switches and push buttons on the mixer operate per manufacturer's specifications. The mixer will be operated without the impeller being submerged in water or other representative fluid or powder. The control that requires testing is the speed control valve. A tachometer will be required.

- Verify that the mixer is in the off position by rotating the speed control valve clockwise until it reaches a stop. This is for safety purposes. If the valve is on, when the mixer is connected to the air supply it will start operating.
- Connect the mixer's air pressure hose to the compressed air supply. Rotate the speed control valve counterclockwise and verify that the mixer speed

**Table 7.9. Calibrated and Noncalibrated Instruments**

| |
|---|
| *There were no calibrated or noncalibrated instruments on this equipment.* |

increases. Measure the mixer speed with a calibrated tachometer and record the measurement in Table 7.10. Record the instrument used to measure the speed in Table 7.11.

- Rotate the Speed Control Valve clockwise and verify that the mixer speed increases. Measure the mixer speed with a calibrated tachometer and record the measurement in Table 7.10. Record the instrument used to measure the speed in Table 7.11.

### *Equipment Operation: Mixer Speed Test*

The objective of speed test is to verify that the measured speeds are within ±10 percent of the manufacturer's specification. The mixer will be operated without the impeller being submerged in water or other representative fluid or powder. A tachometer will be required. Measure the speed of the mixer with a calibrated tachometer. Verify that the measured speeds are within ±10 percent of the manufacturer's specification of 30 rpm minimum and 1,750 rpm maximum and record the results in Table 7.12. Record the instrument used to measure the speed in Table 7.11.

***Mixer speed calculation:***

Mixer minimum speed specification = 30 rpm ± 10%

±10% of 30 = ±3

30 + 3 = 33

30 – 3 = 27

The measured rpm of 29 falls within ±10%.

**Table 7.10. Equipment Control Function Test Results**

| Test Function | Results | Acceptable (Yes/No) |
|---|---|---|
| Speed control valve operation | The mixer speed increases when the speed control valve is rotated counterclockwise. | *Yes* |
| | Initial speed *377.3 rpm*, increased speed *786.4 rpm* | |
| | The mixer speed decreases when the speed control valve is rotated clockwise. | *Yes* |
| | Initial speed *1,184.1 rpm*, decreased speed *865.2 rpm* | |

**Table 7.11. Instrument Used**

| Test Instrument | Identification Number | Calibration Due Date |
|---|---|---|
| *Tachometer* | *64020* | *06/21/97* |

**Table 7.12.** ***Mixer Speed Test Results***

| Item | Specification (rpm ± 10%) | Measured Speed (rpm) | Acceptable (Yes/No) |
|---|---|---|---|
| Minimum speed | 30 | *29* | *Yes* |
| Maximum speed | 1,750 | 1,753 | *Yes* |

Mixer maximum speed specification = 1750 rpm ± 10%

±10% of 1750 = ±175

1750 + 175 = 1925

1750 – 175 = 1575

The measured rpm of 1753 falls within ±10%.

## Performance Qualification

Once it has been established that the equipment is properly installed and functioning within specified operating parameters, it must be shown that the mixer can perform reliably under routine, minimum, and maximum operating conditions (Figure 7.5).

### *Lab Mixer Operation*

The mixer operation test is designed to document the performance of the mixer using water and a colored dye. The water is used for maximum loading conditions. Also, the objective of this test is to verify that the mixer can move the fluid about the container (i.e., mixing). The variable speed of the mixer will be measured and recorded.

The materials and instruments required for this test are a mixing container, test fluid, dye, and a tachometer. Use the following procedure to test mixer operation:

- Fill a mixing container with water to its maximum working volume of 4.5 L and record the amount used in Table 7.13. The impeller should extend to at least 2/3 of the liquid's depth. Position the mixing head off center and in an angular direction. Record the required materials in Table 7.13.

  Total container volume = 6 L

  Mixing material volume = 75% of 6 = 4.5 L

- Add FDC blue #1 dye to the mixing container and observe the mixing action of the fluid. The added solution should be distributed uniformly throughout the container. Record the results in Table 7.14. Record the dye used in Table 7.13.

***Figure 7.5. PQ Elements of a Lab Mixer***

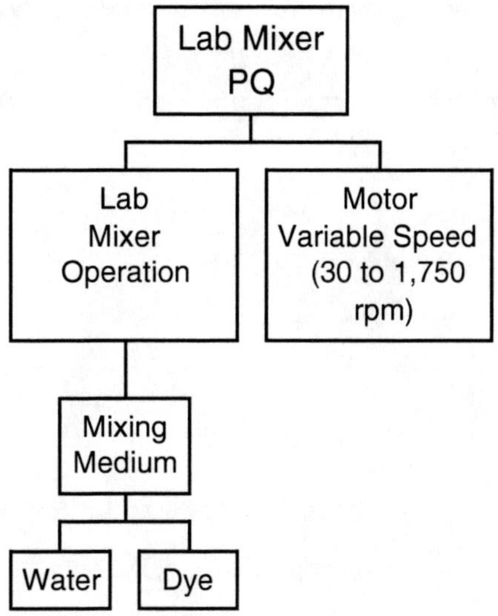


***Table 7.13. Test Materials***

| Item | Specifications |
|---|---|
| Mixing container | *6 L* |
| Test fluid | *Water* |
| Test dye | *FDC Blue #1* |
| Test fluid volume | *4.5 L* |

- Measure the speed of the mixer with a calibrated tachometer. Verify that the measured speeds are within ±10 percent of the manufacturer's specification of 30 rpm minimum and 1,750 rpm maximum and record the results in Table 7.14. Record the instrument used to measure the speed in Table 7.15.

***Table 7.14. Mixer Speed Test Results***

| Test Function | Expected Results | Measured Speed | Acceptable (Yes/No) |
|---|---|---|---|
| Fluid mixing capabilities | The added dye should be distributed uniformly throughout the container. | *The dye was distributed uniformly throughout the container.** | *Yes* |
| Minimum speed | 30 rpm ± 10% | *28 rpm* | *Yes* |
| Maximum speed | 1750 rpm ± 10% | *1749 rpm* | *Yes* |

*The results of this test are qualitative only, and were based on observation.

**Table 7.15. Instrument Used**

| Test Instrument | Identification Number | Calibration Due Date |
|---|---|---|
| *Tachometer* | *64020* | *06/21/97* |

Mixer minimum speed specification = 30 rpm ± 10%

±10% of 30 = ±3

30 + 3 = 33

30 – 3 = 27

The measured rpm of 28 falls within ±10%.

Mixer maximum speed specification = 1,750 rpm ± 10%

±10% of 1,750 = ±175

1,750 + 175 = 1,925

1,750 – 175 = 1,575

The measured rpm of 1,749 falls within ±10%.

## Tapped Density Tester

The validation of a tapped density tester (Figure 7.6) will be described here. Powdered or granular materials of a known weight are placed in a graduated cylinder. The electronic counter is set for the desired number of taps, from 1 to 9,999. The equipment is switched on, and in a few minutes the equipment switches off automatically after the predetermined number of taps. The tapped volume of the material can now be read directly on the scale of the graduated cylinder. By knowing the packed volume and the weight, the tap density is determined.

The validation protocol for a tapped density tester is described in the following sections. This protocol example can be used as a cGMP compliance guide, and it should be tailored to meet individual company requirements. Italicized words in the tables represent examples of information that could be recorded in a validation protocol for a tapped density tester.

### Installation Qualification

An IQ evaluation establishes confidence that the equipment is properly installed (Figure 7.7). The installation must meet the manufacturer's specified guidelines, along with design changes at installation. Also, the supporting electrical utilities must meet all electrical codes. The information required for an IQ evaluation is equipment identification, required documentation, equipment utility requirements,

***Figure 7.6.*** ***Tapped Density Tester (courtesy of Van-Kel Industries, Inc.)***

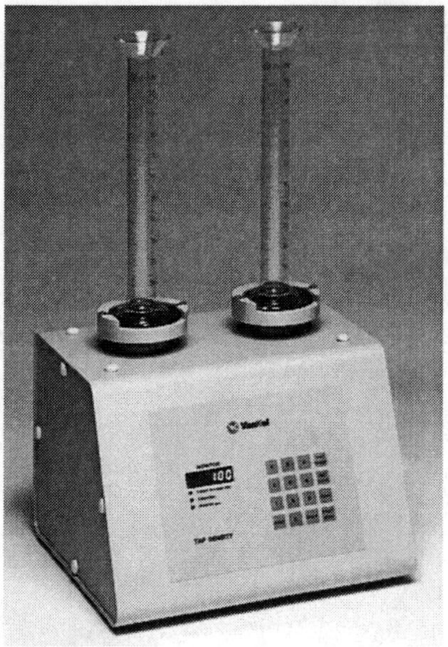

***Figure 7.7.*** ***IQ Elements of a Tapped Density Tester***

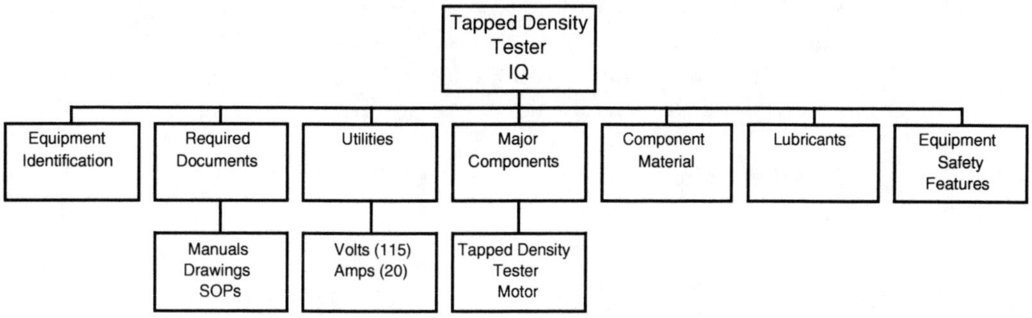


major component specifications, component material, lubricants, and equipment safety features.

## *Equipment Identification*

Record the equipment identification numbers in Table 7.16, along with the following information: equipment manufacturer's purchase order number, model number, serial number, company assigned equipment number, and the location of the equipment.

**Table 7.16. Equipment Identification**

| Required Information | As-Found Conditions |
|---|---|
| Manufacturer | *Any Tapped Density Tester Co.* |
| Purchase order number | *0010792* |
| Model number | *47868M* |
| Serial number | *0219977* |
| Equipment number | *5381* |
| Location | *Room 263 (This equipment is portable.)* |

### *Required Documentation*

Record the equipment manufacturer's operation and maintenance manual and drawings in Table 7.17. Record the standard operating procedures that cover the setup, operation, and cleaning of the tapped density tester in Table 7.18.

### *Equipment Utility Requirements*

Compare the manufacturer's specified volt (V) and amps (A) requirements to their as-found condition at the time of qualification testing and record the results in Table 7.19. Also record the location of the power supply source. Record the instrument used to measure the volts and amps in Table 7.20.

***Volt calculation:***

Volt specification = 115 V ± 10%

±10% of 115 = ±11.5

115 + 11.5 = 126.5

115 – 11.5 = 103.5

The measured volts of 120 falls within ±10%.

**Table 7.17. Manufacturers' Manuals and Drawings**

| Number | Description | Date |
|---|---|---|
| *None* | *Tapped Density Tester Operating Instructions Manual* | *None* |

**Table 7.18. Standard Operating Procedures**

| Number | Description | Date |
|---|---|---|
| *LAB009* | *Setup and Operation of the Tapped Density Tester* | *08/05/94* |
| *LAB010* | *Lab—General House Cleaning Procedures* | *03/20/96* |

**Table 7.19. *Utilities***

| Utility | Specified | Measured Results | Acceptable (Yes/No) |
|---|---|---|---|
| Volts | 115 ± 10% | *120* | *Yes* |
| Amps | 2 | 20 circuit rating | *Yes* |

Power supply source, breaker box *BB23, wire number: 5*

**Table 7.20. *Instrument Used***

| Test Instrument | Identification Number | Calibration Due Date |
|---|---|---|
| *Multimeter* | *ME-025* | *04/19/97* |

***Amp calculation:***

Circuit rating = 20 A

Equipment current draw = 2 A

The circuit rating of 20 amps is greater than the maximum current draw of the equipment.

### *Major Component Specifications*

The component specifications section of the protocol verifies that the tapped density tester components purchased were delivered and installed. Record the major components in Table 7.21.

### *Component Material*

Record the material of each component that contacts the product in Table 7.22.

**Table 7.21. *Major Components***

| Components | As-Found Conditions |
|---|---|
| Tapped density tester motor | Manufacturer: *Any Motor Co.*<br>Model number: *3BE07414*<br>Serial number: *3BE07414*<br>Volts: *115*<br>Amperes *Not specified*<br>Phases: *1*<br>Cycles: *60 Hz*<br>hp: *1*<br>Taps per minute (tpm): *250* |

**Table 7.22. Component Material**

| Component | Material |
|---|---|
| Graduated cylinder | *Glass* |

### *Lubricants*

Record the lubricants used to operate the tester in Table 7.23 and indicate if they make contact with the product.

### *Equipment Safety Features*

There are no safety features on this equipment.

## Operational Qualification

An OQ evaluation should establish that the equipment can operate within specified tolerances and limits (Figure 7.8). The mechanical ranges of the tapped density tester are challenged, along with the basic tapped density tester operations that the operator will use. The tapped density tester will be validated for its operating ability. The information required for the OQ evaluation is calibration of the instrument used to control the tapped density tester, equipment control functions (switches and push buttons) and equipment operation (tapped density tester motor rotation direction and speed tests).

### *Calibration Requirements*

Verify that all critical instruments on the equipment have been logged into the calibration system, have calibration procedures in place, and are currently in calibration at the time of qualification testing. Record all information for the calibrated instruments used to control the tapped density tester in Table 7.24.

**Table 7.23. Lubricants**

| Where Used | Type | Manufacturer | Product Contact (Yes/No) |
|---|---|---|---|
| *The motor gearing was packed at the vendor and requires no further attention.* | | | |
| Is there a preventive maintenance procedure on file? (Yes/No) | | | *Yes* |

**Figure 7.8.** ***OQ Elements of a Trapped Density Tester***

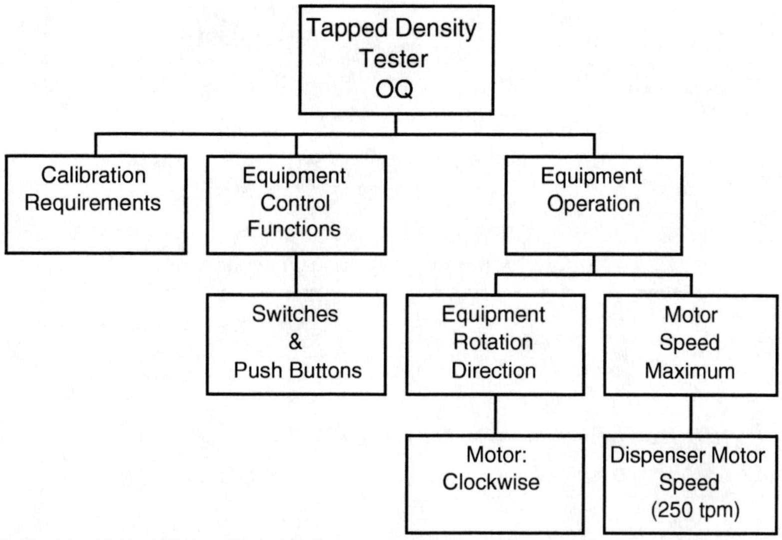


**Table 7.24.** ***Calibrated and Noncalibrated Instruments***

*There were no noncalibrated instruments on the tapped density tester.*

| Calibrated Instrument | As-Found Conditions |
|---|---|
| Identification number | *Test-rpm-00* |
| Type | *Counter, Analog* |
| Manufacturer | *Any Gauge Co.* |
| Model number | *3002 VATS* |
| Serial number | *61916671* |
| Range | *250 tpm* |
| Scale division | *2 tpm* |
| Location | *On the control panel* |
| Use | *Counts the number of taps* |
| Calibration due date | *03/30/97* |
| Critical or not critical | *Not critical to the operation of the tester* |

## *Equipment Control Functions*

The objective of testing equipment control functions is to verify that the push buttons on the tapped density tester operate per manufacturer specifications. The tapped density tester will be operated with the tester empty. Operate each control listed in Table 7.25 and verify their proper operation.

**Table 7.25. *Control Function Test Results***

| Test Function | Expected Results | Acceptable (Yes/No) |
|---|---|---|
| On/off switch | When the on/off switch is pressed, the indicator light goes on and off. | *Yes* |
| Counter setting bar | When the tester is set to 10 taps and the power is turned on, the tester taps 10 times. | *Yes* |
| Counter setting bar | When the counter setting bar reset button is pressed, while the power is still on, the tester taps 10 times. | *Yes* |

### *Equipment Operation: Tapped Density Tester Motor Rotation Direction Test*

The objective of the rotation test is to verify that the tapped density tester's motor rotates in the proper direction. The tapped density tester will be operated with the tester empty. Press the on/off button and observe the direction of rotation of the tapped density tester motor as viewed from behind the motor and record the results in Table 7.26.

### *Equipment Operation: Tapped Density Tester Motor Speed Test*

The objective of the motor speed test is to document the speed of the motor in tpm. The tapped density tester will be empty. A stopwatch will be required. Count the number of taps for one minute and verify that the calculated speed is within ±1 tap of the fixed speed of 250 tpm. Record the results in Table 7.27. Record the instrument used to measure the time in Table 7.28.

**Table 7.26. *Tapped Density Tester Motor Rotation Direction Test Results***

| Item | Expected Results | Results | Acceptable (Yes/No) |
|---|---|---|---|
| Trapped density tester motor rotation direction | Rotation should be clockwise as viewed from behind the motor. | *Clockwise rotation was observed.* | *Yes* |

**Table 7.27. *Tapped Density Tester Speed Test Results***

| Test Function | Speed Specification (tpm ± 1 tap) | Calculated Speed (tpm) | Acceptable (Yes/No) |
|---|---|---|---|
| Number of taps | 250 | *249* | *Yes* |

***Table 7.28. Instrument Used***

| Test Instrument | Identification Number | Calibration Due Date |
|---|---|---|
| *Stopwatch* | *54039* | *02/16/97* |

Tapped density tester motor speed specification = 250 tpm ± 1 tap

±10% of 250 = ±1 tap

250 + 1 = 251

250 – 1 = 249

The measured rpm of 249 falls within ±1 tap.

## Performance Qualification

Once it has been established that the equipment is properly installed and functioning within specified operating parameters, it must be shown that the tapped density tester can operate reliably under routine, minimum, and maximum operating conditions (Figure 7.9).

### *Tapped Density Tester Operation*

The tapped density tester operation test is designed to document the performance of the tapped density tester using a placebo powder. The speed of the tapped density tester will be calculated.

***Figure 7.9. PQ Elements of a Tapped Density Tester***

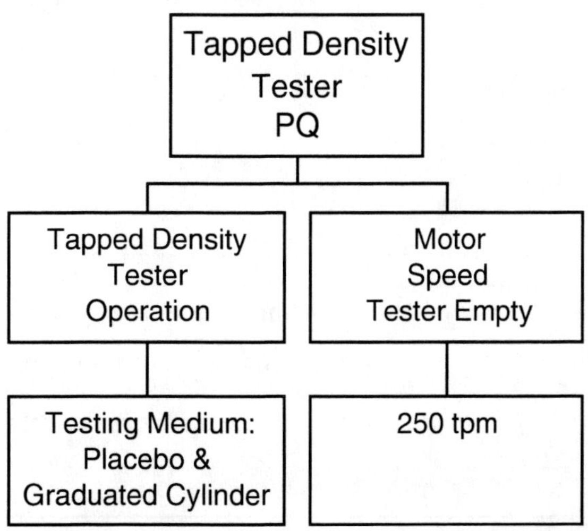

The materials and instruments required for this test are placebo, a graduated cylinder, and a stopwatch (Table 7.29). The test is performed as follows:

- Pour the material to be measured at an even flow rate into a tared 100 ml graduated cylinder, filling the cylinder to between 70 and 90 mL. Important: If the material in the graduated cylinder is not level, calculate the average volume by adding the high and low volume level and dividing by two. Record the results in Table 7.30.

  Placebo average volume = (High Volume + Low Volume) ÷ 2 =
  80 + 82 = 81 mL

- Divide the weight in g into the starting volume in mL to calculate the bulk density and record the results in the following Table 7.29.

  Bulk density = (Starting volume in mL) ÷ g = 81 ÷ 100.248 = 0.808 g/mL

- Place the graduated cylinder containing the placebo on the tapped density tester and secure with the locking collar. Record the instrument used to measure the volume in Table 7.32.

- Setting Number of Tabs:

  a. Press the counter setting bar and open the counter window with the sliding up arrow button.

  b. Hold the counter window open and set the number of taps with the push buttons above the counter window. Tapping should be performed in sets of 100 repetitions until the volume change is less than 1 mL per 100 repetitions. Close the counter window.

  c. Press the on-off button to the on position.

  d. To repeat the same number of taps in step b reset the counter setting bar.

**Table 7.29. Required Materials and Conditions**

| Item | Specifications |
|---|---|
| Placebo | *Potato starch* |
| Graduated cylinder | *100 mL* |
| Placebo average volume | *81 mL* |
| Bulk density | *0.808 g/mL* |
| Trapped density | *0.905 g/mL* |

**Table 7.30. Tapped Density Tester Performance Test Results**

| Number of Taps | Volume Change (mL) | Volume Change Is Less Than 1 mL per 100 (Yes/No) |
|---|---|---|
| 100 | 8.5 | *No* |
| 100 | 0.5 | *No* |

- Divide the weight (g) into the final tapped volume to calculate the tapped density and record the results in the Table 7.30.

  Tapped density = (Final Volume in mL) ÷ g = 72 ÷ 79.558 = 0.905 g/mL

- Count the number of taps for one minute and verify that the calculated speed is within ±1 tap of the fixed speed of 250 tpm. Record the results in Table 7.31. Record the instrument used to measure the time in Table 7.32.

  Tapped density tester speed specification = 250 tpm ± 1 tap

  ±10% of 250 = ±1

  250 + 1 = 251

  250 – 1 = 249

  The calculated tpm of 250 falls within ±10%.

## FRIABILITY TESTER

The validation of a friability tester (Figure 7.10) will be described here. The friability tester consists of two drums or wheels which are rotated by a motor. Tablets are placed inside the drums and rotated at a fixed speed. The testing is used to determine how well tablets will survive the rigors of the coating process, and shipping and handling. It is used to test bonding integrity during the production phase. By using the findings, potential edge crumbling, capping, or break up can be avoided.

The validation protocol for a friability tester is described in the following sections. This protocol example can be used as a cGMP compliance guide, and it should be tailored to meet individual company requirements. Italicized words in the tables represent examples of information that could be recorded in a validation protocol for a friability tester.

**Table 7.31.** ***Tapped Density Tester Speed Test Results***

| Test | Specification (tpm) | Circulated Speed (tpm) | Acceptable (Yes/No) |
|---|---|---|---|
| Number of taps | 250 ± 1 taps | *250* | *Yes* |

**Table 7.32.** ***Instrument Used***

| Test Instrument | Identification Number | Calibration Due Date |
|---|---|---|
| *Graduated cylinder* | *00-770-CTM* | *01/16/97* |
| *Stopwatch* | *54039* | *02/16/97* |

***Figure 7.10. Friability Tester (courtesy of Van-Kel Industries, Inc.)***

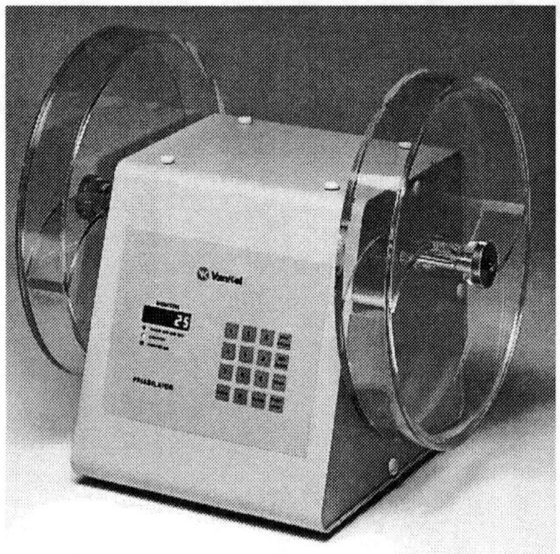

## Installation Qualification

An IQ evaluation establishes confidence that the equipment is properly installed (Figure 7.11). The installation must meet the manufacturer's specified guidelines, along with design changes at installation. Also, the supporting electrical utilities must meet all electrical codes. The information required for an IQ evaluation is equipment identification, required documentation, equipment utility requirements, major component specifications, component material, lubricants, and equipment safety features.

***Figure 7.11. IQ Elements of a Friability Tester***

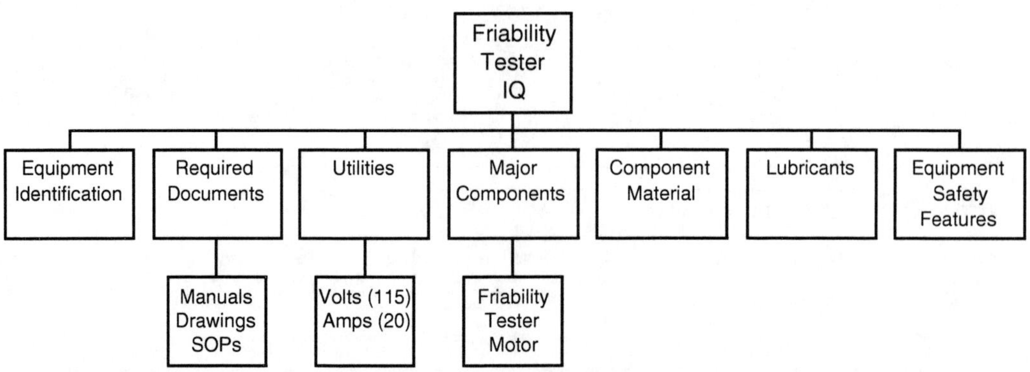

### *Equipment Identification*

Record the equipment identification numbers in Table 7.33, along with the following information: equipment manufacturer's purchase order number, model number, serial number, company assigned equipment number, and the location of the equipment.

### *Required Documentation*

Record the equipment manufacturer's operation and maintenance manual and drawings in Table 7.34. Record SOPs that cover the setup, operation, and cleaning of the friability tester in Table 7.35.

### *Equipment Utility Requirements*

Compare the manufacturer's specified volts (V) and amps (A) requirements to their as-found condition at the time of qualification testing and record the results in Table 7.36. Also record the location of the power supply source. Record the instrument used to measure the volts and amps in Table 7.37.

**Table 7.33. Equipment Identification**

| Required Information | As-Found Conditions |
|---|---|
| Manufacturer | *Any Friability Tester Co.* |
| Purchase order number | *None—expensed item* |
| Model number | *90801* |
| Serial number | *CD4810924* |
| Equipment number | *5910* |
| Location | *45 (This equipment is portable.)* |

**Table 7.34. Manufacturers' Manuals and Drawings**

| Number | Description | Date |
|---|---|---|
| *None* | *Friability Tester Operation & Installation Manual* | *94* |
| *109* | *Electronic Controls Wiring Diagram Drawing* | *None* |

**Table 7.35 Standard Operating Procedures**

| Number | Description | Release Date |
|---|---|---|
| *LAB007* | *Setup and Operation of a Friability Tester* | *01/26/96* |
| *LAB014* | *Friability Tester Cleaning Procedures* | *3/08/96* |

**Table 7.36. *Utilities***

| Utility | Specified | Measured Results | Acceptable (Yes/No) |
|---|---|---|---|
| Volts | 115 ± 10% | *121* | *Yes* |
| Amps | Motor = 0.5 | 20 circuit rating | *Yes* |
| Power supply source, breaker box *BB27, wire number: 5* | | | |

**Table 7.37. *Instrument Used***

| Test Instrument | Identification Number | Calibration Due Date |
|---|---|---|
| *Multimeter* | *ME-025* | *04/19/97* |

***Volt calculation:***

Friability tester motor volt specification = 115 V ± 10%

±10% of 115 = ± 11.5

115 + 11.5 = 126.5

115 – 11.5 = 103.5

The measured volts of 121 fall within the ±10%.

***Amp calculation:***

Circuit rating = 20 A

Equipment current draw = 0.5 A

The circuit amp rating of 20 is greater than the maximum current draw of the equipment.

### *Major Component Specifications*

The component specifications section of the protocol verifies friability tester components purchased were delivered and installed. Record the major components in Table 7.38.

### *Component Material*

Record the material of each component that contacts the product in Table 7.39.

### *Lubricants*

Record the lubricants used to operate the friability tester in Table 7.40 and indicate if they make contact with the product.

**Table 7.38. *Major Components***

| Components | As-Found Conditions |
| --- | --- |
| Friability tester motor | Manufacturer: *Any Motor Co.*<br>Model number: *4T42-CYK4*<br>Serial number: *32057U477*<br>Volts: *115*<br>Amperes: *0.5*<br>Phases: *1*<br>Cycles: *60 Hz*<br>hp: *1/8*<br>rpm: *25* |

**Table 7.39. *Component Material***

| Component | Material |
| --- | --- |
| Drums | *Plastic* |

**Table 7.40. *Lubricants***

| Where Used | Type | Manufacturer | Product Contact (Yes/No) |
| --- | --- | --- | --- |
| *None* | | | |
| Is there a preventive maintenance procedure on file? (Yes/No) | | | *Yes* |

### *Equipment Safety Features*

There are no safety features on this equipment.

## Operational Qualification

An OQ evaluation should establish that the equipment can operate within specified tolerances and limits (Figure 7.12). The mechanical ranges of the friability tester are challenged along with the basic friability tester operations. The friability tester will be validated for its operating ability. The information required for the OQ evaluation is calibration of the instrument used to control the friability tester, equipment control functions (switches and push buttons), and equipment operation (friability tester motor rotation direction, and speed tests).

### *Calibration Requirements*

Verify that all critical instruments on the equipment have been logged into the calibration system, have calibration procedures in place, and are currently in calibration at the time of qualification testing. Record all information for the calibrated instruments used to control the friability tester in Table 7.41.

**Figure 7.12. OQ Elements of a Friability Tester**

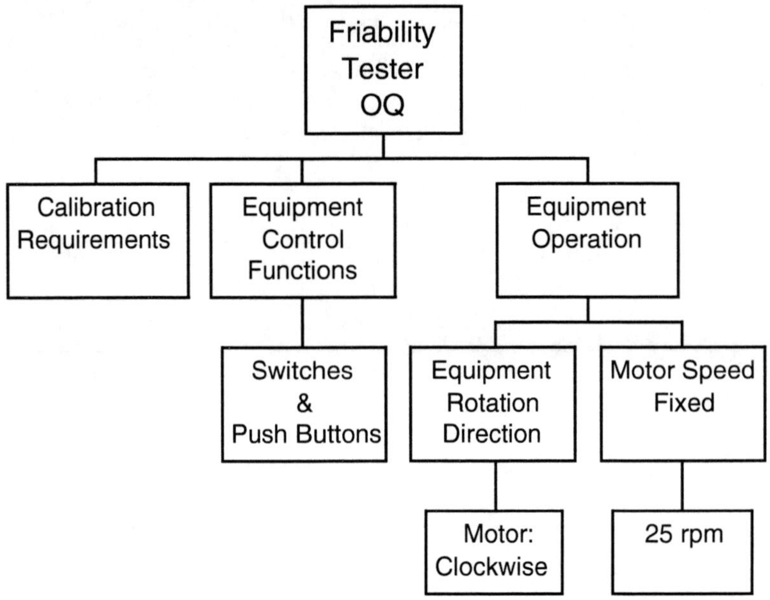


**Table 7.41. Calibrated and Noncalibrated Instruments**

*There were no calibrated or noncalibrated instruments on the friability tester.*

### *Equipment Control Functions*

The objective of testing equipment control functions is to verify that the push buttons on the friability tester operate according to manufacturer specifications. The friability tester will be operated with the tester empty. Operate each control listed in Table 7.42 and verify their proper operation.

**Table 7.42. Control Function Test Results**

| Test Function | Expected Results | Acceptable (Yes/No) |
|---|---|---|
| On/off switch operation | When the power switch is turned on, the indicator light goes on and off. | *Yes* |
| Preset button operation | When the preset buttons are set to 10 revolutions and the power is turned on, the drum rotates 10 times. | *Yes* |
| Reset button operation | When the reset button is pressed, while the power is still on, the number of revolutions returns to 10 and the cycle is repeated. | *Yes* |

### *Equipment Operation: Friability Tester Motor Rotation Direction Test*

The objective of the rotation direction test is to verify that the friability tester motor rotates in the proper direction. The tests are performed with the tester empty. Press the start key and observe the direction of rotation of the friability tester as viewed from the top of the friability tester and record the results in Table 7.43.

### *Equipment Operation: Friability Tester Speed Test*

The objective of tester speed test is to document the speed of the motor with the friability tester empty. A tachometer is required. Measure the speed of the motor with a calibrated tachometer. Verify that the measured speed is within ±10 percent of the fixed speed of 25 rpm. Record the results in Table 7.44. Record the instrument used to measure the speed in Table 7.45.

Motor speed specification = 25 rpm ± 10%

±10% of 25 = ±2.5

25 + 2.5 = 27.5

25 – 2.5 = 22.5

The measured rpm of 25 falls within ±10%.

**Table 7.43.** ***Friability Tester Motor Rotation Direction Test Results***

| Item | Expected Results | Results | Acceptable (Yes/No) |
|---|---|---|---|
| Friability tester motor rotation direction | Rotation should be clockwise as viewed from the top of the friability tester. | *Clockwise rotation was observed.* | *Yes* |

**Table 7.44.** ***Motor Speed Test Results***

| Motor | Speed Specification (rpm ± 10%) | Measured Speed (rpm) | Acceptable (Yes/No) |
|---|---|---|---|
| Friability tester | 25 | *25* | *Yes* |

**Table 7.45.** ***Instrument Used***

| Test Instrument | Identification Number | Calibration Due Date |
|---|---|---|
| *Tachometer* | *64020* | *06/21/97* |

## Performance Qualification

Once it has been established that the equipment is properly installed and functioning within specified operating parameters, it must be shown that the friability tester can operate reliably under routine, minimum, and maximum operating conditions (Figure 7.13).

### *Friability Tester Operation*

The friability tester operation test is designed to document the performance of the friability tester using placebo tablets. The speed of the motor will be measured and recorded.

The materials and instruments required for this test are placebo tablets, a sieve, and a tachometer. Use the following procedure to test friability tester operation:

- Place a sample of 20 (6 g) placebo tablets in a No. 10 sieve and remove any loose dust with the aid of vacuum, or a soft brush. Record the placebo tablets and sieve used in Table 7.46.
- Weigh the tablet sample, initial weight (IW), and record the amount in the Table 7.47. Place the tablet samples in a drum.
- Rotate the drum 100 times.
- Remove any loose dust from the tablets then weigh tablet samples, final weight (FW), and record the amount in Table 7.47.
- Calculate the % friability as follows and record the results in Table 7.47.

$$\frac{\text{inital weight (IW) - final weight (FW)}}{\text{IW}} \times 100 = \%\ \text{friability}$$

% Friability = (IW – FW) ÷ IW = 120 g - 119.05 g ÷ 120 g = 0.792

**Figure 7.13. PQ Elements of a Friability Tester**

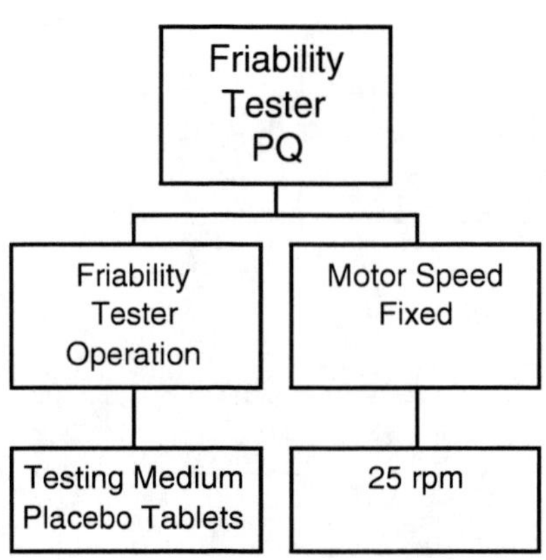

**Table 7.46. Required Materials and Conditions**

| Item | Specifications |
|---|---|
| Placebo tablets | *Placebo blend* |
| Sieve | *20 mesh screen* |

**Table 7.47. Friability Tester Performance Test Results**

| Initial Weight (IW) (g) | Final Weight (FW) (g < 1%) | % Friability | Acceptable (Yes/No) |
|---|---|---|---|
| 120 | 119.05 | *0.792* | *Yes* |

- Measure the speed of the motor with a calibrated tachometer. Verify that the measured speed is within ±10 percent of the fixed speed of 25 rpm. Record the results in Table 7.48. Record the instrument used to measure the speed in Table 7.49.

  Motor speed specification = 25 rpm ± 10%

  ±10% of 25 = ±2.5

  25 + 2.5 = 27.5

  25 – 2.5 = 22.5

  The measured rpm of 24 falls within ±10%.

## INCUBATOR

The validation of an incubator (Figure 7.14) used for incubating microbial testing samples of finished products, room environmental monitoring, water samples,

**Table 7.48. Motor Speed Test Results**

| Motor | Speed Specification (rpm) | Measured Speed (rpm) | Acceptable (Yes/No) |
|---|---|---|---|
| Friability tester | 25 | *24* | *Yes* |

**Table 7.49. Instrument Used**

| Test Instrument | Identification Number | Calibration Due Date |
|---|---|---|
| *Tachometer* | *64020* | *06/21/97* |

**Figure 7.14. Lab Incubator (courtesy of Fisher Scientific)**

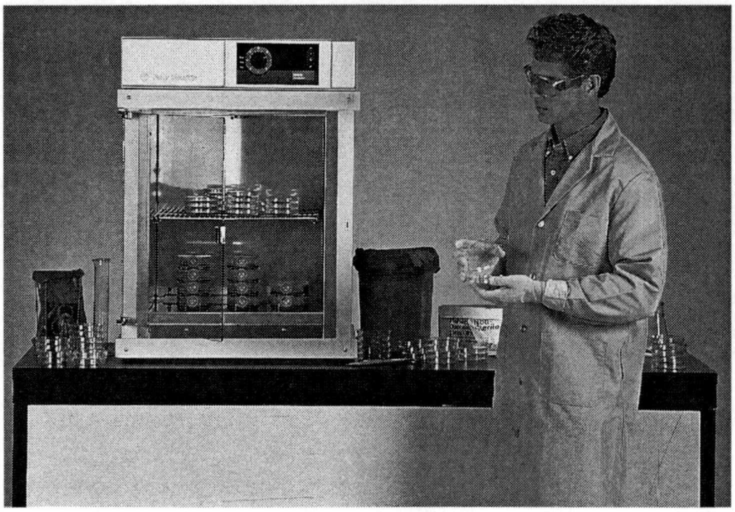

media qualification, and raw materials will be described here. The operating temperature is microprocessor controlled and by using a keypad to increase and decrease the set point temperature. Temperature readouts and control parameters are displayed with LEDs. Three additional LEDs indicate when the temperature is being set, and when the heater power is being applied, or when error conditions are encountered. Fresh air enters through an air intake on the bottom of the incubator, then it is heated in a plenum below the chamber, and finally flows into the incubator chamber in uniform flow patterns. Exhaust air is vented through a port at the top of the incubator.

The validation protocol for an incubator are described in the following sections. This protocol example can be used as a cGMP compliance guide, and it should be tailored to meet individual company requirements. Italicized words in the tables represent examples of information that could be recorded in a validation protocol for an incubator.

## Installation Qualification

The IQ evaluation establishes confidence that the equipment is properly installed (Figure 7.15). The installation must meet the manufacturer's specified guidelines along with design changes at installation. Also, the supporting electrical utilities must meet all electrical codes. The information required for an IQ evaluation is equipment identification, required documentation, equipment utility requirements, major component specifications, component material, lubricants, and equipment safety features.

**Figure 7.15. IQ Elements of an Incubator**

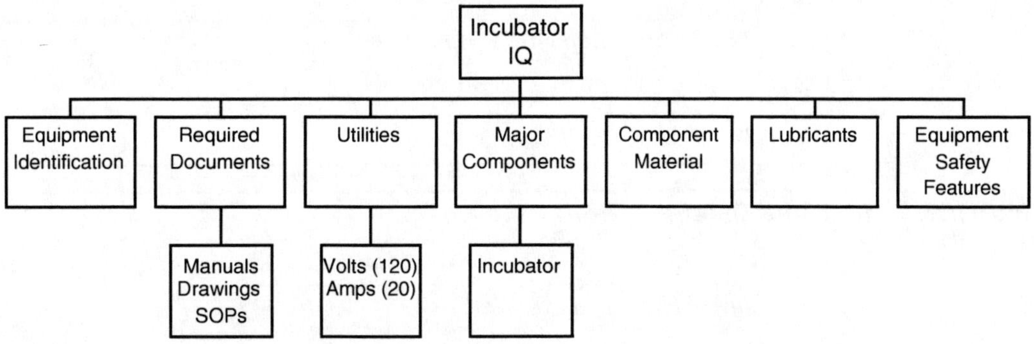


### *Equipment Identification*

Record the equipment identification numbers in Table 7.50, along with the following information: equipment manufacturer's purchase order number, model number, serial number, company assigned equipment number, and the location of the equipment.

### *Required Documentation*

Record the equipment manufacturer's operation and maintenance manual and drawings in Table 7.51. The SOPs that cover the setup, operation, and cleaning of the incubator are recorded in Table 7.52.

### *Equipment Utility Requirements*

Compare the manufacturer's specified volts (V) and amps (A) requirements to their as-found conditions at the time of qualification testing and recorded the results in Table 7.53. Also record the location of the power supply source. Record the instruments used to measure the volts and amps in Table 7.54.

**Table 7.50. Equipment Identification**

| Required Information | As-Found Conditions |
|---|---|
| Manufacturer | *Any Incubator Co.* |
| Purchase order number | *078650* |
| Model number | *D526* |
| Serial number | *010107* |
| Equipment number | *1773* |
| Location | *QC lab (This equipment is portable.)* |

**Table 7.51. Manufacturers' Manuals and Drawings**

| Number | Description | Date |
|---|---|---|
| *D526-11* | *Incubator Operating Instructions Manual* | *1/96* |
| *CRM-1456* | *Remote Sensing Temperature Recorder Manual* | *5/97* |
| *D527* | *Electrical Schematic Drawing* | *2/96* |

**Table 7.52. Standard Operating Procedures**

| Number | Description | Release Date |
|---|---|---|
| *LAB078* | *Incubator Setup and Operation* | *04/23/96* |
| *LAB079* | *Incubator Cleaning Procedure* | *10/12/96* |

**Table 7.53. Utilities**

| Utility | Specified | Measured Results | Acceptable (Yes/No) |
|---|---|---|---|
| Volts | 120 ± 10% | *117* | *Yes* |
| Amps | 7.2 | 20 circuit rating | *Yes* |
| Power supply source, breaker box *BB09, wire number: L1* | | | |

**Table 7.54. Instrument Used**

| Test Instrument | Identification Number | Calibration Due Date |
|---|---|---|
| *Multimeter* | *MU-025* | *04/19/97* |

***Volt calculation:***

Volt specification = 120 V ± 10%

±10% of 120 = ±12

120 + 12 = 132

120 – 12 = 108

The measured volts of 117 falls within ± 10%.

***Amp calculation:***

Circuit rating = 20 A

Equipment current draw = 7.2 A

The circuit rating of 20 amps is greater than the maximum current draw of the equipment.

### *Major Component Specifications*

The component specifications section of the protocol verifies that the incubator components purchased were delivered and installed. Record the major component in the following Table 7.55.

### *Component Material*

Record the material of the component that contacts the product in Table 7.56.

### *Lubricants*

Record the lubricant used to operate the incubator in Table 7.57 and indicate if it makes contact with the product.

### *Equipment Safety Features*

A deviation alarm interrupts the heater power when the incubator temperature differs from the set temperature by more than 3°C. When the temperature exceeds the alarm limit, the alarm indicator LED lights and the display indicates EEE. When the set temperature is changed to more than 3°C below the present incubator temperature, the alarm is tripped. Power is removed from the heaters when the alarm condition occurs.

**Table 7.55. Major Components**

| Components | As-Found Conditions |
|---|---|
| Incubator | Manufacturer: *Any Incubator Co.*<br>Model number: *D526*<br>Serial number: *010107*<br>Temperature range: *30–75°C*<br>Size: *2.5 ft*$^3$ |

**Table 7.56. Component Material**

| Component | Material |
|---|---|
| Incubator chamber | *Stainless steel* |

**Table 7.57. Lubricants**

| Where Used | Type | Manufacturer | Product Contact (Yes/No) |
|---|---|---|---|
| *Door hinges* | APG 90 Oil | Any Oil Co. | *No* |
| Is there a preventive maintenance procedure on file? (Yes/No) | | | *Yes* |

## Operational Qualification

An OQ evaluation should establish that the equipment can operate within specified tolerances and limits (Figure 7.16). The mechanical ranges of the incubator are challenged along with the basic incubator operations. The incubator will be validated for its operating ability, not how well it incubates. Black box software testing was performed. Black box testing evaluates the microprocessor system against the documented functional requirements or intended uses of the equipment. Information required for the OQ evaluation is calibration of the instrument used with the incubator, equipment control functions (switches, push buttons and indicators) and equipment operation (temperature distribution).

### *Calibration Requirements*

Verify that all critical instruments on the equipment have been logged into the calibration system, have calibration procedures in place, and are in calibration at the time of qualification testing. Record all of the necessary information for the calibrated instruments used to control the incubator in Table 7.58.

### *Equipment Control Functions*

The objective of testing equipment control functions was to verify that the switches, push buttons, and indicators on the incubator operated per manufacturer's specifications. Operate each control listed in Table 7.59 and verify its proper operation. Operate each indicator and display in Table 7.60 and verify of their proper operation.

***Figure 7.16. OQ Elements of an Incubator***

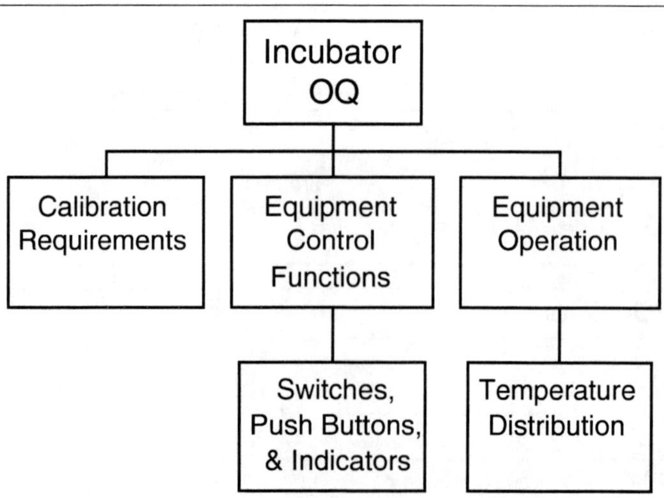

**Table 7.58. Calibrated and Noncalibrated Instruments**

*There were no noncalibrated instruments on the incubator.*

| Calibrated Instrument | As-Found Conditions |
|---|---|
| Identification number | *6567* |
| Type | *Chart recorder, analog* |
| Manufacturer | *Any Chart Recorder Co.* |
| Model number | *CRM-1456* |
| Serial number | *SN8739* |
| Range | *24 h, 20–80°C* |
| Scale division | *15 min, 2°C chart, 1°C digital* |
| Location | *On a bench outside the incubator* |
| Use | *Monitors the temperature of the process* |
| Calibration due date | *10/12/98* |
| Critical or not critical | *Not critical to the operation of the incubator* |

## *Equipment Operation: Temperature Distribution Test*

The objective of the temperature distribution test is to perform temperature mapping of the incubator with the chamber empty. The materials and instruments required for this test are a data recorder and thermocouple wires.

**Table 7.59. Equipment Control Function Test Results**

| Test Function | Expected Results | Acceptable (Yes/No) |
|---|---|---|
| Power on/off switch | When the power on/off switch is pressed, the incubator goes on and off. | *Yes* |
| Menu button operation | When the menu button is pressed, C A L is displayed, then when the set button is pressed, the calibration offset is displayed. | *Yes* |
| ▲ button | When the ▲ button is pressed while holding the set button, the temperature is increased. | *Yes* |
| ▼ button | When the ▼ button is pressed while holding the set button, the temperature is decreased. | *Yes* |
| Set button | When pressing the set button, the set temperature is displayed. | *Yes* |
| | When the set button is used with the ▲ button and ▼ buttons, the set temperature can be changed. | *Yes* |

**Table 7.60. *Equipment Indicator and Display Function Test Results***

| Test Function | Expected Results | Acceptable (Yes/No) |
|---|---|---|
| Heat indicator | When power is being supplied to the heater, the light goes on. | *Yes* |
| Alarm indicator | When the actual chamber temperature exceeds the alarm temperature, the light goes on. | *Yes* |
| Program indicator | When the temperature is being set, the light goes on. | *Yes* |
| Temperature display operation | When programming a temperature, the incubator set temperature is displayed. | *Yes* |
| Set button | When pressing the set button, the set temperature is displayed. | *Yes* |
| | When in the normal operating mode the current incubator temperature is displayed. | *Yes* |

- Calibrate 16 thermocouple wires (TCW) in a range above and below the incubator's setpoint temperature of 60°C.
- Place 5 TCWs at the top, middle, and bottom of the chamber (one in each corner at least 3 in. from the side of the chamber and one in the center). Place TCW 16 next to the data recorder probe. Record the data recorder used in Table 7.62.
- Install a 24 h chart in the chart recorder.
- Program the data recorder to print the average, minimum, and maximum temperature for all 16 TCWs every 10 min.
- When the TCW 16 registers between 55 to 60°C, the temperature recording will need to be started and continuously monitored for 24 h. Record the results in Table 7.61.

**Table 7.61. *Temperature Mapping Results***

| Temperature | Results (24 h) | Chamber Location | Acceptable (Yes/No) |
|---|---|---|---|
| Setpoint | 60.0 °C | — | *Yes* |
| Start | 59.9 °C | — | *Yes* |
| Minimum | 55.6 °C | *TCW 7* | *Yes* |
| Maximum | 61.1 °C | *TCW 13* | *Yes* |
| Average | 58.3 °C | — | *Yes* |

**Table 7.62. *Instrument Used***

| Test Instrument | Identification Number | Calibration Due Date |
|---|---|---|
| *Data recorder* | *DL-066* | *03/26/97* |

## Performance Qualification

Once it has been established that the equipment is properly installed and functioning within specified operating parameters, it must be shown that the incubator can perform reliably under routine operating conditions (Figure 7.17).

### *Incubator Operation: Temperature Distribution Test*

The temperature distribution is designed to perform temperature mapping of the incubator with a load in the chamber. The materials and instruments required for this test are a data recorder and thermocouples.

- Perform this test directly after the empty chamber test and the TCWs will not have to be recalibrated.
- Place 5 TCWs at the top, middle, and bottom of the chamber (one in each corner at least 3 in. from the side of the chamber and one in the center). Place each TCW in a container filled with mineral oil. Place TCW 16 next to the data recorder probe. Record the data recorder used in Table 7.64.
- Install a 24 h chart in the chart recorder.

**Figure 7.17. *PQ Elements of an Incubator***

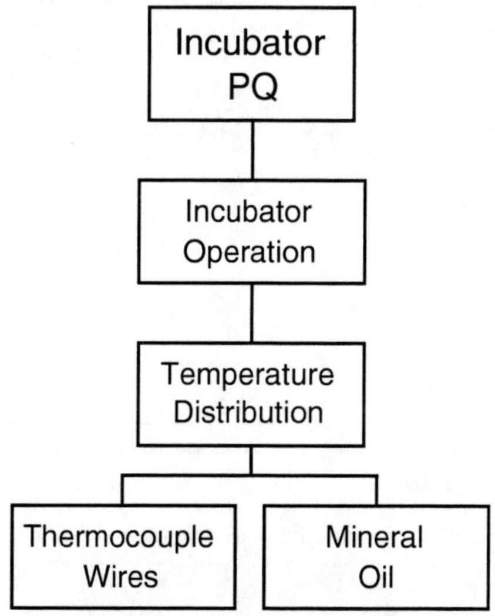

- Program the data recorder to print the average, minimum, and maximum temperature for all 16 TCWs every 10 min.
- When TCW 16 registers between 55 to 60°C, the temperature recording will need to be started and continuously monitored for 24 h. Record the results in Table 7.63.

## Refrigerator/Freezer

The validation of a refrigerator/freezer (Figure 7.18) will be described here used for storing in-process product samples. These materials are environmentally controlled within 2 to 8°C in the refrigerator.

The validation protocol for a refrigerator/freezer is described in the following sections. This protocol example can be used as a cGMP compliance guide, and it should be tailored to meet individual company requirements. Italicized words in the tables represent examples of information that could be recorded in a validation protocol for a refrigerator/freezer.

### Installation Qualification

The IQ evaluation establishes confidence that the equipment is properly installed (Figure 7.19). The installation must meet the manufacturer's specified guidelines, along with design changes at installation. Also, the supporting electrical utilities must meet all electrical codes. The information required for an IQ evaluation is equipment identification, required documentation, equipment utility requirements, major component specifications, component material, lubricants, and equipment safety features.

**Table 7.63. Temperature Mapping Results**

| Temperature | Results (24 h) | Chamber Location | Acceptable (Yes/No) |
|---|---|---|---|
| Setpoint | 60.0°C | — | *Yes* |
| Start | 59.8°C | — | *Yes* |
| Minimum | 55.4°C | *TCW 2* | *Yes* |
| Maximum | 60.1°C | *TCW 12* | *Yes* |
| Average | 57.7°C | — | *Yes* |

**Table 7.64. Instrument Used**

| Test Instrument | Identification Number | Calibration Due Date |
|---|---|---|
| *Data recorder* | *DL-066* | *03/26/97* |

***Figure 7.18. Scientific Refrigerator/Freezer (courtesy of Kelvinator Scientific)***

***Figure 7.19. IQ Elements of a Refrigerator/Freezer***

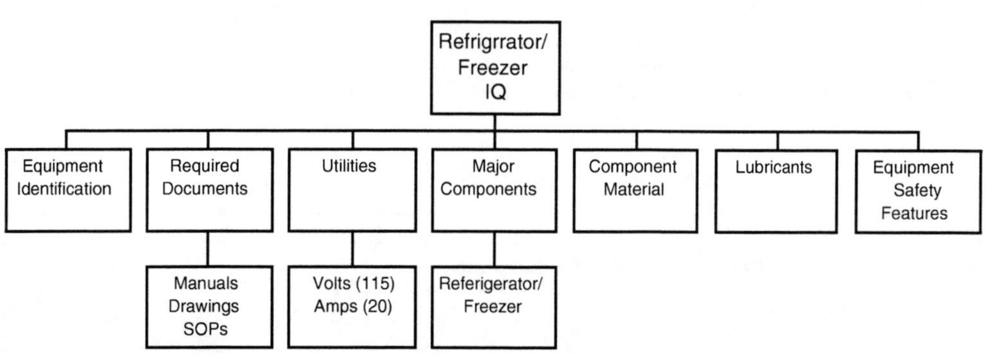

### *Equipment Identification*

Record the equipment identification numbers in Table 7.65, along with the following information: equipment manufacturer's purchase order number, model number, serial number, company assigned equipment number, and the location of the equipment.

### *Required Documentation*

Record the equipment manufacturer's operation and maintenance manual and drawings in Table 7.66. The SOPs that cover the setup, operation, and cleaning of the refrigerator/freezer are recorded in Table 7.67.

### *Equipment Utility Requirements*

Compare the manufacturer's specified volts (V) and amps (A) requirements to their as-found condition at the time of qualification testing and recorded the results in Table 7.68. Also record the location of the power supply source. Record the instruments used to measure the volts and amps in Table 7.69.

***Volt calculation:***

Volt specification = 115 V ± 10%

±10% of 115 = ±11.5

**Table 7.65. Equipment Identification**

| Required Information | As-Found Conditions |
|---|---|
| Manufacturer | *Any Refrigerator/Freezer Co.* |
| Purchase order number | *066231* |
| Model number | *2-SR03TB* |
| Serial number | *81187045* |
| Equipment number | *4233* |
| Location | *Room 455* |

**Table 7.66 Manufacturers' Manuals and Drawings**

| Number | Description | Date |
|---|---|---|
| *None* | *Refrigerator/Freezer Operating Instructions Manual* | *None* |

**Table 7.67. Standard Operating Procedures**

| Number | Description | Release Date |
|---|---|---|
| *LAB120* | *Refrigerator/Freezer Operation of Microbiology* | *03/01/96* |
| *LAB121* | *Refrigerator/Freezer Cleaning Procedure* | *09/26/97* |

**Table 7.68. *Utilities***

| Utility | Specified | Measured Results | Acceptable (Yes/No) |
|---|---|---|---|
| Volts | 115 ± 10% | *114* | *Yes* |
| Amps | 2.5 | 20 circuit rating | *Yes* |
| Power supply source, breaker box *BB28, wire number: 23* | | | |

**Table 7.69. *Instrument Used***

| Test Instrument | Identification Number | Calibration Due Date |
|---|---|---|
| *Multimeter* | *MU-025* | *04/19/97* |

115 + 11.5 = 126.5

115 – 11.5 = 103.5

The measured volts of 114 falls within ±10%.

***Amp calculation:***

Circuit rating = 20 A

Equipment current draw = 2.5 A

The circuit rating of 20 amps is greater than the maximum current draw of the equipment.

## *Major Component Specifications*

The component specifications section of the protocol verifies that the refrigerator/freezer components purchased were delivered and installed. Record the major component in the Table 7.70.

**Table 7.70. *Major Components***

| Components | As-Found Conditions |
|---|---|
| Refrigerator/freezer | Manufacturer: *Any Refrigerator/Freezer Co.*<br>Volts: *115*<br>Amperes: *2.5*<br>Volts: *115*<br>Phases: *1*<br>Cycles: *60 Hz*<br>Temperature range: *-15 to +15°C*<br>Size: *26" wide × 27.5" deep × 60" high = 8 ft*$^3$ |

### *Component Material*

Record the material of the component that contacts the product in Table 7.71.

### *Lubricants*

Record the lubricant used to operate the refrigerator/freezer in Table 7.72 and indicate if it makes contact with the product.

### *Equipment Safety Features*

There are no safety features on this equipment.

## Operational Qualification

An OQ evaluation should establish that the equipment can operate within specified tolerances and limits (Figure 7.20). The mechanical ranges of the refrigerator/

**Table 7.71. Component Material**

| Component | Material |
|---|---|
| Refrigerator/freezer chamber | *Plastic* |

**Table 7.72. Lubricants**

| Where Used | Type | Manufacturer | Product Contact (Yes/No) |
|---|---|---|---|
| *Refrigerant* | *R-22/24 oz Charge* | *Any Refrigerant Co.* | *No* |
| Is there a preventive maintenance procedure on file? (Yes/No) | | | *Yes* |

**Figure 7.20. OQ Elements of a Refrigerator/Freezer**

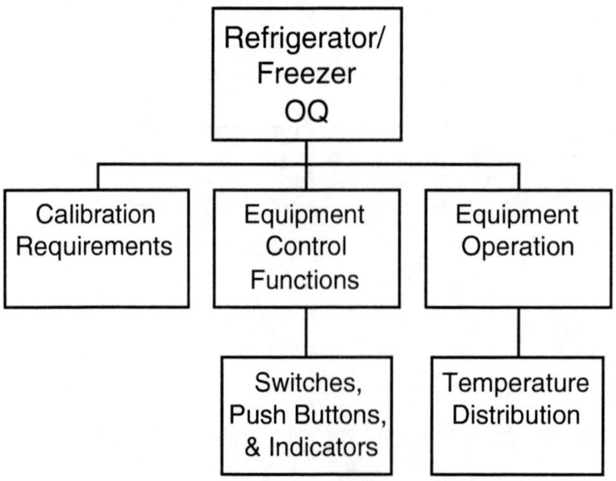

freezer are challenged along with the basic refrigerator/freezer operations that the operator will use. The refrigerator/freezer will be validated for its operating ability, not how well it freezes. Black box software testing was performed. Black box software testing evaluates the microprocessor system against the documented functional requirements, or intended uses of the equipment. Information required for the OQ evaluation is calibration of the instrument used with the refrigerator/freezer, equipment control functions (switches, push buttons and indicators) and equipment operation (temperature distribution).

### *Calibration Requirements*

Verify that all critical instruments on the equipment have been logged into the calibration system, have calibration procedures in place, and are in calibration at the time of qualification testing. Record all of the necessary information for the calibrated instruments used to control the refrigerator/freezer in Table 7.73.

### *Equipment Control Functions*

The objective of testing equipment control functions was to verify that the switches, push buttons and indicators on the refrigerator/freezer operated per manufacturer's specifications. Set the refrigerator to five and the freezer to three, then operate each control listed in Table 7.74 and verify their proper operation.

### *Equipment Operation: Temperature Distribution Test*

The objective of the temperature distribution test is to perform temperature mapping of the refrigerator/freezer with the chamber empty. A data recorder and thermocouples are required.

- Calibrate 16 thermocouple wires (TCW) in a range above and below the refrigerator/freezer's setpoint temperature of 5°C.
- Place 3 TCWs on each of the 5 shelves in the chamber (one in each corner at least 1 in. from the side of the chamber and one in the center). Place TCW 16 next to the data recorder probe. Record the data recorder used in Table 7.76.
- Install a 24 h chart in the chart recorder.
- Program the data recorder to print the average, minimum and maximum temperature for all 16 TCWs every 10 min.
- When TCW 16 registers between 2 and 8°C, the temperature recording will need to be started and continuously monitored for 24 h. Record the results in Table 7.75.

## Performance Qualification

Once it has been established that the equipment is properly installed and functioning within specified operating parameters, it must be shown that the refrigerator/freezer can perform reliably under routine operating conditions (Figure 7.21).

**Table 7.73. *Calibrated and Noncalibrated Instruments***

*There were no noncalibrated instruments on the refrigerator/freezer.*

| Calibrated Instrument | As-Found Conditions |
|---|---|
| Identification number | *100-TID* |
| Type | *Temperature gauge, digital* |
| Manufacturer | *Any Gauge Co.* |
| Model number | *0008-296* |
| Serial number | *13311970* |
| Range | *0–100°C* |
| Scale division | *1°C* |
| Location | *On top of the refrigerator* |
| Use | *Display chamber temperature* |
| Calibration due date | *08/28/97* |
| Critical or not critical | *Critical to the operation of the refrigerator/freezer* |
| **Calibrated Instrument** | **As-Found Conditions** |
| Identification number | *210-CR* |
| Type | *Chart recorder, analog* |
| Manufacturer | *Any Chart Recorder Co.* |
| Model number | *L-136-302-LA* |
| Serial number | *149152* |
| Range | *-15 to +15°F* |
| Scale division | *1°C* |
| Location | *Control panel* |
| Use | *Records the chamber temperature* |
| Calibration due date | *02/27/97* |
| Critical or not critical | *Critical to the operation of the refrigerator/freezer* |

**Table 7.74. *Equipment Control Function Test Results***

| Test Function | Expected Results | Acceptable (Yes/No) |
|---|---|---|
| Refrigerator control knob | When the refrigerator control knob is rotated to a higher number, the refrigerator temperature goes down. | *Yes* |
| Freezer control knob | When the refrigerator control knob is rotated to a lower number, the refrigerator temperature goes up. | *Yes* |
| Freezer control knob | When the freezer control knob is rotated to a higher number, the freezer temperature goes down. | *Yes* |
| | When the freezer control knob is rotated to a lower number, the freezer temperature goes up. | *Yes* |

***Table 7.75. Temperature Mapping Results***

| Time (h:min) | Chart Recorder (Temp. °C) | TCW 16 (Temp. °C) | Average Chamber (Temp. °C) | Acceptable (Yes/No) |
|---|---|---|---|---|
| 17:20 | 5 | 2.6 | 2.8 | Yes |
| 18:30 | 6 | 5.6 | 5.7 | Yes |
| 07:40 | 4 | 2.9 | 3.2 | Yes |
| 08:40 | 6 | 7.4 | 7.4 | Yes |
| 11:50 | 3 | 2.1 | 2.4 | Yes |
| 12:50 | 7 | 6.9 | 6.9 | Yes |
| 14:00 | 6 | 6.7 | 6.7 | Yes |
| 15:40 | 6 | 6.0 | 5.9 | Yes |
| 17:00 | 7 | 7.7 | 7.7 | Yes |

***Table 7.76 Instrument Used***

| Test Instrument | Identification Number | Calibration Due Date |
|---|---|---|
| *Data recorder* | *DL-066* | *03/26/97* |

***Figure 7.21. PQ Elements of a Refrigerator/Freezer***

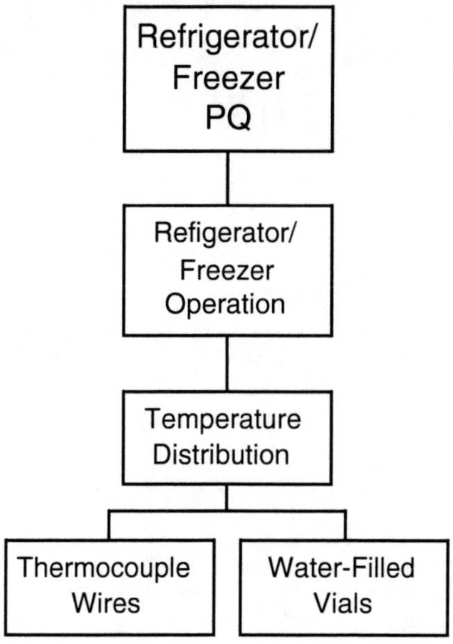


## *Temperature Distribution Test*

The objective of refrigerator/freezer operation is designed to perform temperature mapping of the refrigerator/freezer with a load in the chamber. The materials and instruments required for this test are a data recorder and thermocouples. Use the following procedure to test temperature distribution operation:

- Perform this test directly after the empty chamber test and the TCWs will not have to be recalibrated.
- Place 3 TCWs on each of the 5 shelves in the chamber (one in each corner at least 1 in. from the side of the chamber and one in the center). Place each TCW in water filled vials. Place TCW 16 next to the data recorder probe. Record the data recorder used in Table 7.78.
- Install a 24 h chart in the chart recorder.
- Program the data recorder to print the average, minimum, and maximum temperature for all 16 TCWs every 10 min.
- When TCW 16 registers between 2 and 8°C, the temperature recording will need to be started and continuously monitored for 24 h. Record the results in Table 7.77.

***Table 7.77. Temperature Mapping Results***

| Time (h:min) | Chart Recorder (Temp. °C) | TCW 16 (Temp. °C) | Average Chamber (Temp. °C) | Acceptable (Yes/No) |
|---|---|---|---|---|
| 13:20 | 4 | 6.6 | 5.4 | Yes |
| 14:30 | 5 | 3.3 | 4.6 | Yes |
| 15:50 | 3 | 4.9 | 2.7 | Yes |
| 16:30 | 5 | 2.2 | 5.3 | Yes |
| 17:40 | 6 | 6.7 | 5.2 | Yes |
| 19:00 | 7 | 5.9 | 4.8 | Yes |
| 08:20 | 6 | 3.1 | 4.9 | Yes |
| 10:50 | 6 | 4.4 | 6.0 | Yes |
| 13:10 | 5 | 5.2 | 4.7 | Yes |

***Table 7.78. Instrument Used***

| Test Instrument | Identification Number | Calibration Due Date |
|---|---|---|
| *Data recorder* | *DL-066* | *03/26/97* |

# Appendix A

# Acronyms

| | |
|---|---|
| A, Amp | Ampere |
| AC | Alternating Current |
| AQL | Acceptable Quality Level |
| °C | Degrees Celsius |
| cap/hr | Capsules per Hour |
| cm | Centimeter |
| cfm | Cubic Feet Per Minute |
| cc | Cubic Centimeter |
| cGMP | Current Good Manufacturing Practice |
| °F | Degrees Fahrenheit |
| Dia. | Diameter |
| FW | Final Weight |
| GMP | Good Manufacturing Practice |
| g | Gram |
| hp | Horsepower |
| hr | Hour |
| Hz | Hertz |
| in. | Inch |
| in-lb | Inch Pounds |
| in. W.C. | Inches Water Column |
| IW | Initial Weight |
| IQ | Installation Qualification |
| kg | Kilogram |
| lbs. | Pounds |
| LED | Light Emitting Diode |
| L | Liter |
| LOD | Loss on Drying |
| mg | Milligram |
| mL | Milliliter |
| mm | Millimeter |
| mm/ss | Minutes/Seconds |
| min. | Minute |
| PC | Personal Computer |
| OQ | Operational Qualification |
| P&ID | Piping and Instrumentation Diagram |
| PLC | Programmable Logic Controller |
| PQ | Performance Qualification |
| psig | Pounds per Square Inch Gauge |

| | |
|---|---|
| rpm | Revolutions per Minute |
| sec. | Second |
| SOP | Standard Operating Procedure |
| Sc | Strongcobb Units |
| tpm | Tablets per Minute |
| TCW | Thermal Couple Wires |
| V | Voltage |
| $H_2O$ | Water |
| W.C. | Water Column |

# Appendix B

# Glossary

**Black Box Software Testing:** Evaluates the microprocessor system against the documented functional requirements or intended uses of the system.

**Installation Qualification:** Documented verification that the equipment is installed in a proper way and that all of the components are installed in an environment suitable for their intended use.

**Operational Qualification:** Documented verification that the equipment performs as expected throughout its entire operating ranges.

**Performance Qualification:** Documented verification that the equipment demonstrates effectiveness and repeatability and consistently produces an acceptable product.

**Protocol:** Protocols are used to document the qualification testing of processing equipment. A written procedure stating how validation will be conducted, including test parameters, production equipment, and decision points on what constitutes acceptable test results.

**Qualification:** Qualification is generally related to equipment and is used to determine whether the equipment operates as it was designed to in a reproducible manner.

**Validation:** Established documented evidence that provides a high degree of assurance that a specific process will consistently produce a product meeting its predetermined specifications and quality attributes.

# Appendix C

# Recommended Reading


Baker, G. S., and Rhodes, C. T. 1996. *Modern Pharmaceutics*. New York: Marcel Dekker, Inc.

Berry, I. R., and Harpaz, D. 1997. *Validation of Bulk Pharmaceutical Chemicals*. Buffalo Grove, Ill., USA: Interpharm Press, Inc.

Berry, I. R., and Nash, R. A. 1993. *Pharmaceutical Process Validation*. New York: Marcel Dekker, Inc.

DeSain, C. 1993a. *Drug, Device and Diagnostic Manufacturing*. Buffalo Grove, Ill., USA: Interpharm Press, Inc.

DeSain, C. 1993b. *Documentation Basics That Support Good Manufacturing Practices*. Cleveland: Advanstar Communications.

Juran, J. M., and Gryna, F. M. 1988. *Juran's Quality Control Handbook*. New York: McGraw-Hill, Inc.

Lachman, L., Lieberman, H. A., and Kanig, J. L. 1986. *The Theory and Practice of Industrial Pharmacy*. Philadelphia: Lea & Febiger.

Stokes, T., Branning, R. C., Chapman, K. G., Hamblock, H., and Trill, A. J. 1994. *Good Computer Validation Practices: Common Sense Implementation*. Buffalo Grove, Ill., USA: Interpharm Press, Inc.

Willing, S. H., and Stoker, J. R. 1992. *Good Manufacturing Practices for Pharmaceuticals*. New York: Marcel Dekker, Inc.

Wingate, G. 1997. *Validating Automated Manufacturing and Laboratory Applications: Putting Principles into Practice*. Buffalo Grove, Ill., USA: Interpharm Press, Inc.

# Index